W.-M. Ni L.A. Peletier J. Serrin
Editors

Nonlinear Diffusion Equations and Their Equilibrium States I

Proceedings of a Microprogram held August 25-September 12, 1986

With 12 Illustrations

Springer-Verlag
New York Berlin Heidelberg London Paris Tokyo

W.-M. Ni
James Serrin
School of Mathematics
University of Minnesota
Minneapolis, MN 55455

L.A. Peletier
Department of Mathematics and Computer Science
University of Leiden
The Netherlands

Mathematical Sciences Research Institute
1000 Centennial Drive
Berkeley, CA 94720
USA

The Mathematical Sciences Research Institute wishes to acknowledge support by the National Science Foundation.

Mathematics Subject Classification (1980): 35JXX, 35KXX

Library of Congress Cataloging-in-Publication Data
Nonlinear diffusion equations and their equilibrium states :
proceedings from a conference held August 25–September 12, 1986 / W.-M. Ni, L. Peletier, J. Serrin, editors.
p. cm. — (Mathematical Sciences Research Institute publications ; 12–13)
Includes bibliographies.

ISBN-13: 978-1-4613-9607-9 **e-ISBN-13: 978-1-4613-9605-5**
DOI:10.1007/ 978-1-4613-9605-5

1. Differential equations, Partial—Congresses. 2. Differential equations, Nonlinear—Congresses. 3. Diffusion—Mathematical models—Congresses. I. Nim W.-M. (Wei-Ming) II. Peletier, L. (Lambertus) III. Serrin, J. (James), 1926– . IV. Series.
QA377.N645 1988
515.3'53—dc19 86–12370

Softcover reprint of the hardcover 1st edition 1988

Camera-ready text prepared by the Mathematical Sciences Research Institute using PC $\mathrm{T_EX}$.

9 8 7 6 5 4 3 2 1

Preface

In recent years considerable interest has been focused on nonlinear diffusion problems, the archetypical equation for these being

$$u_t = \Delta u + f(u).$$

Here Δ denotes the n-dimensional Laplacian, the solution $u = u(x,t)$ is defined over some space-time domain of the form $\Omega \times [0,T]$, and $f(u)$ is a given real function whose form is determined by various physical and mathematical applications. These applications have become more varied and widespread as problem after problem has been shown to lead to an equation of this type or to its time-independent counterpart, the elliptic equation of equilibrium

$$\Delta u + f(u) = 0.$$

Particular cases arise, for example, in population genetics, the physics of nuclear stability, phase transitions between liquids and gases, flows in porous media, the Lend-Emden equation of astrophysics, various simplified combustion models, and in determining metrics which realize given scalar or Gaussian curvatures. In the latter direction, for example, the problem of finding conformal metrics with prescribed curvature leads to a ground state problem involving critical exponents. Thus not only analysts, but geometers as well, can find common ground in the present work.

The corresponding mathematical problem is to determine how the structure of the nonlinear function $f(u)$ influences the behavior of the solution. Since the function $f(u)$ in almost all applications seriously departs from linearity it might be thought that the basic problem would be nearly intractable, but with increasing experience, gained from recent research, certain important patterns have emerged. At the same time, new directions of investigation, which might have appeared impossibly difficult only a few years ago, are now open and inviting.

The papers which comprise the two volumes of this work attest to this growing level of interest in nonlinear diffusion equations. Their content and subject matter is broadly directed, but nevertheless can be roughly classified into several individual categories of particular interest:

Ground state problems, critical exponent problems and critical dimension problems associated with the elliptic equation $\Delta u + f(u) = 0$.

Singularity theory for solutions in both the elliptic and parabolic cases.

Asymptotic behavior of solutions in both space and time.

Blow-up phenomena associated with the non-continuability of solutions as time increases, as well as compact support questions when the operator arises in the theory of porous media.

Studies of similarity solutions for elliptic and parabolic problems, leading to a number of fascinating questions in the theory of ordinary differential equations.

Finally, the underlying Laplace operator in the problems discussed above can be replaced by more generalized variational operators of quasi-linear type, or by degenerate elliptic operators, leading to still further ramifications of the basic theory.

The above list is by no means comprehensive, including neither uniqueness nor convexity questions, nor related work in critical point theory or the use of specialized phase plane analysis.

In summary, the present volumes bring together for the intended reader a wealth of stimulating ideas in an important and rapidly expanding branch of differential equations and applied analysis.

The organizers of the Microprogram are pleased to thank the Mathematical Sciences Research Institute for the opportunity to present this work, both as lectures at the original Microprogram at the Institute in 1986 as well as in the form of the papers included here. We also wish to thank the Office of Naval Research, Department of Defense, for their generous financial support of the Microprogram itself.

Our special and particular thanks also go to Professor Irving Kaplansky, who worked capably and imaginatively and with ever-present patience and good humor to ensure the success of the project from the moment of its inception, to Arlene Baxter who coordinated the editorial effort for the many papers involved, and to David Mostardi who carried out the immense effort of transcribing the original manuscripts into this final published version.

To all these, and to the authors of the papers included here, the organizing committee conveys its thanks.

James Serrin, Chairman
Wei-Ming Ni
Lambertus A. Peletier

Nonlinear Diffusion Equations and their Equilibrium States

TABLE OF CONTENTS – VOLUME 1

Nonlinear Diffusion Equations and their Equilibrium States

TABLE OF CONTENTS – VOLUME 2

On the Initial Growth of the Interfaces in Nonlinear Diffusion-Convection Processes

L. ALVAREZ, J. I. DIAZ AND R. KERSNER

Abstract. We study the qualitative behavior of the fronts or interfaces generated by the solutions of the equation

$$u_t = (u^m)_{xx} + b(u^\lambda)_x,$$

where $m, \lambda > 0$ and b is a real number, $b > 0$. In particular we focus our attention on the waiting time phenomenon and give "necessary" and sufficient conditions on the initial data $u_0(x)$ in order to have such a property. Since the convection term in the equation introduces an asymmetry, a separated study of the left and right fronts is needed. The results depend in a fundamental way on the values of m and λ. In particular, the different answer with respect to the case of pure diffusion ($b = 0$) occurs for $0 < \lambda < 1$, where the left front does not exist and the right one may already be reversing.

1. Introduction.

The nonlinear diffusion-convection processes referred to in the title of this paper are those described by the equation

$$u_t = (u^m)_{xx} + b(u^\lambda)_x, \tag{1}$$

where $m > 0$, $\lambda > 0$ and $b \in \mathbb{R}$, $b > 0$.

This equation arises as a model for a number of different physical phenomena. For instance, when u denotes unsaturated soil-moisture content, the equation describes the infiltration of water in a homogenous porous medium and some natural conditions in this context are $m > 1$ and $\lambda > 0$ (Bear [**3**], Phillip [**15**]). The equation also occurs in the study of the flow of a thin viscous film over an inclined bed for the specific exponents $m = 3$ and $\lambda = 4$ (Buckmaster [**5**]). By analogy with the classical equation from statistical mechanics (Chandrasekhar [**6**]), equation (1) is often referred to as the nonlinear Fokker-Planck equation. Equation (1) is also used in connection with transport of thermal energy in plasma (then $0 < m < 1$ and $\lambda = 1$, Rosenau and Kamin [**16**]). Finally, the equation has additional interest as a generalization of the well-known equation of Burgers approximating the associated hyperbolic conservation law equation.

It is a well-known fact that nonnegative solutions u of (1) may give rise to interfaces (or free boundaries) separating regions where $u > 0$ from ones where $u = 0$. These fronts are relevant in the physical problems modelled and their occurrence is essentially due to slow diffusion ($m > 1$) or to convective phenomena dominating over diffusion ($\lambda < m$) (See, e.g. Gilding [**10,12**] and Diaz-Kersner [**8**]). Another kind of front (the time of extinction of the solution) is intrinsic to fast diffusion ($m < 1$) and will not be considered here (Berryman-Holland [**4**]). Note finally that we cannot expect, in general, to have classical solutions of (1), and that discontinuities of the gradient of solutions take place on the interfaces.

The existence, uniqueness and regularity of weak solutions of the Cauchy problem, the Cauchy-Dirichlet problem and the first boundary-value problem for (1) was given by Diaz-Kersner [**7**] for $m \geq 1$ and $\lambda > 0$ and later extended by Gilding [**11**] to any $m > 0$ (see the references in these articles for earlier works).

The main goal of this work is to study the initial growth of the interfaces

$$\varsigma_-(t) = \inf\{x : u(x,t) > 0\}$$
$$\varsigma_+(t) = \sup\{x : u(x,t) > 0\}$$

in relation to different values of m and λ. For the sake of simplicity, we restrict the discussion to the Cauchy problem for (1). However, we remark that the initial growth of ς_- and ς_+ will depend only on the behavior of the initial data $u_0(x)$ near the boundary of its support $[\varsigma_-(0), \varsigma_+(0)]$. Thus our results may be extended to solutions of other initial boundary-value problems associated with (1) (the extension to the study of "interior" fronts is possible as well).

To be explicit, we shall focus our attention on continuous nonnegative weak solutions of the problem

$$(CP) \qquad \begin{cases} u_t = (u^m)_{xx} + b(u^\lambda)_x & \text{for } (x,t) \in \mathsf{R} \times \mathsf{R}^+ \\ u(x,0) = u_0(x) & \text{for } x \in \mathsf{R}, \end{cases}$$

where u_0 is a given continuous nonnegative function on R which, for simplicity, is assumed to satisfy

$$(2) \qquad u_0(x) > 0 \text{ on } (a_-, a_+) \text{ and } u_0(x) = 0 \text{ on } \mathsf{R} - (a_-, a_+).$$

In analogy with the case of nonconvective flows in porous media ($b \equiv 0, m > 1$) the interfaces may be stationary until a certain finite time, called the "waiting time", of the interface (some discussions of this property for the porous media equation can be found in Aronson [**2**], Knerr [**14**] and Vazquez [**17**]). In the following, we shall place special emphasis on giving "neccesary and sufficient" conditions on u_0 in order to have a waiting time. In contrast with the case $b \equiv 0$, a separate study of ς_- and ς_+ is needed because the convective term introduces an inherent asymmetry into the problem.

As we will show, the initial growth of the interfaces is different in each one of the following regions of (λ, m) parameter space.

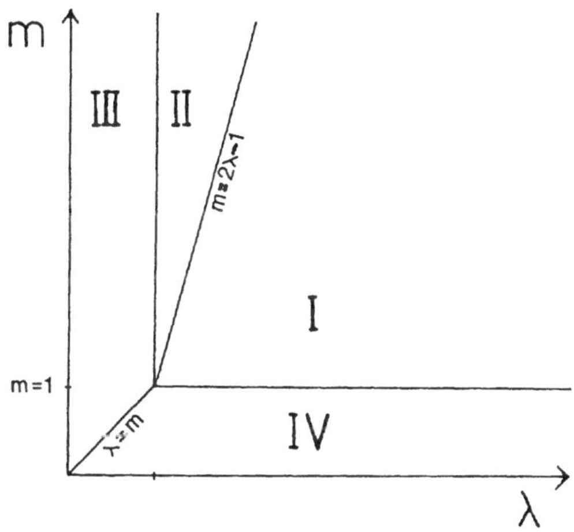

Figure 1

It is a curious fact that the interplay between diffusion and convection may be completely different in other contexts. For instance, the asymptotic behavior of solutions of (CP) was studied by Grundy [**13**], where a different decomposition of the (λ, m) parameter space occurs.

In order to describe our results we remark that the behavior of the interface $\varsigma_\mp(t)$ depends only on the values of m and λ as well as on the local behavior of the initial data $u_0(x)$ near $a_\mp$. For the sake of simplicity we shall use the notation $u_0(x) \sim |x - a_\mp|^\alpha$ to indicate that $u_0(x) \leq C\,|x - a_\mp|^\alpha$ and $u_0(x) \geq C\,|x - a_\mp|^{\tilde{\alpha}}$ for x near $a_\mp$, where C, $\tilde{C} > 0$ and $\tilde{\alpha}$ is some number $< \alpha$.

As a final general remark we point out that the conclusions of this paper also hold upon replacing pointwise comparison conditions on u_0 by more general assumptions indicating how the mass $M_\pm(x)$ of u_0 grows near $a_\mp$ (here $M_\pm(x) = \pm\int_x^{a_\mp} u_0(s)ds$). For instance, conditions assuring the existence of a waiting time can be formulated in terms of the relation $\limsup_{x\to a_\mp} M_\mp(x)\,|x-a_\mp|^{-\alpha-1} = 0$ (see Alvarez-Diaz [**1**]; a pioneer work in this direction is Vazquez [**17**], where the case of nonconvection was examined). We also point out that our proof of the nonexistence of a waiting time always leads to growth estimates on $\varsigma_\pm(t)$.

In Section 2 we study the interfaces for (λ, m) in the region I, defined by $\{(\lambda,m) : \lambda \geq \frac{1}{2}(m+1), m > 1\}$. This case corresponds to a slow diffusion dominating convection, in the sense that the results are of the same nature as in the equation without convection; the interface $\varsigma_\pm(t)$ has a waiting time if and only if $u_0(x) \sim |x-a_\mp|^{2/(m-1)}$. The presence of the convection term only leads to a natural displacement of the interfaces compared to the case without convection (such a property occurs for any value of λ when there is some interface).

Section 3 is devoted to the case in which (λ, m) belongs to the region II defined by $\{(\lambda,m) : 1 < \lambda < \frac{1}{2}(m+1)\}$. Some differences compared to the case of pure diffusion appear, namely the criterion for the existence of a waiting time for $\varsigma_-(t)$ is weaker ($u_0(x) \sim |x-a_-|^{1/(\lambda-1)}$) while for the front $\varsigma_+(t)$ a stronger criterion is needed: $u_0(x) \leq C_0\,|x-a_+|^{1/(m-\lambda)}$ for x near a_+ and $C_0 = [b(m-\lambda)/m]^{1/(m-\lambda)}$. We also show that this last condition is "necessary" for the existence of a waiting time. We would describe this by saying that convection already dominates over diffusion but in a weak way, because many other properties of $\varsigma_\mp(t)$, for small t, remain unchanged: $\varsigma_-(t)$ is finite and nonincreasing, $\varsigma_+(t)$ is finite and nondecreasing, etc.

The region III, defined by $\{(\lambda,m) : \lambda \leq 1 \text{ and } \lambda < m\}$, is examined in Section 4 and reflects the case in which there is a great contrast with pure diffusion phenomena (especially when $0 < \lambda < 1$) because then the interface $\varsigma_-(t)$ does not exist (Diaz-Kersner [**8,9**]). Here we show that convection dominates strongly over diffusion; namely, it is enough to know that $u_0(x) \leq C\,|x-a_+|^{1/(m-\lambda)}$ for some $C < C_0$ to conclude that $\varsigma_+(t)$ is initially a reversing front and that $\varsigma_+(t) \leq a_+ - \bar{C}t^{(m-\lambda)/(m+1-2\lambda)}$ for some suitable $\bar{C} > 0$ and any t small. Moreover, if $u_0(x) \geq C\,|x-a_+|^{1/(m-\lambda)}$ for some $C > C_0$ then $\varsigma_+(t)$ is initially a progressing front and $\varsigma_+(t) \geq a_+ + \underline{C}t^{(m-\lambda)/(m+1-2\lambda)}$ for some suitable $\underline{C} > 0$ and any t small. Moreover, if $u_0(x) \geq C\,|x-a_+|^{1/(m-\lambda)}$ for some suitable $C > C_0$ then $\varsigma_+(t)$ is initially a progressing front and $\varsigma_+(t) \geq a_+ + \underline{C}t^{(m-\lambda)/(m+1-2\lambda)}$ for some suitable

$\underline{C} > 0$ and any t small.

We finish the introduction with two remarks. The first is that the region IV, defined by $\{(\lambda, m) : m \leq 1 \text{ and } \lambda \geq m\}$, corresponds to a fast or linear diffusion with a weak convection. In this case none of the fronts exist (Gilding [**12**]), hence this region is not of interest to us. Second, we can consider a more general formulation of the equation,

$$u_t = (u^m)_{xx} + f(u)_x, \tag{3}$$

where $m > 0$ and f is a continuous real function. This program will be developed elsewhere. As an illustration, consider the function $f(s) = \mu s + bs^\lambda$ with $\mu, b, \lambda > 0$. Making the change of variables ($\bar{x} = x$, $\bar{t} = t - \mu x$) it is easy to see that the function $v(\bar{x}, \bar{t}) = u(t, x)$ will satisfy the equation (1). Thus a waiting time phenomenon for v means that u follows the characteristics of the hyperbolic conservation law $u_t = f(u)_x$ during some finite time (see figure 2). Obviously, a systematic study of the different possibilities can be carried out in terms of the values of λ, m and the assumptions on the behavior of u_0 near the boundary of its support.

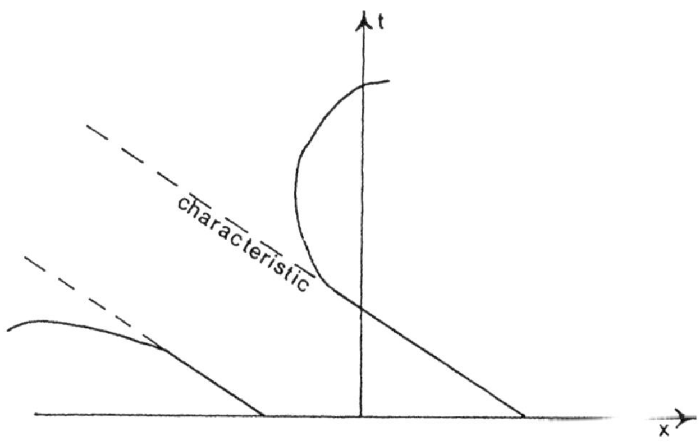

Figure 2

2. Region I: Slow diffusion dominating convection.

We shall assume in this section that $\lambda \geq \frac{1}{2}(m+1)$ and $m > 1$. Under these assumptions it turns out that the initial behavior of the interfaces is the same as for the porous medium equation ($b \equiv 0$). We start by recalling a well-known result

THEOREM 1 Gilding [**10**]. *Let u be a weak solution of the Cauchy problem (CP) and suppose that*

$$u_0(x) \leq C\,|x - x_0|^{2/(m-1)} \tag{4}$$

for some constants $C > 0$, $\delta > 0$ and for x such that $|x - x_0| \leq \delta$. Then there exists a finite time $t^* > 0$ such that $u(x_0, t) = 0$ for any $t \in [0, t^*)$.

IDEA OF THE PROOF: Without loss of generality we may choose $x_0 = 0$. Then, it is enough to show that the separable function

$$\bar{u}(x,t) = (Ax^2)^{1/(m-1)}\{\tau_0/(\tau_0 - t)\}^{\alpha}$$

is a supersolution on the set $(x,t) \in [-\delta, \delta] \times [0, \tau_0)$, for some positive constants A, α and τ_0.

□

Concerning the nonexistence of a waiting time, we have

THEOREM 2. *Let u be a weak solution of (CP) and suppose that*

$$u_0(x) \geq C\,|x - a_{\mp}|^{\gamma}, \tag{5}$$

for some $\gamma \in (0, 2/(m-1))$, $C > 0$, $\delta > 0$ and for x respectively such that $0 \leq x - a_- \leq \delta$ or $0 \leq a_+ - x \leq \delta$. Then there exist positive constants τ and $\bar{c}$ such that

$$\varsigma_-(t) \leq a_- - \bar{c}t^{1/(2-\gamma(m-1))} \tag{6}$$

for any $t \in [0, \tau]$, and (if in addition $\gamma(m-1) \geq 1$)

$$\varsigma_+(t) \geq a_+ - \bar{c}t^{1/(2-\gamma(m-1))} \tag{7}$$

for any $t \in [0, \tau]$. In particular, $u(a_{\mp}, t) > 0$ for any $t > 0$.

In order to prove Theorem 2 we will define a family of auxiliary functions depending on two parameters K and $\bar{x}$ in the following way:

$$\underline{u}(x, t; K, \bar{x}) = \left[K^2 t - K M_1 (1 - M_2 t)(x - \bar{x})\right]_+^{1/(m-1)}, \tag{8}$$

where $[a]_+ = \max\{a, 0\}$ and M_1 and M_2 will be chosen later. The following lemma shows that $\underline{u}$ is a subsolution of the equation on sets of the form $[-\delta, \infty) \times [0, \tau]$ for some suitable $\tau > 0$.

LEMMA 1. *Let M and δ be given positive constants. Then there exist M_1, M_2 and $\tau > 0$ such that $\underline{u}$ is a subsolution of (1) in the region $(x,t) \in [-\delta, +\infty) \times [0,\tau]$ for every $K \in (0,M)$ and $\bar{x} \in [-\delta, 0]$.*

PROOF: It is easy to see that if u satisfies (1) then the function $v = m/(m-1)u^{m-1}$ satisfies the equation

$$\mathcal{L}(v) = -v_t + (m-1)vv_{xx} + (v_x)^2 + q(v^p)_x = 0,$$

where

$$q = \frac{bm\lambda}{m+\lambda-2}\left(\frac{m-1}{\lambda}\right)^{\frac{m+\lambda-2}{m-1}}, \quad p = 1 + \frac{\lambda-1}{m-1}.$$

Now, take $\underline{v}(x,t;K,\bar{x}) = \frac{m}{m-1}\left[K^2 t - KM_1(1-M_2 t)(x-\bar{x})\right]_+$. A direct computation gives

$$\mathcal{L}(\underline{v}) = \frac{m}{m-1}K\left[K\left\{\frac{mM_1^2}{m-1}(1-M_2 t)^2 - 1\right\} - M_1 M_2 (x-\bar{x}) + \right.$$
$$\left. + pM_1\left(\frac{m}{m-1}\right)^{p-1} q(M_2 t - 1)\left\{K^2 t - KM_1(1-M_2 t)(x-\bar{x})\right\}^{p-1}\right]_+ .$$

In order to prove that $\mathcal{L}(\underline{v}) \geq 0$ in $[-\delta, +\infty) \times [0,\tau]$ for some $\tau > 0$ we study the following two cases separately.

Case 1: $x \leq -K$. Since $K \in (0,M)$, $\bar{x} \in [-\delta, 0]$ and $\lambda \geq \frac{1}{2}(m+1)$ then $\underline{v}$ is uniformly bounded in $[-\delta, +\infty) \times [0,\tau]$ and moreover $p - 1 \geq \frac{1}{2}$. Now, let

$$\tau = \min\{1/2M_1, 1\} \quad \text{and} \quad M_1 \geq 2\left(\frac{m-1}{m}\right)^{1/2}.$$

Then, there exists $N = N(M,\delta) > 0$ such that

$$\mathcal{L}(\underline{v}) \geq \frac{m}{m-1}KM_1|x|\left(M_2 - N(1+M_1)^{1/2}\right),$$

and so, if we take $M_2 = N(1+M_1)^{1/2}$ we have $\mathcal{L}(\underline{v}) \geq 0$.

Case 2: $x \geq -K$. Since $\underline{v} \equiv 0$ if $x \geq 2K/M_1$ it suffices to assume $x < 2K/M_1$. Then

$$\mathcal{L}(\underline{v}) \geq \frac{m}{m-1}K^2\left[\frac{m}{m-1}\frac{M_1^2}{4} - 1 - 2N(1+M_1)^{1/2} + M_1 N(1+M_1)^{1/2}\right].$$

Therefore, if M_1 is big enough we conclude that $\mathcal{L}(\underline{v}) \geq 0$.

□

PROOF OF THEOREM 2: We shall first prove (7). Without loss of generality, we may assume $a_+ = 0$, $1 < \gamma(m-1) < 2$ and $u_0(x) = C|x|^\gamma$ for $x \in [-\delta, 0]$. We shall compare the function $\underline{u}$, defined by (8), and u in a region $[-\delta, \infty) \times [0, \tau]$ when the parameter $\bar{x}$ belongs to $(-\delta, 0)$ and τ is given in Lemma 1. First of all, we remark that by the continuity of u, there exists a value $\theta > 0$ such that $u(-\delta, t) \geq \theta$ for every $t \in [0, \tau]$ (recall (2)). The initial inequality $\underline{u}(x, 0; K, \bar{x}) \leq C|x|^\gamma$ is verified if and only if

$$[-M_1 K(x-\bar{x})]_+ \leq C^{m-1}|x|^{\gamma(m-1)}. \tag{9}$$

Since $\gamma(m-1) > 1$, by a convexity argument, (9) is satisfied if we choose

$$K = K(\bar{x}) = \frac{\gamma(m-1)C^{m-1}}{M_1}|\bar{x}|^{\gamma(m-1)-1}.$$

Next, choosing $\bar{x}$ small enough we have

$$\underline{u}(-\delta, t; K(\bar{x}), \bar{x}) \leq \theta \leq u(-\delta, t) \quad \text{for any } t \in [0, \tau]$$

and

$$\underline{u}(x, 0; K(\bar{x}), \bar{x}) \leq u_0(x) \qquad \text{if } x \in [-\delta, +\infty).$$

Then, by the comparison principle, we conclude that

$$\underline{u}(x, t; K(\bar{x}), \bar{x}) \leq u(x, t) \text{ in } [-\delta, +\infty) \times [0, \tau].$$

A direct computation shows that $\underline{u}(0, t; K(\bar{x}), \bar{x}) > 0$ if

$$t > t(\bar{x}) = \frac{M_1^2|\bar{x}|^{2-\gamma(m-1)}}{\gamma(m-1)C^{m-1} + M_1^2 M_2|\bar{x}|^{2-\gamma(m-1)}}. \tag{10}$$

Then for $t_0 \in (0, \tau)$ we choose $\bar{x} = -\beta t_0^{1/(2-\gamma(m-1))}$ with $\beta > 0$ large enough, and derive from (10) that

$$\varsigma_+(t_0) \geq \left(\frac{\gamma(m-1)C^{m-1}\beta^{\gamma(m-1)-1}}{M_1^2(1-M_2 t_0)} - \beta\right) t_0^{1/(2-\gamma(m-1))}; \tag{11}$$

this proves the inequality (7) because t_0 is arbitrary. By a well-known result (Gilding [**10**, Theorem 3]) we have, in fact, $u(a_+, t) > 0$ for any $t > 0$.

The proof of the inequality (6) is similar. As a matter of fact, the study of the left interface $\varsigma_-(t)$ is easier due to the sign $(b > 0)$ of the convection term (see Remark 1). In particular the proof that $u(t, a_-) > 0$ given in Knerr [**14**] remains true without any significant change.

□

REMARK 1: The assumptions (4) and (5) are the same as in the the proof of the corresponding results for the porous media equation, $b \equiv 0$ (see, for instance, Knerr [**14**]). (We point out that the proof of Theorem 2 is different from the one given in Knerr [**14**] and that the estimates (6) and (7) seem to be new in the literature.) As already mentioned, in region I the diffusion dominates the convection. Nevertheless, it is clear that the presence of the convection terms modifies the behavior of the solution (and so of the fronts) compared to the solution of the pure diffusion equation, *independent of the value of* λ. In particular, it is shown in Alvarez-Diaz [**1**] that if we denote by w the solution of the Cauchy problem (CP) without convection $(b \equiv 0)$ and by $\xi_-(t)$ and $\xi_+(t)$ the left and right fronts generated by w, then

$$\varsigma_-(t) \leq \xi_-(t) \quad \text{and} \quad \xi_+(t) \leq \varsigma_+(T). \tag{12}$$

Notice that (12) makes sense only when $m > 1$ because otherwise $\xi_-(t) = -\infty$ and $\xi_+(t) = +\infty$.

3. Region II: Convection weakly dominating diffusion.

We shall now assume that $1 < \lambda < \frac{1}{2}(m+1)$. As we shall show, in this case convection dominates diffusion weakly and therefore the influence of convection is different for each front, making it necessary to study $\varsigma_-(t)$ and $\varsigma_+(t)$ separately.

We start by studying the left interface $\varsigma_-(t)$. A sufficient condition for the existence of a waiting time is given by the following result.

THEOREM 3. *Let u be a weak solution of (CP) and assume that*

$$u_0(x) \leq C\,|x - a_-|^{1/(\lambda-1)} \text{ if } 0 \leq x - a_- \leq \delta \tag{13}$$

for some positive constants C and δ. Then there exists a finite time $t^* > 0$ such that $u(a_-, t) = 0$ for any $t \in [0, t^*]$.

PROOF: Without loss of generality we may choose $a_- = 0$ and $\delta \leq 1$. As in Theorem 1, we shall derive our conclusion from the construction of a suitable supersolution. We define

$$\bar{u}(x,t) = \left(\frac{1}{K_1 - K_2 t}\right)^{1/(m-1)} |x|^{1/(\lambda-1)} \tag{14}$$

with

$$K_1 = \min\left\{1, \quad C^{1-m}, \quad M^{1-m}\delta^{(m-1)/(\lambda-1)}\right\}, \quad M = \|u_0\|_{L^\infty(\mathbb{R})},$$
$$K_2 = (m-1)\left[\frac{m(m-\lambda+1)}{(\lambda-1)^2} + b\frac{\lambda}{\lambda-1}\right].$$

From the choice of K_2, $\lambda < m$ and $m + 2 - 2\lambda > 1$, it is not difficult to verify that $\underline{u}$ satisfies the inequality

$$\bar{u}_t - (\bar{u}^m)_{xx} - b(\bar{u}^\lambda)_x \geq 0.$$

Moreover, from the definition of K_1 we have

$$\begin{aligned} &\bar{u}(x,0) \geq C|x|^{1/(\lambda-1)} && \text{if } x \in [-\delta,\delta] \\ &\bar{u}(\delta,t) \geq M && \text{if } t \in [0,t^*], \end{aligned}$$

with $t^* = K_1/K_2$. Then, applying the comparison principle to the region $(-\delta,\delta) \times [0,t^*)$ we conclude that $0 \leq u \leq \bar{u}$ and the result follows.

□

The nonexistence of any waiting time for the front ς_- can also be proved when the opposite inequality in (13) holds.

THEOREM 4. *Let u be a weak solution of (CP) and suppose that*

$$u_0(x) \geq C\,|x - a_-|^\gamma \ \text{ if } 0 \leq x - a_- \leq \delta \tag{15}$$

for some $\gamma \in (0, 1/(\lambda-1))$, $C > 0$ and $\delta > 0$. Then there exist positive constants τ and $\bar{c}$ such that

$$\xi_-(t) \leq a_- - \bar{c}t^{1/1-\gamma(\lambda-1)}$$

for every $t \in [0,\tau]$. In particular, $u(a_-,t) > 0$ for any $t > 0$.

PROOF: As before, it suffices to take $a_- = 0$ and to construct a suitable subsolution, which we shall choose as the following "traveling wave solution" depending on two parameters $K > 0$ and $\bar{x} \in (0,\delta)$:

$$v(x,t;K,\bar{x}) = \mu_K([x + Kt - \bar{x}]_+) \tag{17}$$

where μ_K is defined by

(18) $$y = m\int_0^{\mu_K(y)} \frac{s^{m-2}}{K - bs^{\lambda-1}}\,ds.$$

Given τ small enough, we shall compare u and v on the region $(-\infty,\delta)\times[0,\tau)$. From the continuity of u and (2) there exists $\tau > 0$ such that $u(\delta,t) \geq \theta$ for any $t \in [0,\tau]$. Then we shall choose $\bar{x}$ and K such that

(19) $$v(\delta,t;K,\bar{x}) \leq \theta \leq u(\delta,t) \quad \text{for } t \in [0,\tau]$$

and

(20) $$v(x,0;K,\bar{x}) \leq u_0(x) \qquad \text{for } x \in (-\infty,\delta).$$

In order to have (20) we take $K = K(\bar{x})$ and since

$$\sup_{x\in\mathbb{R}} v(x,0;K,\bar{x}) = bK^{1/(\lambda-1)},$$

using (15) it is enough to choose

(21) $$K = K(\bar{x}) = (C/b)^{(\lambda-1)}\bar{x}^{\gamma(\lambda-1)}.$$

On the other hand, by taking $\bar{x}$ small enough it is easy to see that (19) holds. Then, by the comparison principle we conclude that $v \leq u$ in $(-\infty,\delta)\times[0,\tau)$. Now a direct computation shows that $v(0,t;K(\bar{x}),\bar{x}) > 0$ if $t > t(\bar{x}) = (b/c)^{(\lambda-1)}\bar{x}^{1-\gamma(\lambda-1)}$. Since $\gamma < 1/(\lambda-1)$, it is clear that $t(\bar{x}) \to 0$ when $\bar{x} \to 0^+$. Now let $t_0 \in [0,\tau)$ and choose $\bar{x} = \beta t_0^{1/(1-\gamma(\lambda-1))}$ with $\beta > 0$ small. Then we have

$$\varsigma_-(t_0) \leq a_- - \left[\beta - (b/c)^{(\lambda-1)}\beta^{1-\gamma(\lambda-1)}\right] t_0^{1/(1-\gamma(\lambda-1))}.$$

Then, if β is large enough, we have $\beta > (b/c)^{(\lambda-1)}\beta^{1-\gamma(\lambda-1)}$. This proves the estimate (16). The fact that $u(a_-,t) > 0$ for any $t > 0$ again follows from the initial positivity of $u(a_-,t)$ and Theorem 2 of Gilding [**12**].

□

We shall now study the initial behavior of the right front $\varsigma_+(t)$. We start by giving a sufficient condition for the waiting time property.

THEOREM 5. *Let u be a weak solution of (CP) and suppose that*

$$u_0(x) \le C\,|x-a_+|^{1/(m-\lambda)} \quad \text{if } 0 \le a_+ - x \le \delta \tag{22}$$

for some positive constants δ and $C > 0$, $C < C_0$, where

$$C_0 = \left(\frac{b(m-\lambda)}{m}\right)^{1/(m-\lambda)}. \tag{23}$$

Then there exists a finite time $\tau^* > 0$ such that $u(a_+,t) = 0$ for every $t \in [0,\tau^*]$.

PROOF: We first remark that the function $z(x) = C_0[a_+ - x]_+^{1/(m-\lambda)}$ is a stationary solution of the equation. From (22) and the continuity of u we deduce that there exists $\tau^* > 0$ such that $u(a_+ - \delta, t) \le z(-\delta)$ for any $t \in [0,\tau^*]$. The conclusion follows from the comparison principle applied to the solutions u and z in the region $(a_+ - \delta, +\infty) \times [0,\tau^*]$.

□

The optimality of the assumption (22) is given by the following result, showing the expanding nature of $\varsigma_+(t)$ under an opposite hypothesis on u_0.

THEOREM 6. *Let u be a weak solution of (CP) and suppose that*

$$u_0(x) \ge C\,|x-a_+|^{1/(m-\lambda)} \quad \text{if } 0 \le a_+ - x \le \delta \tag{24}$$

for some $\delta > 0$ and $C > C_0$, C_0 given in (23). Then there exist positive constants τ and $\bar{c}$ such that

$$\varsigma_+(t) \ge a_+ + \bar{c}t^{(m-\lambda)/(m-2\lambda+1)}$$

for every $t \in [0,\tau]$. In particular, $u(a_+,t) > 0$ for every $t > 0$.

PROOF: Without loss of generality we may assume $a_+ = 0$. We define a traveling wave solution, depending on two parameters, in the following way:

$$w(x,t : K,\bar{x}) = \nu_K([Kt - x + \bar{x}]_+) \tag{26}$$

where ν_K is defined by

$$y = m\int_0^{\nu_K(y)} \frac{s^{m-2}}{K + bs^{\lambda-1}}\,ds. \tag{27}$$

Given τ small enough, we shall compare u and w on the region $(-\delta,\infty)\times[0,\tau)$. By the comparison principle it suffices to have

$$w(x,0;K,\bar{x})\leq u_0(x) \text{ for } x\in(-\delta,+\infty) \tag{28}$$

and

$$w(-\delta,t;K,\bar{x})\leq u(-\delta,t) \text{ for } t\in[0,\tau]. \tag{29}$$

As for (28) we remark that, by (24), it is equivalent to the following condition:

$$[\bar{x}-x]_+\leq m\int_0^{C|x|^{1/(m-\lambda)}}\frac{s^{(m-2)}}{K+bs^{(\lambda-1)}}\,ds \tag{30}$$

for $x\in(-\delta,\infty)$. But if $\alpha>0$ we have

$$\begin{aligned} m\int_0^{C|x|^{1/(m-\lambda)}} & \frac{s^{(m-2)}}{K+bs^{(\lambda-1)}}\,ds \\ &\geq m\int_\alpha^{C|x|^{1/(m-\lambda)}}\frac{s^{(m-2)}}{\frac{K}{\alpha^{(\lambda-1)}}s^{(\lambda-1)}+bs^{(\lambda-1)}}\,ds \\ &= \frac{m}{(m-\lambda)\left(\frac{K}{\alpha^{(\lambda-1)}}+b\right)}\left(C^{m-\lambda}|x|-\alpha^{m-\lambda}\right). \end{aligned}$$

Since $C>C_0$, there exists $\varepsilon>0$ such that

$$\frac{mC^{m-\lambda}}{(m-\lambda)(\varepsilon+b)}=1.$$

Then, choosing

$$K=K(\bar{x})=\varepsilon\left(\frac{(m-\lambda)(\varepsilon+b)}{m}\right)^{\frac{(\lambda-1)}{(m-\lambda)}}|\bar{x}|^{\frac{(\lambda-1)}{(m-\lambda)}}$$

and $\alpha=(K/\varepsilon)^{1/(\lambda-1)}$ we deduce (30). Finally, condition (29) is satisfied for $\bar{x}$ small enough (use the same argument as in the proof of Theorem 4). Then $w\leq u$ in $(-\delta,\infty)\times[0,\tau)$. A direct computation shows that $w(0,t;K(\bar{x}),\bar{x})>0$ if

$$t>t(\bar{x})=\frac{1}{\varepsilon}\left(\frac{m}{(m-\lambda)(\varepsilon+b)}\right)^{\frac{(\lambda-1)}{(m-\lambda)}}|\bar{x}|^{1-(\lambda-1)/(m-\lambda)}.$$

Since $1 > (\lambda-1)/(m-\lambda)$, then $t(\bar{x}) \to 0$ as $\bar{x} \to 0^-$. On the other hand, from the inequality $w \le u$ in $(-\delta,\infty)\times[0,\tau)$ we deduce that

$$\varsigma_+(t_0) \ge -|\bar{x}| + K(\bar{x})t_0 \text{ for any } t_0 \in [0,\tau).$$

Then, choosing $\bar{x} = -\beta t_0^{(m-\lambda)/(m-2\lambda+1)}$ with β small, we have that

$$\varsigma_+(t_0) \ge \left[-\beta + \varepsilon\left(\frac{(m-1)(\varepsilon+b)\beta}{m}\right)^{(\lambda-1)/(m-1)}\right] t_0^{(m-\lambda)/(m-2\lambda+1)}$$

Since $(\lambda-1)/(m-\lambda) < 1$ we conclude that the coefficient of $t_0^{(m-\lambda)/(m-2\lambda+1)}$ is positive if we choose β big enough and, since t_0 is arbitrary in $[0,\tau)$, (25) is proved. As before, this also proves that $u(a_+,t) > 0$ for any $t > 0$.

□

REMARK 2: When (λ, m) is in the region II the asymmetry caused by the convection is quite clear. So, in that case the previous results and the inequalities

$$\frac{1}{\lambda-1} \ge \frac{2}{m-1} \ge \frac{1}{m-\lambda}$$

show that, since the convection plays an important role, the condition on u_0 to have a waiting time in the interface $\varsigma_-(t)$ is stronger than for $\varsigma_+(t)$ and also stronger than $\varsigma_-(t)$ with (λ, m) belonging to the region I.

REMARK 3: Some other qualitative properties on ς_- and ς_+ are well-known when that $\lambda > 1$ and $m > 1$. Thus in Gilding [**10,12**] it is shown that $\varsigma_-(t)$ is monotone and nonincreasing and $\varsigma_-(t) \searrow -\infty$ as $t \nearrow +\infty$, $\varsigma_+(t)$ is monotone nondecreasing and $\varsigma_+(t) \nearrow +\infty$ if $\lambda \ge m$ or $\varsigma_+(t) \nearrow A_+$ if $\lambda < m$ for some real number A_+, as $t \nearrow +\infty$. He also proves that both interfaces satisfy the equation

$$\varsigma'_\pm(t) = \left\{-[(u^m)_x + bu^\lambda]/u\right\}(\varsigma_\pm(t),t)$$

in a certain sense.

4. Region III: Convection strongly dominating over diffusion.

In this last section we shall assume that $\lambda \le 1$ and $\lambda < m$. The first important difference compared to the cases in which (λ, m) belongs to regions I or II appears already for $\lambda = 1$. Indeed, in that case it is well-known that $m > 1$ implies the existence of the interfaces $\varsigma_-(t)$ and $\varsigma_+(t)$ (see Diaz-Kersner [**8**]), nevertheless, the following result shows that in this case $\varsigma_-(t)$ can never exhibit a waiting time.

THEOREM 7. *Let $m > 1$, $\lambda = 1$, and u be a weak solution of (CP). Then $\varsigma_-(t) \leq a_- - bt$ for all $t \geq 0$.*

PROOF: We introduce the transformation $v(x,t) = u(x - bt, t)$. Then it is easy to see that v satisfies the equation $v_t = (v^m)_\alpha (v^m)_{xx}$ and $v(x,0) = u_0(x)$. Moreover if $\xi_-(t)$ is the left face generated by v, i.e. $\xi_-(t) = \inf\{x : v(x,t) > 0\}$, then we have $\varsigma_-(t) = \xi_-(t) - bt$. Since the fronts for the porous media equation are nonincreasing the result follows at once.

□

When $\lambda < 1$ it turns out that the convection dominates diffusion in such a strong way that, in fact, the left interface $\varsigma_-(t)$ does not exist, i.e. $\inf\{x : u(x,t) > 0\} = -\infty$ for any $t > 0$. This result was first proved in Diaz-Kersner [8] for $m \geq 1$, and later, for any $m > 0$, in Gilding [**12**].

The behavior of $\varsigma_+(t)$ is completely different than that of $\varsigma_-(t)$ when (λ, m) is in the region III. Thus, this front does exist for any value of (λ, m) in that region (see Diaz-Kersner [8] for $\lambda < 1 \leq m$ and Gilding [**12**] for the general case $\lambda < m$). Besides, the stationary solution $z(x) = C_0[a_+ - x]_+^{1/(m-\lambda)}$, C_0 given by (23), shows that (CP) admits solutions with an infinite waiting time. The following result gives a stronger result ensuring that, under a suitable assumption on u_0, $\varsigma_+(t)$ is in fact a "reversing front" near $t = 0$ (the property $\varsigma_+(t) \searrow -\infty$ as $t \nearrow +\infty$ without any additional assumption on the initial value u_0, was first proved in Diaz-Kersner [8], see also Gilding [**12**,Theorem 5]).

THEOREM 8. *Let u be a weak solution of (CP) and suppose that*

$$u_0(x) \leq C\,|x - a_+|^{1/(m-\lambda)} \text{ if } 0 \leq a_+ - x \leq \delta, \tag{31}$$

for some positive constants δ and C, $C < C_0$, where C_0 is given by (23). Then there exist positive constants $\bar{c}$ and τ such that

$$\varsigma_+(t) \leq a_+ - \bar{c}t^{(m-\lambda)/(m-2\lambda+1)} \tag{32}$$

for all $t \in [0, \tau]$.

PROOF: We assume $a_+ = 0$ and introduce the following traveling wave solution

$$v(x,t;K) = \alpha_K([-x - Kt]_+) \tag{33}$$

where α_K is defined by

$$y = m\int_0^{\alpha_K(y)} \frac{s^{m-2}}{bs^{\lambda-1}-K}\,ds, \tag{34}$$

and K is a positive constant to be chosen such that $K < b$ if $\lambda = 1$.

In order to compare v with u in the region $(-\delta,+\infty)\times[0,\tau)$, with τ suitably chosen, we notice that the condition

$$v(x,0;K) \geq u_0(x) \text{ for } x \in (-\delta,+\infty) \tag{35}$$

is verified if

$$|x| \geq m\int_0^{C_*|x|^{1/(m-\lambda)}} \frac{s^{m-2}}{bs^{\lambda-1}-K}\,ds$$

for $-\delta < x \leq 0$ for any $C_* \in [C, C_0)$. But if $x \in (-\delta, 0]$ we have

$$\begin{aligned} m\int_0^{C_*|x|^{1/(m-\lambda)}} &\frac{s^{m-2}}{bs^{\lambda-1}-K}\,ds \\ &\leq m\int_0^{C_*|x|^{1/(m-\lambda)}} \frac{s^{m-2}}{bs^{\lambda-1}-K\left(\frac{s}{C_*|x|^{1/(m-\lambda)}}\right)^{(\lambda-1)}}\,ds \\ &= \frac{mC_*^{m-\lambda}}{(m-\lambda)\left(b-\frac{K}{C_*^{(\lambda-1)}}\delta^{(\lambda-1)/(m-\lambda)}\right)}|x|. \end{aligned}$$

Since $C_* < C_0$, there exist positive constants ε and ε' such that

$$1-\frac{mC_*^{(m-\lambda)}}{(m-\lambda)(b-\varepsilon)} = \varepsilon'.$$

Thus (35) holds if we take

$$K = \varepsilon C_*^{(\lambda-1)}\delta^{(\lambda-1)/(m-\lambda)}. \tag{36}$$

On the other hand, by continuity, we deduce from (31) that for any $C_* \in (C, C_0)$ there exists a $\tau^* > 0$ such that

$$u(-\delta,t) \leq C_*\delta^{1/(m-\lambda)} \quad \text{if} \quad t \in [0,\tau_*].$$

Then, choosing $\tau \leq \tau^*$ the condition

$$v(-\delta,t;K) \geq u(-\delta,t) \text{ for } t \in [0,\tau) \tag{37}$$

holds if

$$\alpha_K\left([\delta - Kt]_+\right) \geq C_* \delta^{1/(m-\lambda)}. \tag{38}$$

From the definition of α_K we deduce that (38) holds if $\delta - K\tau \geq (1-\varepsilon')\delta$, that is,

$$\tau \leq \varepsilon' \frac{\delta}{K} = \frac{\varepsilon'}{\varepsilon C_*^{\lambda-1}} \delta^{(m-2\lambda+1)/(m-\lambda)}. \tag{39}$$

Then, for τ small enough we deduce from the comparison principle that $v \geq u$ in $(-\delta, +\infty) \times [0, \tau)$. This shows that

$$\varsigma_+(t_0) \leq -\varepsilon C_*^{\lambda-1} \delta^{(\lambda-1)/(m-\lambda)} t_0 \quad \text{if } t_0 \in [0, \tau).$$

Obviously, the same arguments may be applied to any $\delta_0 \in (0, \delta)$. Thus, in particular, by choosing

$$\delta_0 = \left(\frac{\varepsilon C_*^{\lambda-1}}{\varepsilon'} t_0\right)^{(m-\lambda)/(m-2\lambda+1)}$$

we obtain the estimate (32) and the proof is finished.

□

Our last result shows that even when $\varsigma_+(t) \searrow -\infty$ as $t \nearrow +\infty$, the front $\varsigma_+(t)$ may expand initially under an assumption on u_0 which is the opposite to that in the above theorem.

THEOREM 9. *Let u be a weak solution of (CP) and suppose that*

$$u_0(x) \geq C\,|x - a_+|^{1/(m-\lambda)} \quad \text{if } 0 \leq a_+ - x \leq \delta, \tag{40}$$

for some positive constants δ and C, $C > C_0$, where C_0 is given by (23). Then there exist positive constant $\bar{c}$ and τ such that

$$\varsigma_+(t) \geq a_+ + \bar{c}t^{(m-\lambda)/(m-2\lambda+1)} \tag{41}$$

for all $t \in [0, \tau]$.

PROOF: As before, we can assume without loss of generality that $a_+ = 0$. We introduce the traveling wave solution defined in Theorem 6 but now for $\bar{x} = 0$, i.e.,

$$w(x, t; K) = \nu_K\left([Kt - x]_+\right)$$

where ν_K is defined by (27). To compare u with w we first remark that the condition

$$w(x,0;K) \leq u_0(x) \quad \text{for } x \in (-\delta,+\infty) \tag{42}$$

holds if

$$|x| \leq m \int_0^{C_*|x|^{1/(m-\lambda)}} \frac{s^{m-2}}{bs^{\lambda-1}-K}\,ds$$

for $-\delta < x \leq 0$ and any $C_* \in (C_0, C)$. But for $x \in (-\delta, 0]$, arguing as in Theorem 8 we have

$$m \int_0^{C_*|x|^{1/(m-\lambda)}} \frac{s^{m-2}}{bs^{\lambda-1}-K}\,ds \geq \frac{mC_*^{m-\lambda}}{(m-\lambda)\left(b+\frac{K}{C_*^{(\lambda-1)}}\delta^{(\lambda-1)/(m-\lambda)}\right)}|x|.$$

Since $C_* > C_0$, there exist positive constants ε and ε' such that

$$\frac{mC_*^{(m-\lambda)}}{(m-\lambda)(b+\varepsilon)} - 1 = \varepsilon.$$

Thus, taking

$$K = \varepsilon C_*^{(\lambda-1)}\delta^{(\lambda-1)/(m-\lambda)}.$$

condition (42) holds. On the other hand, from (40) and the continuity of u we deduce that for any $C_* \in (C, C_0)$ there exists a $\tau^* > 0$ such that

$$u(-\delta,t) \geq C_*\delta^{1/(m-\lambda)}.$$

Then, choosing $\tau \leq \tau^*$, the condition

$$w(-\delta,t;K) \leq u(-\delta,t) \quad \text{for } \; t \in [0,\tau) \tag{43}$$

holds if

$$\nu_K(K\tau+\delta) \leq C_*\delta^{1/(m-\lambda)}. \tag{44}$$

As before, (44) is verified when

$$\tau \leq \frac{\varepsilon'}{\varepsilon C_*^{(\lambda-1)}}\delta^{(m-2\lambda+1)/(m-\lambda)}. \tag{45}$$

It now follows from the comparison principle that $w \leq u$ in the region $(-\delta,\infty)\times[0,\tau]$ for τ small enough. This shows that

$$\varsigma_+(t_0) \geq \varepsilon C_*^{(\lambda-1)}\delta^{(\lambda-1)/(m-\lambda)}t_0 \quad \text{if } t_0 \in [0,\tau).$$

and the same conclusion holds if δ is replaced by any $\delta_0 \in [0, \delta]$. Then, choosing

$$\delta_0 = \left(\frac{\varepsilon}{\varepsilon'} C_*^{(\lambda-1)} t_0\right)^{(m-\lambda)/(m-2\lambda+1)}$$

we obtain the estimate (41) and this establishes the result.

□

Keywords. nonlinear diffusion-convection, degenerate second-order parabolic equation, interfaces, free boundaries.
1980 *Mathematics subject classifications*: 35K55, 35B99

The research of Prof. Diaz was partially sponsored by the CAICYT (Spain) project 3308/83.

The research of Prof. Kersner was sponsored by the AKA (Hungarian Academy of Sciences) project 13-86-261.

Departamento de Matematica Aplicada, Universidad Complutense de Madrid, 28040, SPAIN

Departamento de Matematica Aplicada, Universidad Complutense de Madrid, 28040, SPAIN

Computer and Automation Institute, Hungarian Academy of Sciences, Budapest, HUNGARY

References

1. L. Alvarez and J. I. Diaz, in preparation.
2. D. G. Aronson, *Regularity properties of flows through porous media: a counterexample*, SIAM J. Appl. Math. **19** (1970), 299–307.
3. J. Bear, "Dynamics of Fluids in Porous Media," American Elsevier Publishing Company, New York, 1972.
4. J. G. Berryman and C. J. Holland, *Stability of the separable solution for fast diffusion*, Arch. Rat. Mech. Anal. **74** (1980), 379–388.
5. J. Buckmaster, *Viscous sheets advancing over dry beds*, J. Fluid Mech. **81** (1977), 735–756.
6. S. Chandrasekhar, *Stochastic problems in physics and astronomy*, Rev. Modern Phys. **15** (1943), 1–89.
7. J. I. Diaz and R. Kersner, *On a nonlinear degenerate parabolic equation in infiltration or evaporation through a porous medium*, J. Diff. Eqs. **69** (1987), 368–403. (A preliminary version appears as Technical Summary Report #2502 of the MRC, University of Wisconsin, Madison, 1983).
8. J. I. Diaz and R. Kersner, *Nonexistence d'une des frontières libres dans une équation dégénérée en théorie de la filtration*, C. R. Acad. Sci. Paris **296** (1983), 505–508.
9. J. I. Diaz and R. Kersner, *On the behavior and cases of nonexistence of the free boundary in a semibounded porous medium*, to appear in J. Math. Anal. and Applic.
10. B. H. Gilding, *Properties of solutions of an equation in the theory of infiltration*, Arch. Rat. Mech. Anal. **65** (1977), 203–225.
11. B. H. Gilding, *Improved theory for a nonlinear degenerate parabolic equation*, Twente Univ. of Technology Dept. of Applied Math. Memorandum 587 (1986).
12. B. H. Gilding, *The occurrence of interfaces in nonlinear diffusion-advection processes*, Twente Univ. of Technology Dept. of Applied Math. Memorandum 595, to appear in Arch. Rat. Mech. Anal. (1987).
13. R. E. Grundy, *Asymptotic solution of a model nonlinear convective diffusion equation*, IMA Journal of Appl. Math. **31** (1983), 121–137.
14. B. F. Knerr, *The porous medium equation in one dimension*, Trans. Amer. Math. Soc. **234** (1977), 381–415.
15. J. R. Phillip, *Evaporation and moisture and heat fields in the soil*, J. of Meteorology **14** (1957), 354–366.
16. P. Rosenau and S. Kamin, *Thermal waves in an absorbing and convective medium*, Physica **8D** (1983), 273–283.
17. J. L. Vazquez, *The interface of one-dimensional flows in porous media*, Trans. Amer. Math. Soc. **285** (1984), 717–737.

Large Time Asymptotics for the Porous Media Equation

SIGURD ANGENENT

§1. Introduction.

In this note I compute asymptotic expansions for certain solutions of the degenerate parabolic partial differential equation

$$u_t = uu_{xx} + \gamma(u_x)^2. \tag{1.1}$$

Here $\gamma > -\frac{1}{2}$ is a real parameter.

For $\gamma > 0$ this equation may be rewritten as the porous media equation. Indeed $U = u^\gamma$ satisfies

$$U_t = \tfrac{1}{m}(U^m)_{xx}$$

where $m = 1 + \gamma^{-1}$.

For $\gamma < 0$, (1.1) may be rewritten as

$$V_t = V^n V_{xx}$$

where $V = u^{1+\gamma}$ and $n = (1+\gamma)^{-1}$.

For all $\gamma > -\frac{1}{2}$, (1.1) has a *similarity solution*, given by

$$u(x,t) = \frac{1}{4\gamma+1} t^{-\frac{1}{2\gamma+1}} U\left(x \cdot t^{-\frac{\gamma}{2\gamma+1}}\right) \tag{1.2}$$

in which $U(\xi) = \max(0, \frac{1}{2}(1-\xi^2))$. The two curves

$$x = \pm t^{\frac{\gamma}{2\gamma+1}}$$

are called the *free boundaries* of the solution. The form of the solution (1.2) suggests that we look for "nearby" solutions of the form

$$u(x,t) = U\left(\frac{x - M(t)}{R(t)}, t\right) \tag{1.3}$$

where $U(\xi,t)$ is defined on the strip

$$Q_T = \{(\xi,t) : -1 \le \xi \le 1, 1 \le t \le T\}$$

(we shall find it convenient to take $t = 1$ as the origin on the time axis).

More specifically we shall look for triples (U, M, R) which satisfy

(a) $U, U_t, U_\xi, U_{\xi\xi}, U_{\xi\xi\xi}$ *are Hölder continuous on* Q_T, *and* $U > 0$ *on* $\overset{o}{Q}_T$.

(b) u *defined by (1.3) satisfies the equation (i.e. (1.1)) on the domain* $\{(x,t) : |x - M(t)| < R(t), 1 < t < T\}$.

(c) $U(\pm 1, t) \equiv 0$ *and* $U_\xi(\pm 1, t) \ne 0$ *for all* $t \in [1, T]$.

(d) $M(t)$ *and* $R(t)$ *are* C^1 *for* $1 \le t \le T$.

In the case $\gamma > 0$, when (1.1) can be reduced to the porous media equation, any u given by (1.3) with (U, M, R) as above, is a "weak solution" in the usual sense. For $\gamma \leqq 0$ weak solutions have not (yet) been defined so we take the above as definition of the solution.

In [**1**,**3**] we showed that solutions in this sense exist locally in time for "nice" initial data. More precisely:

THEOREM A. *Let* $M_0 \in \mathbb{R}$, $R_0 > 0$ *and* $U_0(\xi) \in C^{3+\alpha}([-1, +1])$ *be given. If* $U_0(\pm 1) = 0$, $U_0'(\pm 1) \ne 0$ *and* $U_0(\xi) > 0$ *for* $-1 < \xi < +1$, *then there is a* $T > 1$ *and a solution* (U, M, R) *of (a,b,c,d) which also satisfies the initial condition*

$$M(1) = M_0, \quad R(1) = R_0$$
$$U(\xi, 1) = U_0(\xi) \quad (-1 \le \xi \le +1)$$

In addition we showed in [**3**] that the $M(\cdot)$ and $R(\cdot)$ and $U(\xi, \cdot)$ are real analytic functions of time for $1 < t \le T$. Furthermore, when $\gamma > 0$ the $C^{3+\alpha}$ may be replaced by $C^{2+\alpha}$.

Our main result in this note is:

THEOREM B. *Let* M_0, $R_0 > 0$ *be given and let* $U_0 \in C^{3+\alpha}([-1, +1])$ *be* $C^{3+\alpha}$ *close to* $(1 - \xi^2)$. *Then the solution of Theorem A exists for all time (i.e.* $T = +\infty$*) and* $R(t)$ *has the asymptotic expansion*

$$R(t) \sim t^{\frac{\gamma}{2\gamma+1}} \left(R_\infty + \sum_\alpha r_\alpha(\log t) t^{\frac{\alpha}{2\gamma+1}} \right)$$

as $t \to \infty$. *Here* R_∞ *is a constant, the* r_α *are polynomials in* $\log t$ *and the sum is taken over all* α *of the form*

$$\alpha = -\lambda_{n_1} - \lambda_{n_2} - \lambda_{n_3} - \cdots - \lambda_{n_k}$$

with $n_1, \ldots, n_k \geq 2$ and $\lambda_n = \frac{1}{2}n(n + 2\gamma - 1)$. For $\gamma > 0$, $C^{3+\alpha}$ is to be replaced by $C^{2+\alpha}$.

We also obtain similar expansions for $M(t)$ and $U(\xi, t)$ as $t \to \infty$ (see section 6). Furthermore we find that, if γ is an irrational number, then the polynomials $r_\alpha(\log t)$ are actually constants.

Briefly, our method is to produce a local semiflow on a Banach space, related to the equation (1.1). This semiflow has a fixed point V corresponding to the similarity solution. The fixed point turns out to be exponentially attracting, so that orbits close to V exist for all time. Using Laplace transform techniques we then obtain asymptotic expansions for orbits close to V, and these expansions can be turned into expansions for (U, M, R).

In [5] Zeldovitch also computed asymptotic expansions for porous media type equations. The computations in [5] are purely formal, however. The numbers λ_k are computed, but not the numbers α. The logarithmic terms in the expansion are also missing in [5]. This is to be expected since Zeldovitch in fact computes asymptotic expansions for a linearized equation, and the logarithmic terms are due to nonlinear phenomena ("resonance," see sections 7 and 8 of this note).

Finally, J. L. Vasquez has informed me that he and D. G. Aronson have proved that for $\gamma > 0$ any solution (in the weak sense) gets $C^{2+\alpha}$ close to $(1 - \xi^2)$, after rescaling, and provided that the initial data has compact support. This means that Theorem B can be applied to all such solutions.

ACKNOWLEDGEMENTS: I would like to thank S. Kamin for pointing out the work [5] of Zeldovitch, and for taking the trouble of translating the Russian text.

§2. A semiflow with a fixed point.

If $u(x,t) = U(\frac{x - M(t)}{R(t)}, t)$ satisfies (1.1) then we get the following equation for U:

$$U_t = R^{-2} \cdot \{UU_{\xi\xi} + \gamma(U_\xi)^2 + R(M_t + \xi R_t) \cdot U_\xi\}.$$

the condition "$U = 0$ at $\xi = \pm 1$" gives us equations for M_t and R_t. We get the following system for the triple (U, M, R):

$$
\begin{aligned}
U_t &= R^{-2}\{UU_{\xi\xi} + \gamma P(U_\xi)\cdot U_\xi\} \\
R_t &= -\frac{\gamma}{2R}\left(U_\xi(1,t) - U_\xi(-1,t)\right) \\
M_t &= -\frac{\gamma}{2R}\left(U_\xi(1,t) + U_\xi(-1,t)\right)
\end{aligned} \tag{2.1}
$$

Here we have used $P(f)$ to denote

$$P(f) = f(\xi) - f(1)\frac{1+\xi}{2} - f(-1)\frac{1-\xi}{2}.$$

The system (2.1) governs the evolution of (U, M, R).

We claim that

$$A = R\cdot\int_{-1}^{+1} U^\gamma d\xi \tag{2.2}$$

is an integral of the equations (2.1). Indeed, for $\gamma > 0$, A is the total mass associated with a solution of the porous media equation. In [**2,3**] we showed that the solution (U, M, R) depends analytically on parameters (such as γ). Hence, by analytic continuation we see that A is also time independent if $-\frac{1}{2} < \gamma \leq 0$. Of course, a direct calculation of A_t will also show that $A_t \equiv 0$.

Anyhow, we can use this to eliminate M and R, and get a single equation for U alone:

$$U_t = A^{-2}\left(\int_{-1}^{+1} U^\gamma d\xi\right)^2 \left(UU_{\xi\xi} + \gamma P(U_\xi)U_\xi\right) \tag{2.3}$$

Finally, because we want the similarlity solution to become a fixed point (or a stationary solution of the equation) we introduce

$$
\begin{aligned}
v &= t^{\frac{1}{2\gamma+1}}U \\
\tau &= \log(t)/(1+2\gamma).
\end{aligned}
$$

The new variable v satisfies

$$v_\tau = \frac{2\gamma+1}{A^2}\left(\int_{-1}^{+1} v^\gamma d\xi\right)^2 \left(vv_{\xi\xi} + \gamma P(v_\xi)v_\xi\right) + v. \tag{2.4}$$

This equation has $K(1-\xi^2) = V$ as a stationary solution, where

$$2K^{2\gamma+1} = \frac{A^2}{2\gamma+1}\left(\int_{-1}^{+1}(1-\xi^2)^\gamma d\xi\right)^2$$

(the integral can be computed using Euler's Γ function).

We shall study the initial value problem for (2.4) and use a linearization procedure to obtain asymptotic expansions of the solution as $\tau \to \infty$.

§3. A new space coordinate.

The equation (2.4) is degenerate at $\xi = \pm 1$, since our boundary conditions require $v = 0$ at $\xi = \pm 1$. In fact near $\xi = 1$ the highest order term in (2.4) will behave like $(1-\xi)v_{\xi\xi}$. If we choose a coordinate y such that $y^2 = 1 - \xi + \ldots$ (near $\xi = 1$) then the highest order term becomes v_{yy}, i.e. nondegenerate. A coordinate which has this behavior at $\xi = 1$ and at -1 is given by

$$\xi = \cos y.$$

From here on we regard any function of ξ as a 2π-periodic and even function of y.

Define the Banach spaces

$$E_s = \left\{ v \in h^s(\mathbb{R}) : v(y) \equiv v(-y) \equiv v(y + 2\pi) \text{and } v(0) = v(\pi) = 0 \right\}$$

for any non-integer $s > 0$. Here $h^s(\mathbb{R})$ denotes the little Hölder space of exponent s. For $s > 2$ define $\mathcal{O}_s$ to be the subset of E_s containing those v for which

$$v(y) > 0, \text{ if } 0 < y < \pi$$
$$v_{yy}(0) > 0, \quad v_{yy}(\pi) > 0.$$

Then $\mathcal{O}_s$ is an open subset of E_s. On $\mathcal{O}_s$ we define

$$F(v) = \frac{2\gamma + 1}{A^2} \left(\int_{-1}^{+1} v^\gamma d\xi \right)^2 (v v_{\xi\xi} + \gamma P(v_\xi) v_\xi) + v. \tag{3.1}$$

Just as in [**3**] one now verifies that F is a real analytic mapping from $\mathcal{O}_s$ to E_{s-2}, and that if $\lfloor s \rfloor - 2 + \gamma > 0$, then the equation (2.4), rewritten as $v_t = F(v)$ generates a real analytic local semiflow on $\mathcal{O}_s$ ($\lfloor s \rfloor$ denotes the integer part of s). We denote this semiflow by $\{\varphi_\tau\}_{\tau \geq 0}$. Thus, for each $v_0 \in \mathcal{O}_s$ there is a $\tau(v_0) > 0$ such that $v(\tau) = \varphi_\tau(v_0)$ $(0 \leq \tau < \tau_0)$ is the maximal solution of (2.4).

§4. A linearized equation and its spectrum.

Recall that we have a fixed point $V = K(1-\xi^2)$ of the semiflow $\{\varphi_\tau\}_{\tau\geq 0}$, which corresponds to the similarity solution (1.2). The Fréchet derivative $D\varphi_\tau(V) : E_s \to E_s$ may be computed by the following recipe. For $w_0 \in E_s$ we have $w_1 = D\varphi_\tau(V)\cdot w_0$ if and only if $w_1 = w(\tau)$ where $w : [0,\tau] \to E_s$ is the solution of

$$\begin{aligned} w'(\tau) &= Lw(\tau) \\ w(0) &= w_0 \end{aligned}$$

and $L = DF(V) : E_s \to E_{s-2}$ is the Fréchet derivative of F at V. At a formal level this should be clear. That this description of $D\varphi_\tau(V)$ actually holds follows from our results in [**2**,**3**] (indeed, in [**3**] we show that $L \in MR(E_s, E_{s-2})$ so that we may apply Theorem 2.2 of [**2**]).

Thus we have to compute $DF(V)$. Substituting $v = V + w$ in (3.1) and computing the terms of first order in w leads to

$$\begin{aligned} L(w) = \frac{2\gamma+1}{A^2}\left(\int V^\gamma d\xi\right)^2 (Vw_{\xi\xi} + \gamma V_\xi P(V_\xi) + V_{\xi\xi}w) + w \\ + \frac{2\gamma+1}{A^2} 2\gamma \int V^\gamma d\xi \int V^{\gamma-1} w d\xi \, (VV_{\xi\xi} + \gamma V_\xi P(V_\xi)) \end{aligned}$$

(we have already used $P(V_\xi) = 0$). Further simplifications lead to

$$Lw = \tfrac{1}{2}(1-\xi^2)w_{\xi\xi} - \gamma\xi P(w_\xi) - 2\gamma \frac{\int V^{\gamma-1} w\, d\xi}{\int V^\gamma\, d\xi}\cdot V$$

(all integrals are to be taken over the interval $(-1,+1)$).

Next we compute the spectrum of L. Observe that if X is the vector space of all polynomials $p(\xi)$ which vanish at $\xi = \pm 1$ (i.e. are divisible by $(1-\xi^2)$) then L maps X into X. In fact if X_k denotes all polynomials in X of degree $\leq k$ then we still have $L(X_k) \subset X_k$. Choose a sequence of polynomials $p_2, p_3, p_4, \ldots$ in X such that $\{p_2, \ldots, p_k\}$ is a basis of X_k (e.g. $p_k = \xi^k - \xi^{k-2}$ will do). Then the matrix of L with respect to this system $\{p_n\}$ will be upper triangular. A short computation shows that

$$L_{p_n} = -\tfrac{n}{2}(n + 2\gamma - 1)\cdot p_n + \text{ lower order terms}$$

holds for $n \geq 2$.

It follows that the numbers $-\lambda_n = -\frac{n}{2}(n+2\gamma-1)$ for $n \geq 2$ are eigenvalues of the operator L on X. Since the λ_n are pairwise different these eigenvalues are all simple, and to each $n \geq 2$ there corresponds a polynomial $q_n \in X$ of degree n such that

$$Lq_n = -\lambda_n q_n = -\frac{n}{2}(n+2\gamma-1)q_n.$$

We claim that the $\{-\lambda_n\}_{n\geq 2}$ exhaust the whole spectrum of L. To see this, observe that the inclusion $E_s \subset E_{s-2}$ is compact so that L has only point spectrum, and all eigenvalues of L are isolated and of finite algebraic multiplicity. Let μ be an eigenvalue of L and assume $\mu \neq -\lambda_n$ for all $n \geq 2$. For such a μ we can find a splitting of the space $E_s : E_s = M \oplus N$ where M and N are two closed subspaces. One of these spaces (say M) contains the eigenvector belonging to μ, and the other (N) will contain (at least) the other eigenfunctions $\{q_n\}_{n\geq 2}$ (the projection on M with kernel N is given by the residue of the resolvent of L at μ). But now we have a contradiction: the span of the q_n is dense in E_s but they lie in the closed subspace N.

Our conclusion is that the spectrum of L consists of the simple eigenvalues $\{-\lambda_n\}$ for $n \geq 2$.

§5. Linearized stability.

Since $\sigma(L) = \{-\frac{n}{2}(n+2\gamma-1)\}_{n\geq 2}$ and $D\varphi_\tau(V)$ is given by $\exp(\tau L)$ we see that

$$\sigma(D\varphi_\tau(V)) = \left\{\exp(-\frac{1}{2}n(n+2\gamma-1)\tau)\right\}_{n\geq 2}.$$

By our assumption $\gamma > -\frac{1}{2}$ this implies that the spectral radius of $D\varphi_\tau(V)$ satisfies:

$$r(D\varphi_\tau(V)) = e^{-\lambda_2\tau} = e^{-\frac{1}{2}(2\gamma+1)\tau} < 1.$$

Therefore $v = V$ is a hyperbolic attracting fixed point of the semiflow φ_τ on E_s. Thus there is a neighborhood $\mathcal{N}$ in E_s of V such that for all $v \in \mathcal{N}$ one has

$$\|\varphi_\tau(v) - V\|_{E_s} \leq Ce^{-\varepsilon\tau} \quad (0 \leq \tau < \infty)$$

for some $C > 0$ and $0 < \varepsilon < \lambda_2$.

To construct such a neighborhood one chooses a $\tau_* > 0$ for which

$$\|D\varphi_{\tau_*}(V)\| < e^{-\varepsilon\tau_*}.$$

In view of the formula $r = \lim_{\tau\to\infty} \|D\varphi_\tau(V)\|^{1/\tau}$ for the spectral radius, this can be done. By continuity there is a ball $B = \{v : \|V - v\| < \delta\}$ such that for all $v \in B$ we still have $\|D\varphi_{\tau_*}(v)\| \leqq \exp(-\varepsilon\tau_*)$. Also by continuity there is a neighborhood $\mathcal{N}$ of V such that $\varphi_\tau(\mathcal{N}) \subset B$ for $0 \le \tau \le \tau_*$. Moreover we can choose $\mathcal{N}$ so small that $\|D\varphi_\tau(v)\| \leqq Ce^{-\varepsilon\tau_*}$ for all $v \in \mathcal{N}$ and $0 \le \tau \le \tau_*$. Finally, if $\mathcal{N}$ is a ball in the norm $\|\cdot\|_{E_s}$ around V then we have $\varphi_{\tau_*}(\mathcal{N}) \subset \mathcal{N}$, because φ_{τ_*} is a contraction on B.

It follows from this construction that for all $v \in \mathcal{N}$ and $\tau > 0$ one has $\|D\varphi_\tau(v)\| \le Ce^{-\varepsilon\tau}$, which implies the exponential stability of $v = V$.

Let $v(\tau)$ be $\varphi_\tau(v)$. Then it follows from $v'(\tau + \sigma) = D\varphi_\sigma(v(\tau)) \cdot v'(\tau)$ that $\|v'(\tau)\|_{E_s}$ also decays exponentially (just let $\sigma \nearrow \infty$). Repeated differentiation with respect to τ shows that all higher derivatives of $v(t)$ decay exponentially (with the same rate, $e^{-\varepsilon\tau}$).

§6. Asymptotic expansions.

Let $v \in E_s$ be close to V, i.e. let v belong to the neighborhood $\mathcal{N}$ defined above. Consider $w(\tau) = \varphi_\tau(v) - V$. Then $w(\tau)$ is a solution of

$$w'(\tau) = F(V + w(\tau)) - F(V) = F(V + w(\tau)).$$

Since F is a real analytic we may expand this in a power series:

$$w'(\tau) = Lw(\tau) + \sum_{k\ge 2} L_k(w(\tau), \dots, w(\tau)). \tag{6.1}$$

where $L_k : E_s \times \dots \times E_s \to E_{s-2}$ is a bounded k-linear mapping. We shall assume that the neighborhood $\mathcal{N}$ is so small that the series in (6.1) converges uniformly on $\mathcal{N}$. We shall also write the higher-order terms as

$$N(w) = \sum_{k\ge 2} L_k(w, \dots, w) = F(V + w) - Lw.$$

To obtain asymptotic expansion of w we compute its Laplace transform; more precisely we show that its Laplace transform is a meromorphic function. The location and nature of its poles will then tell us what kind of terms occur in the expansion of $w(\tau)$.

It will be convenient to study

$$W(\tau) = \eta(\tau)\cdot w(\tau) \text{ if } \tau \ge 0$$
$$0 \text{ if } \tau \le 0$$

where $\eta \in C^\infty(\mathrm{R})$ is zero for $\tau \le 1$ and $\eta(\tau) = 1$ for $\tau \ge 2$. Clearly W and w have the same asymptotic behavior. Since $W - w$ has compact support their Laplace transforms differ by an entire function.

For W we have the equation

$$W'(\tau) - LW(\tau) = N(W(\tau)) + f(\tau) \tag{6.2}$$

where $f(\tau) = N(w(\tau)) - N(W(\tau)) + \varphi'(\tau)w(\tau)$. The regularity theory in [**2,3**] implies that w is real analytic for $\tau > 0$, so that $f(\tau)$ is a C^∞ function of compact support, and its Laplace transform is an entire function with values in E_{s-2}.

If we denote the Laplace transform of a function $g(t)$ by $\tilde{g}(\lambda)$, so that

$$\tilde{g}(\lambda) = \int_0^\infty e^{-\lambda t} g(t)\, dt,$$

then (6.2) implies

$$(\lambda - L)\tilde{W}(\lambda) = N(W)^\sim + \tilde{f}(\lambda)$$

i.e.,

$$\tilde{W}(\lambda) = (\lambda - L)^{-1}\left[N(W)^\sim + \tilde{f}(\lambda)\right] \tag{6.3}$$

Let the set $\mathcal{A}$ be defined by

$$\mathcal{A} = \{-(\lambda_{k+1} + \lambda_{k_2} + \cdots + \lambda_{k_j}) \mid k_1, \ldots, k_j \ge 2\}.$$

For any Banach space X and real number $a > 0$ we define $\mathcal{M}_a(X)$ to be the set of X valued functions which are meromorphic on the region $\{\lambda \in C : \operatorname{Re}\lambda > -a\}$, whose poles lie in $\mathcal{A}$ and which also satisfy the following estimates:

$$\lim_{\substack{z\to\infty \\ \operatorname{Re} z \ge -a+\varepsilon}} |z|^m \|f(z)\|_X = 0$$

for any $\varepsilon > 0$ and $m \geq 0$.

Suppose $g_0 \in C^\infty(\mathbb{R}, X)$ has support in $\mathbb{R}_+$ and all derivatives of g_0 decay exponentially:

$$\lim_{\tau\to\infty} e^{(a-\varepsilon)\tau} \left\| g_0^{(m)}(\tau) \right\|_X = 0$$

for all $m \geq 0$ and $\varepsilon < 0$. Moreover, let $p_1(\tau), \ldots, p_k(\tau)$ be X-valued polynomials in τ. Then the Laplace transform of

$$g(\tau) = g_0(\tau) + \left(p_1(\tau)e^{\alpha_1\tau} + \cdots + p_k(\tau)e^{a_k\tau}\right) \cdot \eta(\tau) \tag{6.4}$$

(where the α_j belong to $\mathcal{A}$) belongs to $\mathcal{M}_a$. Conversely, any function g with $\tilde{g} \in \mathcal{M}_a$ is of this form. These are well-known facts from the theory of Fourier-Laplace transforms.

The following lemma follows directly from this description of $\mathcal{M}_a$.

LEMMA. *Let $\tilde{W}$ belong to $\mathcal{M}_a(E_s)$. If $a < \lambda_2$ then $N(W)^\sim \in \mathcal{M}_{2a}(E_{s-2})$. If $a \geq \lambda_2$ then $N(W)^\sim \in \mathcal{M}_{\alpha+\lambda_2}(E_{s-2})$.*

PROOF: Assume W has the form (6.4) (i.e. $W = g$). Then for $\varepsilon < \min(a, \lambda_2)$ we have

$$\left\| g^{(n)}(\tau) \right\|_{E_s} \leqq C_n e^{-\varepsilon\tau}$$

for $n = 0, 1, 2, \ldots$. Consider the (finite) Taylor's series of N at $v = 0$:

$$N(v) = L_2(v) + \cdots + L_\ell(v) + R_\ell(v),$$

where ℓ is chosen so large that $\ell\varepsilon > \max(2a, a + \lambda_2)$. Then $R_\ell(g(\tau))$ is C^∞ and decays exponentially. In fact we have for each $n \geq 0$:

$$(d/d\tau)^n R_\ell(g(\tau)) = \sum c_{\ell_1\ldots\ell_k} R_\ell^{(k)}(g(\tau)) g^{(\ell_1)} \ldots g^{(\ell_k)}$$

for some integers $c_{\ell_1\ldots\ell_k}$. Using this formula one proves that all derivatives of $R_\ell(g(\tau))$ decay like $e^{\ell\varepsilon\tau}$, so that $R_\ell(g(\tau))^\sim$ belongs to $\mathcal{M}_b(E_{s-2})$ where $b = \max(2a, a + \lambda_2)$.

Next consider $L_m(g(\tau))$ with $m \leq \ell$ and assume that $a > \lambda_2$. Inserting (6.4) in $L_m(g(\tau))$ and using the m-linearity of L_m we find that L_m is a finite sum of terms of the form

$$\eta(\tau)^n \cdot L_m\left(g_0(\tau), \ldots, g_0(\tau), p_{j_1} e^{a_{j_1}\tau}, \ldots, p_{j_n} e^{a_{j_n}\tau}\right)$$

with $n \leqq m$. In case $n < m$, so that one factor $g_0(\tau)$ is present one finds that this term decays faster than $\exp((\varepsilon - a - \lambda_2)\tau)$ for any $\varepsilon > 0$. In case

$n = m$ one obtains a term which is of the form $\eta^n\cdot$ polynomial $\cdot e^{\beta\tau}$, with $\beta \in \mathcal{A}$. Therefore $L_m(g(\tau))^\sim \in \mathcal{M}_{a+\lambda_2}(E_{s-2})$.

The case $a \le \lambda_2$ can be dealt with in a similar way.

We return to (6.3).

Since we know that $\tilde{W}$ belongs to $\mathcal{M}_\varepsilon(E_s)$ for some $\varepsilon > 0$ the lemma implies that $N(W)^\sim$ belongs to $M_{2\varepsilon}(E_{s-2})$ if $\varepsilon \le \lambda_2$, and therefore $(\lambda - L)^{-1}[N(W)^\sim + \tilde{f}]$ belongs to $\mathcal{M}_{2\varepsilon}(E_s)$. So, if $\tilde{W}$ belongs to $\mathcal{M}_\varepsilon(E_s)$ and $\varepsilon \le \lambda_2$, then $\tilde{W}$ also belongs to $\mathcal{M}_{2\varepsilon}(E_s)$. Repeating this argument we find that $\tilde{W}$ belongs to $\mathcal{M}_{2^n\varepsilon}(E_s)$ where n is the least integer such that $2^n\varepsilon > \lambda_s$. Again using the lemma and relation (6.3) we see that $\tilde{W} \in \mathcal{M}_a(E_s) \Rightarrow \tilde{W} \in M_{a+\lambda_2}(E_s)$. The final conclusion is that $\tilde{W}$ belongs to $\mathcal{M}_a(E_s)$ for arbitrary large $a > 0$. In other words, the Laplace transform $\tilde{W}(\lambda)$ of $W(\tau)$ is a meromorphic E_s valued function whose poles lie in the set $\mathcal{A}$.

Therefore $W(\tau)$ and $w(\tau)$ have an asymptotic expansion of the form

$$w(\tau) \sim \sum_{\alpha\in\mathcal{A}} p_\alpha(\tau)e^{\alpha\tau}$$

where the $p_\alpha(\tau)$ are E_s valued polynomials in τ.

If we trace the substitutions which led us from u (in (1.1)) to w, then we see that U of (2.1) has the expansion

$$U(t) \sim t^{-\frac{1}{2\gamma+1}}\left(V + \sum_{\alpha\in\mathcal{A}} p_\alpha(\log t, \xi)t^{\frac{\alpha}{2\gamma+1}}\right). \tag{6.5}$$

Substituting this into (2.2) gives an expansion for $R(t)$:

$$R(t) \sim \tau^{\frac{\gamma}{2\gamma+1}}\left(R_0 + \sum_{\alpha\in\mathcal{A}} r_\alpha(\log t)\cdot t^{\frac{\alpha}{2\gamma+1}}\right), \tag{6.6}$$

where the r_α are polynomials, and R_0 can be computed in terms of A. Finally, using these expansions and (2.1) again we get an expansion for M_t which upon integration yields

$$M(t) \sim M_\infty + t^{\frac{\gamma}{2\gamma+1}}\left(M_0 + \sum_{\alpha\in\mathcal{A}} m_\alpha(\log t)t^{\frac{\alpha}{2\gamma+1}}\right). \tag{6.7}$$

This last expansion can be improved, as we shall see in the next section.

§7. Computing the coefficients (resonance).

We compute the principal parts of $\tilde{W}$ at its poles. Its first pole is $\lambda = -\lambda_2$. Since $\tilde{W}$ is holomorphic on $\operatorname{Re}\lambda > -\lambda_2$, arguments as in the lemma of Section 6 show that $N(W)^\sim$ is holomorphic on $\operatorname{Re}\lambda > -\lambda_2 - \varepsilon$ for some $\varepsilon > 0$. The function $\tilde{f}(\lambda)$ is entire, so that $\tilde{W} = (\lambda - L)^{-1}(N(W)^\sim + \tilde{f}(\lambda))$ has at most a simple pole at $\lambda = \lambda_2$. Its residue at $-\lambda_2$ is exactly the eigenfunction q_2 of L, up to a scalar multiple.

Since $q_2(\xi) = c\cdot(1-\xi^2)$ is a multiple of V we see that the first unspecified term in (6.5) is

$$p_{\lambda_2}(\log t, \xi) = c\cdot(1-\xi^2)$$

for some $c \in \mathbb{R}$. If one now does the calculation that led to (6.7) one finds that the second term M_0 vanishes.

The fact that $\tilde{W}(\lambda)$ has a simple pole at λ_2 implies that the polynomial $p_{\lambda_2}(\tau)$ is actually a constant.

If we go on to compute the next term we find that there are three possibilities. The largest number in $\mathcal{A}\backslash\{-\lambda_2\}$ is either $-2\lambda_2$ or $-\lambda_3$.

If $-2\lambda_2 > -\lambda_3$, then $N(W)^\sim$ will have a simple pole at $-2\lambda_2$, and since $(\lambda - L)^{-1}$ is regular at $-2\lambda_2$ so will $\tilde{W}$. In this case the next term in the expansion of $U(t,\xi)$ will be

$$t^{\frac{\gamma}{2\gamma+1}} f(\xi)\cdot t^{-\frac{2\lambda_2}{2\gamma+1}}$$

where $f(\xi)$ can, in principle, be computed.

If $-2\lambda_2 < -\lambda_3$, then $N(W)^\sim$ will be regular at $\lambda = -\lambda_3$, but $(\lambda - L)^{-1}$ has a simple pole, so that $\tilde{W}(\lambda)$ will have a simple pole at $-\lambda_3$. The residue of $\tilde{W}$ at $-\lambda_3$ is a multiple of the eigenfunction q_3 of L. Since $q_3(\xi) = \xi(1-\xi^2)$ we get

$$c\cdot t^{\frac{\gamma-\lambda_3}{2\gamma+1}}\xi(1-\xi^2)$$

as the second term in the expansion (6.5).

The third possibility is that $2\lambda_2 = \lambda_3$ (this happens when $\gamma = 1$). In this case $N(W)^\sim$ has a simple pole at $-2\lambda_2$, so that $(\lambda - L)^{-1}(N(W)^\sim + \tilde{f}(\lambda))$ will have a double pole at $-2\lambda_2 = -\lambda_3$. When this happens the second term in (6.5) will have the form

$$t^{\frac{\gamma-\lambda_3}{2\gamma+1}}(f(\xi) + g(\xi)\log t),$$

where $f(\xi)$ and $g(\xi)$ are again in principle computable.

Clearly one can compute all other terms by continuing this procedure. One sees that $\tilde{W}(\lambda)$ can only have a double pole if one of the eigenvalues λ_k can be written as

$$\lambda_k = \lambda_{n_1} + \cdots + \lambda_{n_j}$$

for certain $n_1, \ldots, n_j \geq 2$. This phenomenon is called *resonance.*

Using the explicit form $\lambda_k = \frac{1}{2}k(k+2\gamma-1)$ one can show that resonance occurs if and only if γ is rational.

Thus, if γ is irrational all the polynomials in the expansions (6.5), (6.6) and (6.7) are independent of time.

§8. Final comments.

We have computed asymptotic expansions of solutions of (1.1) by studying the asymptotic behavior of orbits near an attracting fixed point of a local semiflow related to (1.1).

If the semiflow were defined on a finite dimensional space ($\mathbb{R}^n$) classical results from the theory of ordinary differential equations tell us what to expect. If $-\lambda_1, \ldots, \lambda_N$ are the eigenvalues of the linearized equation then any solution may be written as a convergent power series in the variables $e^{-\lambda_1 t}, \ldots, e^{-\lambda_N t}$, provided no resonances occur. In the presence of resonance polynomials in t occur. These results, which seem to be due to Poincaré, are discussed in Arnold's book [**4**, Chapter 5].

For our infinite dimensional problem the same phenomena occur, on the level of asymptotic expansions. It would be interesting to see if the asymptotic series (6.5), (6.6) and (6.7) are actually convergent (e.g. in case γ is irrational so that no resonances occur).

Department of Mathematics, University of Wisconsin, Van Vleck Hall, 480 Lincoln Dr., Madison WI 53706

This work was supported by a NATO Science Fellowship, and by the Netherlands Organization for the Advancement of Pure Research (ZWO).

REFERENCES

1. S. B. Angenent, *Analyticity of the interface of the porous media equation*, Proc. AMS, to appear.
2. S. B. Angenent, *Abstract parabolic initial value problems*, Preprint 22, Leiden University (1986).
3. S. B. Angenent, *Local existence and regularity for a class of degenerate parabolic equations*, to appear in Math. Annalen.
4. V. I. Arnold, "Geometrical Methods in the Theory of Ordinary Differential Equations," Grundlehren 250, Springer-Verlag, New York, 1983.
5. Zeldovitch, *On the asymptotic properties of self-similar solutions of the equations of non-stationary filtration*, Akademia Nauk SSSR, Doklady **118, no. 4** (1958), 671–674 (in Russian).

Regularity of Flows in Porous Media: A Survey

D. G. ARONSON

§0. Introduction.

Much of the early development of the classical theory of linear partial differential equations was guided by the very detailed body of knowledge accumulated over the years concerning the three model equations: Laplace's equation, the wave equation, and the equation of heat conduction. Indeed, many of us were indoctrinated at the outset of our research careers with maxims such as: "Whatever is true for Laplace's equation is also true, (*sotto voce*) with appropriate modifications, for any elliptic equation." In this lecture I want to describe some of the results of a continuing project, involving a fairly large number of analyists, concerning a model equation for a class of nonlinear diffusion problems, the so-called porous medium equation. Although the theory is certainly not complete, its general outlines are quite clear and a coherent summary is possible.

§1. Formulation.

Consider an ideal gas flowing isentropically in a homogenous porous medium. The flow is governed by three laws:

Equation of state: $p = p_0\rho^\alpha$, where $p = p(x,t)$ is the pressure of the gas at the point $x \in \mathrm{R}^d$ $(d \geq 1)$ at time $t \in \mathrm{R}$, $\rho = \rho(x,t)$ is the density of the gas, and $p_0 \in \mathrm{R}^+$, $\alpha \in [1,\infty)$ are constants.

Conservation of Mass: $\gamma\frac{\partial\rho}{\partial t} + \operatorname{div}(\rho \underset{\sim}{v}) = 0$, where $\underset{\sim}{v} = \underset{\sim}{v}(x,t)$ is the velocity vector and $\gamma \in \mathrm{R}^+$ is the porosity of the medium (i.e., the volume fraction available to the gas).

At this point, for ordinary gas flow we would write down the law of conservation of momentum and thus obtain the usual Navier-Stokes equations.

However, porous medium flow is characterized by an empirically derived relationship between the pressure gradient and the velocity vector known as

Darcy's Law [**D**]: $\nu \underset{\sim}{v} = -\mu \operatorname{grad} p$. Here $\nu \in \mathsf{R}^+$ is the viscosity of the gas and $\mu \in \mathsf{R}^+$ is the permeability of the medium.

If we eliminate p and $\underset{\sim}{v}$ from these equations and rescale so as to eliminate all unnecessary constants we obtain the *porous medium equation*

$$\frac{\partial u}{\partial t} = \Delta(u^m), \tag{1}$$

where $m = 1+\alpha \geq 2$. Since u is a scaled density it is appropriate to assume that

$$u \geq 0$$

and I shall do so throughout this lecture. Because of Darcy's Law, the pressure

$$v \equiv \frac{m}{m-1} u^{m-1}$$

plays a very important role in the theory. The evolution of v is governed by the *pressure equation*

$$\frac{\partial v}{\partial t} = (m-1)v\,\Delta v + |\operatorname{grad} v|^2 . \tag{2}$$

The porous medium equation for various values of $m \geq 1$ also arises in other applications. For example:

(i) the linear heat conduction equation for $m = 1$.
(ii) the theory of ionized gases at very high temperature [**ZR**] for $m > 1$.
(iii) radiative transfer theory [**LP**], where for $m = 7$ it is a special case of the Marshak wave equation.
(iv) boundary layer theory [**O**].

In view of this there is no particular reason to assume that $m \geq 2$ and I will assume from now on that

$$m > 1.$$

There is a very interesting body of work on the so-called fast diffusion case $m < 1$ (cf. [**BH,HP**]), but I will not have time to discuss it here.

The porous medium equation is the simplest example of a general class of equations of the form

$$\frac{\partial u}{\partial t} = \Delta\varphi(u) + \operatorname{grad}\psi(u) + \rho(u)$$

with $\varphi(0) = \varphi'(0) = 0$ and φ increasing near zero. In addition to the nonlinear diffusion term $\Delta\varphi$ these equations may include an advection term grad ψ and a reaction term ρ. Equations of this general form arise in various applications, e.g., in the study of ground water flow [**BP**] and in population dynamics [**GM**]. I will concentrate on the porous medium equation since its theory is very well developed and since it serves as a paradigm for the study of more general equations.

Rewrite equation (1) in the form

$$\frac{\partial u}{\partial t} = \operatorname{div}(mu^{m-1} \operatorname{grad} u).$$

In terms of the usual Fickian diffusion theory, mu^{m-1} is the diffusion coefficient. If u is bounded away from zero, then the porous medium equation is a standard quasilinear parabolic equation which can be treated by the standard methods found in [**LSU**]. However, the porous medium equation is degenerate in the neighborhood of any point where u vanishes. It is this nonlinear degeneracy which gives the theory of porous medium flow its distinctive character. Specifically, for porous medium flow there is a *finite speed of propagation of disturbances from rest.* If, for example, $u(\cdot, 0)$ has compact support, then the support of $u(\cdot, t)$ is also compact for all $t \in \mathbb{R}^+$. The lateral boundary of supp u is called the *interface.* The finite speed of propagation is a nonlinear effect since for the linear heat conduction equation ($m = 1$) any initial positive temperature is felt throughout the space immediately.

A preview of much of the theory of porous medium flow can be obtained by looking at some explicit solutions of (1). These are all selfsimilar solutions and the most important one was found by G. I. Barenblatt [**B**]. Roughly speaking, the Barenblatt solution is the solution of (1) in $\mathbb{R}^d \times \mathbb{R}^+$ whose initial datum is a mass M concentrated at the origin. It is therefore the analog of the fundamental solution of the equation of heat conduction. In terms of pressure the Barenblatt solution is

$$v(x, t; M) = \frac{\beta}{2t}\{r^2(t) - |x|^2\}_+,$$

where $\{\cdot\}_+ = \max(0,\cdot)$, $\beta = \{2+(m-1)d\}^{-1}$,

$$r(t) = \sqrt{\frac{A}{B}}t^{\beta},$$

$B = (m-1)\beta/2m$, and A is a rather complicated function of d, m and M. Details of this and other matters beyond the scope of this lecture can be found in [**Ar4**]. The surface $|x| = r(t)$ is the interface (or free boundary). Note that the Barenblatt solution is not a classical solution of (1) in $\mathbb{R}^d \times \mathbb{R}^+$. Although v is clearly a C^∞ function on the set $\{(x,t) \in \mathbb{R}^d \times \mathbb{R}^+ : |x| < r(t)\}$, globally it is only Lipschitz continuous since $\frac{\partial v}{\partial t}$ and $\operatorname{grad} v$ have bounded jump discontinuities across the interface.

§2. Existence and Uniqueness.

In view of the Barenblatt solution, as well as other examples, it is clear that it is unreasonable to expect to find classical solutions to the porous medium equation unless we require u to be bounded away from zero. To avoid this restriction, we introduce an appropriate notion of generalized solution.

DEFINITION. *$u = u(x,t)$ is said to be a* continuous weak solution *of (1) in $S_T \equiv \mathbb{R}^d \times (0,T]$ for some $T > 0$ if*

(i) *u is continuous and nonnegative in S_T*
(ii) *u satisfies*

$$\iint_{\mathbb{R}^d \times (\tau_1,\tau_2)} u^m \Delta\psi + u\frac{\partial \psi}{\partial t} = \int_{\mathbb{R}^d} u\psi \Big|_{t=\tau_1}^{t=\tau_2}$$

for all τ_i such that $0 < \tau_1 < \tau_2 \le T$ and for all $\psi \in C^{2,1}(S_T)$ such that $\psi(\cdot,t)$ has compact support for each $t \in [\tau_1,\tau_2]$.

Note that this definition is local in time since it says nothing about the behavior at $t = 0$. However, every continuous weak solution of (1) has initial values in the following sense.

THEOREM [**AC1**]. *If u is a continuous weak solution of (1) in S_T then there exists a* unique *nonnegative Borel measure ρ on $\mathbb{R}^d$ such that*

$$\lim_{t\downarrow 0} \int_{\mathbb{R}^d} u(x,t)\psi(x)dx = \int_{\mathbb{R}^d} \psi(x)\rho(dx)$$

for all test functions $\psi \in C_0(\mathbb{R}^d)$. *Moreover, there exists a constant* $C = C(d,m) \in \mathbb{R}^+$ *such that*

$$\int_{B_r(0)} \rho(dx) \leq C\left\{r^{\kappa/(m-1)}T^{-1/(m-1)} + T^{d/2}u^{\kappa/2}(0,T)\right\} \tag{3}$$

for all $r \in \mathbb{R}^+$, *where* $B_r(0) = \{x \in \mathbb{R}^d : |x| < r\}$ *and*

$$\kappa = 2 + d(m-1).$$

I will call ρ the *initial trace* of u. Roughly speaking, the growth condition (3) says that, on average, u must be $O(|x|^{2/(m-1)})$ for $|x|$ large.

For any nonnegative Borel measure ρ on $\mathbb{R}^d$ define

$$|||\rho||| \equiv \sup_{r\geq 1} r^{-\kappa/(m-1)}\rho(B_r(0)).$$

The basic existence and uniqueness result is the following:

THEOREM [**BCP**,**DK**]. *Let* ρ *be a nonnegative Borel measure on* $\mathbb{R}^d$ *with* $|||\rho||| < \infty$. *Define*

$$\ell = \lim_{r\to\infty} \sup_{R\geq r} R^{-\kappa/(m-1)}\rho(B_R(0)).$$

There exists a unique continuous weak solution of (1) on S_{T^*} *with initial trace* ρ, *where* $T^* = c(d,m)/\ell^{m-1}$.

Note that $T^* = \infty$ if and only if $\ell = 0$, i.e., if ρ does not put too much mass at infinity. For simplicity, I will assume that $T^* = \infty$. Moreover, although much of the theory can be done locally (at the expense of some technical effort) I will avoid this and work in the context of a solution to an initial value problem

$$\begin{aligned} \frac{\partial u}{\partial t} &= \Delta(u^m) \text{ in } \mathbb{R}^d \times \mathbb{R}^+ \\ u(\cdot,0) &= u_0 \in L^1(\mathbb{R}^d). \end{aligned} \tag{4}$$

This problem has a unique continuous weak solution and I will simply refer to it as "the solution."

§3. Basic Estimates.

A priori estimates are the basic tools for studying any partial differential equation. For the porous medium equation, however, we are only interested in estimates which are independent of the lower bound for u. The first of these is the

Maximum Principle. If $u_0 \in [0, N]$ in $\mathbb{R}^d$ then the solution u of (4) satisfies $u \in [0, N]$ in $\mathbb{R}^d \times \mathbb{R}^+$.

The maximum principle is a consequence of the more fundamental comparison principle which states that if $u_0 < \hat{u}_0$ in $\mathbb{R}^d$ then the solutions u and $\hat{u}$ of the corresponding initial value problems (4) satisfy $u < \hat{u}$ in $\mathbb{R}^d \times \mathbb{R}^+$.

Having found a bound for the solution, the next step usually is to look for a gradient bound. As we shall see, in general, there is no such bound for solutions to (1) despite the fact that they are C^∞-functions on the interior of their supports [**OKC**]. In fact, one of the most curious aspects of porous medium flow is that the basic tool for analyzing it is a one-sided bound on the Laplacian of the pressure.

Semiconvexity [**Ar3,AB**]. $\Delta v \geq -\frac{k}{t}$ in $\mathcal{D}'(\mathbb{R}^d)$, where $k = (m-1+\frac{2}{d})^{-1}$.

This result is sharp since equality holds for the Barenblatt solution. An immediate consequence of semiconvexity and the nonnegativity of v is

$$\frac{\partial v}{\partial t} \geq -\frac{(m-1)k}{t} v \quad \text{in } \mathcal{D}'(\mathbb{R}^d). \tag{5}$$

The proof of semiconvexity for positive solutions is easy. Let $p = \Delta v$. Then

$$\mathcal{L}(p) \equiv \frac{\partial p}{\partial t} - (m-1)v\,\Delta p - 2m \operatorname{grad} v \cdot \operatorname{grad} p - k^{-1}p^2 \geq 0 = \mathcal{L}(-\frac{k}{t})$$

and the estimate follows by comparison. The extension to nonnegative solutions requires a more technical approximation argument.

Two important elementary consequences of semiconvexity and (5) are:

(i) If $v(x_0, t_0) > 0$ then $v(x_0, t) > 0$ for all $t > t_0$.

(ii) For $x \in \mathbb{R}$ (i.e., for $d = 1$), $v_x + \frac{kx}{t}$ is a nondecreasing function of x and therefore has lateral limits everywhere.

It is natural to ask if there is a corresponding upper bound for Δv. The answer is "no" in general. For $d = 1$ there is an explicit counterexample [**Ar2**]. Consider the initial value problem

$$v_t = (m-1)vv_{xx} + v_x^2 \ \text{ in } \mathbb{R} \times \mathbb{R}^+$$
$$v(x,0) = \sin^2 x \ \text{ in } \mathbb{R}.$$

There exist $c = c(m) \in \mathbb{R}^+$ and $\tau = \tau(m) \in \mathbb{R}^+$ such that for all $p \in \mathbb{Z}$

$$v_{xx}(p\pi, t) = \frac{c}{\tau - t} \quad \text{on } [0, \tau).$$

Thus v_{xx} is not bounded above even though $v(x,0)$ is a real analytic function. I will return to the question of an upper bound for v_{xx} later on.

§4. Regularity for $d = 1$.

In the case $d = 1$ an estimate for v_x (velocity estimate) was known before the semiconvexity estimate was found. It was proved by an adaptation of the classical Bernstein method [**Ar1**]. Here I will give a velocity estimate which is derived from semiconvexity:

$$|v_x(x,t)|^2 \le \frac{2}{(m+1)t} \|v(\cdot,t)\|_{L^\infty(\mathbb{R})}.$$

To prove this fix $y \in \mathbb{R}$ and $t \in \mathbb{R}^+$, and consider the function

$$\varphi(x) \equiv v(x+y,t) + \frac{x^2}{2(m-1)t}.$$

The velocity estimate follows from the fact that φ is continuous, nonnegative, and convex.

From the estimates for $|v_x|$ one can derive the Hölder continuity of v with respect to t [**Kr**,**G**]. However, it turns out that v_t is actually bounded.

THEOREM [**Ar1**,**Be**,**AC2**]. *For $d = 1$ the pressure is globally Lipschitz continuous.*

The Barenblatt solution show that this is the best possible global result. To get more refined regularity results we must first discuss the interface in some detail. To fix ideas, consider the initial value problem

$$u_t = (u^m)_{xx} \text{ in } \mathbb{R} \times \mathbb{R}^+$$
$$u(\cdot,0) = u_0 \text{ in } \mathbb{R},$$

where

$$u_0 \begin{cases} \equiv 0 & \text{on } \mathsf{R}^+ \\ > 0 & \text{for all sufficiently small } x < 0. \end{cases}$$

As usual, I assume $T^* = \infty$.

The function

$$\varsigma(t) \equiv \sup\{x \in \mathsf{R} : u(x,t) > 0\}$$

exists on R^+ and satisfies $\varsigma(0) = 0$. Moreover, $\varsigma(t)$ is nondecreasing and $\varsigma(t) \in \text{Lip}(\mathsf{R}^+)$. The curve $x = \varsigma(t)$ is called the (right-hand) interface. We expect the interface to move with the local velocity of the gas. By Darcy's Law we should therefore have $\dot{\varsigma} = -v_x(\varsigma-, t)$. However, this is not quite true and the correct global result is

$$D^+\varsigma(t) = -v_x(\varsigma(t)-, t) \text{ in } \mathsf{R}^+,$$

where D^+ denotes the right-hand derivative [**K**].

The interface does not necessarily begin to move at $t = 0$, although it must ultimately move. Specifically, there is a $t^* \in [0, \infty)$, called the *waiting time*, such that

(i) $\varsigma(t) \equiv 0$ on $[0, t^*]$

(ii) $\varsigma(t)$ is strictly increasing on (t^*, ∞).

Note that (ii) means that once the interface begins to move it never stops. It is possible, however, to have $t^* > 0$. For example, if $v_0(x) \leq cx^2$ on $(-\delta, 0)$ for some $c \in \mathsf{R}^+$ and $\delta \in \mathsf{R}^+$ then $t^* > 0$. There is a necessary and sufficient condition for $t^* > 0$. Moreoever, one can estimate t^* from above in terms of "local" properties of v_0 and from below in terms of "global" properties of v_0 [**V**]. For our purposes the fundamental properties of the waiting time are given by the following

THEOREM [**CF1**]. *$\varsigma \in C^1(t^*, \infty)$,*

$$\dot{\varsigma} = -v_x(\varsigma-, t) \text{ on } (t^*, \infty),$$

and vv_{xx} is continuous in a neighborhood of $x = \varsigma(t)$ for $t \in (t^*, \infty)$ with

$$\lim_{\substack{(x,t)\to(\varsigma(\tau),\tau) \\ t,\tau > t^*}} (vv_{xx})(x,t) = 0. \tag{6}$$

Suppose that $t^* > 0$. Then $\varsigma \in C^1$ on $(0, t^*) \cup (t^*, \infty)$ and $\varsigma \in C^1(\mathsf{R}^+)$ if and only if $\dot{\varsigma}(t^*)$ exists. Since $D^-\varsigma(t^*) = 0$ it follows that $\varsigma \in C^1(\mathsf{R}^+)$ if and only if $D^+\varsigma(t^*) = 0$. Actually both $D^+\varsigma(t^*) = 0$ and $D^+\varsigma(t^*) > 0$ can occur. For example,

$$D^+\varsigma(t^*) = 0 \quad \text{if} \quad v_0(x) = O(x^2) \text{ as } x \uparrow 0$$

and

$$D^+\varsigma(t^*) > 0 \quad \text{if} \quad v_0(x) = o(x^2) \text{ as } x \uparrow 0.$$

More refined criteria for both regularity and irregularity of ς can be found in [**ACV**].

It is reasonable to guess that once the interface starts to move there is a local relaxation of the flow and everything becomes quite smooth. This is exactly what happens.

THEOREM [**AV**]. *v is locally a C^∞-function on its support at any point of a moving interface. Moreover, any moving interface is a C^∞-curve.*

The first step in proving this result is to derive an upper bound for v_{xx}. Recall that we already have a lower bound and know that an upper bound cannot exist unless $t > t^*$.

Consider $t_0 > t^*$. Then

$$\dot{\varsigma}(t_0) = -v_x(\varsigma(t_0)-, t_0) = a > 0. \tag{7}$$

We bound v_{xx} from above in a neighborhood $R_{\delta\eta}$ of $(\varsigma(t_0), t_0)$ shown in Figure 1 by constructing a barrier for it of the form

$$\varphi(x,t) \equiv \frac{\alpha}{\varsigma(t) - x} + \frac{\beta}{\varsigma^*(t) - x},$$

where α and β are positive constants, and $\varsigma^* > \varsigma$ in a neighborhood of t_0. Specifically, let $p = v_{xx}$. Then

$$\Lambda(p) = p_t - (m-1)vp_{xx} - 2mv_xp_x - (m-1)p^2 = 0.$$

Using (7) and suitably adjusting the dimensions of $R_{\delta\eta}$ one proves the existence of an $\alpha_0 = \alpha_0(d, m, \|v\|_{L^\infty}) \in \mathrm{R}^+$ and $\beta = \beta(d, m, \|v\|_{L^\infty}) \in \mathrm{R}^+$ such that

$$\Lambda(\varphi) \geq 0 \text{ in } R_{\delta\eta}$$

for all $\alpha \in (0, \alpha_0)$. From (6) and (7) we conclude that

$$v_{xx}(x,t) = o\left(\frac{1}{\varsigma(t) - x}\right) \text{ as } x \uparrow \varsigma(t), \tag{8}$$

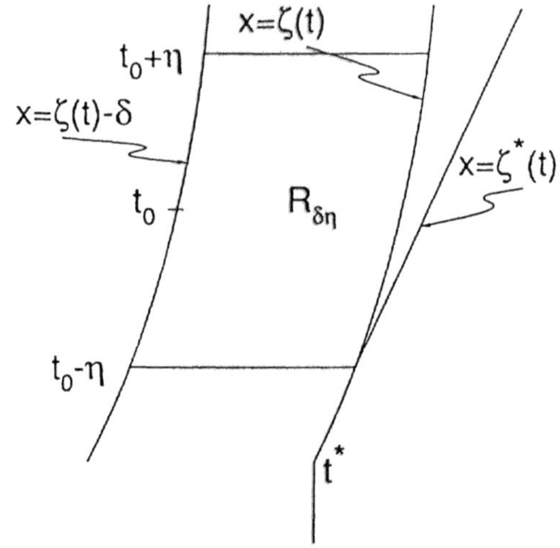

Figure 1

and this enables us to show that $\varphi \geq p$ on the parabolic boundary of $R_{\delta\eta}$. Thus, by the comparison principle,

$$v_{xx}(x,t) \leq \frac{\alpha}{\varsigma(t)-x} + \frac{\beta}{\varsigma^*(t)-x} \quad \text{in } R_{\delta\eta},$$

where β is fixed and $\alpha \in (0, \alpha_0)$ is arbitrary. Now let $\alpha \downarrow 0$ to obtain

$$v_{xx}(x,t) \leq \frac{\beta}{\varsigma^*(t)-x}.$$

Therefore we have a finite upper bound for v_{xx} in $R_{\delta\eta}$ as long as $t \geq t_0 - \eta/2$.

A similar, but technically more difficult, argument permits us to estimate $v^{(j)} \equiv (\frac{\partial}{\partial x})^j v$ for all $j \geq 3$. The main difficulty comes from the fact that we no longer have an a priori estimate analogous to (8) and indeed must begin by proving the weaker estimate

$$v^{(j)}(x,t) \equiv O\Big(\frac{1}{\varsigma(t)-x}\Big) \text{ as } x \uparrow \varsigma(t).$$

Once we have estimates for all the pure spatial derivatives $v^{(j)}$ we can estimate all of the remaining derivatives of v by suitably differentiating the pressure equation (2), and all of the derivatives of ς by differentiating $\dot{\varsigma} = -v_x(\varsigma, t)$. Details can be found in [**AV**].

The fact that $v \in C^\infty$ after the waiting time was first proved by a totally different method (weighted integral estimates) by Höllig and Kreiss [**HK**].

They assume that v_0 has compact support and show that v is globally C^∞ on its support for t larger than the maximal waiting time. In these same circumstances, Angenent [**A**] has shown that the interface is an analytic curve.

§4. Regularity and Irregularity for $d > 1$.

In constrast to the virtually complete regularity theory for flows in one-dimensional porous media, there remain important open questions concerning flows in $\mathbb{R}^d$. Moreover, the theory for $\mathbb{R}^d$ is essentially different from the theory in $\mathbb{R}$ since there is no velocity bound in $\mathbb{R}^d$.

The basic global regularity result for porous medium flow in $\mathbb{R}^d$ is the following

THEOREM [**CF2**]. *v and the interface are Hölder continuous in $\mathbb{R}^d \times \mathbb{R}^+$.*

This result does not give any estimates for the Hölder exponent and leaves many questions unanswered. Qualitatively, however, it is the best possible global result.

Recall that for $d = 1$ there is a bound for $|v_x|$. For $d > 1$, in general, $|\operatorname{grad} v|$ is not bounded. This is shown by a selfsimilar solution which was first found numerically by Graveleau [**Gr**] in about 1972 and whose existence was proved recently [**AG**]. The general problem is to solve (1) starting from a radially symmetric distribution of gas which is supported *outside* a ball. The question of interest is "how does the solution behave at the instant when the gas first reaches the center of the ball?" I will call this the *focusing problem.* Graveleau suggested looking for a selfsimilar solution which fills all of $\mathbb{R}^d$ at $t = 0$ of the form

$$\tilde{v}(x,t) = \frac{r^2}{-t}\varphi(\eta)$$

with $\eta = tr^{-\alpha}$, where $r = |x|$.

To find $\tilde{v}$ we must find the function φ and the constant α. To this end write

$$\tilde{v}(x,t) = r^{2-\alpha}\frac{\varphi(\eta)}{-\eta},$$

and require that

$$\varphi(0) = 0 \text{ and } \varphi'(0) = -1.$$

Then for $r > 0$

$$\lim_{t\uparrow 0} \tilde{v}(x,t) = r^{2-\alpha}.$$

φ satisfies a degenerate second order nonlinear ordinary differential equation which depends on the parameter α. The Graveleau solution is obtained when $\alpha = \alpha^* \equiv \alpha^*(d,m)$ is chosen so that there is an orbit connecting certain critical points (heteroclinic orbit). One shows that α^* is unique and that there exists an $A = A(d,m) \in \mathrm{R}^-$ such that

$$\varphi \begin{cases} > 0 & \text{on } (A,0) \\ \equiv 0 & \text{on } (-\infty, A]. \end{cases}$$

Thus

$$\tilde{v}(x,t) \equiv 0 \text{ for } r \le (t/A)^{1/\alpha^*}$$

(cf. [**AG,Ar4**]).

For $d = 1$, $\alpha^*(1,m) \equiv 1$ and the Graveleau solution consists of a pair of colliding linear pressure waves: $\tilde{v}(x,t) = \gamma(\gamma t \pm x)_+$. For $d > 1$, one can show that $\alpha^*(d,m) \in (1,2)$. However, the value of α^* can only be found numerically. For example,

$$\alpha^*(2,2) = 1.167778795\ldots$$

so that

$$\tilde{v}(x,0) = |x|^{0.832221205\ldots}.$$

I conjecture that an appropriately scaled d-dimensional Graveleau solution will be the leading term of any focusing solution provided that the focusing is genuinely d-dimensional.

On the positive side, there is the following

THEOREM [**CVW**]. *If $v(\cdot,0)$ has compact support then there exists a $\tilde{T} \in \mathrm{R}^+$ such that $v \in \mathrm{Lip}(\mathrm{R}^d \times (\tilde{T},\infty))$.*

To determine $\tilde{T}$, suppose that supp $v(\cdot,0) \subset B_r(0)$. As t increases, supp $v(\cdot,t)$ spreads out and eventually covers all of R^d. Then $\tilde{T}$ is defined to be the first time for which

$$\mathrm{Cl}\, B_r(0) \subset \operatorname{supp} v(\cdot,t).$$

That $\tilde{T} \in \mathrm{R}^+$ is assured by the following consequence of the Aleksandrov reflection principle. Suppose supp $u(\cdot,0) \subset B_R(0)$. Then for any $x_0 \in \mathrm{R}^d$ with $|x_0| \ge R$, $u(\cdot,t)$ is nondecreasing on the ray $\{x = \lambda x_0 : \lambda > 1\}$. Roughly speaking, supp $u(\cdot,t)$ cannot have any "pockets." Moreover, the diameter of supp $u(\cdot,t)$ grows like $t^{1/\kappa}$.

The best local regularity result presently known is the following

THEOREM [**CW**]. *Near any point on a moving interface, v is a $C^{1+\alpha}$ function on its support.*

The natural conjecture is that $v \in C^{\infty}$ on its support near a moving interface. This is an open problem.

School of Mathematics, University of Minnesota, Minneapolis MN 55455

REFERENCES

[**A**] S. Angenent, *Analyticity of the interface of the porous medium equation after the waiting time*, Proc. Amer. Math. Soc., to appear.

[**Ar1**] D. G. Aronson, *Regularity properties of flows through porous media*, SIAM J. Appl. Math. **17** (1969), 461–467.

[**Ar2**] D. G. Aronson, *Regularity properties of flows through porous media: a counterexample*, SIAM J. Appl. Math. **19** (1970), 299–307.

[**Ar3**] D. G. Aronson, *Regularity properties of flows through porous media: the interface*, Arch. Rat. Mech. Anal. **37** (1970), 1–10.

[**Ar4**] D. G. Aronson, *The porous medium equation*, in "Nonlinear Diffusion Problems," (A. Fasano and M. Primicerio, eds.), Springer Lecture Notes in Math, vol. 1224, Springer-Verlag, 1986.

[**AB**] D. G. Aronson and Ph. Benilan, *Regularité des solutions de l'equation des milieux poreux dans* $\mathbb{R}^N$, C.R. Acad. Sci. Paris **288** (1979), 103–105.

[**AC1**] D. G. Aronson and L. A. Caffarelli, *The initial trace of a solution of the porous medium equation*, Trans. Amer. Math. Soc. **280** (1983), 351–366.

[**AC2**] D. G. Aronson and L. A. Caffarelli, *Optimal regularity for one-dimensional porous medium flow*, Revista Matemática Iberoam. **2** (1986), 357–366.

[**ACV**] D. G. Aronson, L. A. Caffarelli and J. L. Vasquez, *Interfaces with a corner point in one-dimensional porous medium flow*, Comm. Pure Appl. Math. **38** (1985), 375–404.

[**AG**] D. G. Aronson and J. Graveleau, in preparation.

[**AV**] D. G. Aronson and J. L. Vasquez, *Eventual* C^∞*-regularity and concavity for flows in one-dimensional porous media*, Arch. Rat. Mech. Anal. **99** (1987), 329–348.

[**B**] G. I. Barenblatt, *On some unsteady motions of a liquid or a gas in a porous medium*, Prikl. Mat. Meh. **16** (1952), 67–78.

[**Be**] Ph. Benilan, *A strong regularity* L^p *for solutions of the porous media equation*, in "Contributions to Nonlinear Partial Differential Equations," (C. Bardos, A. Damlamian, J. I. Diaz and J. Hernandez, eds.), Research Notes in Math., vol. 89, Pitman, London, 1983, pp. 39–58.

[**BCP**] Ph. Benilan, M. G. Crandall, and M. Pierre, *Solutions of the porous medium equation in* $\mathbb{R}^N$ *under optimal conditions on initial values*, Indiana Univ. Math. J. **33** (1984), 51–87.

[**BH**] J. G. Berryman and C. J. Holland, *Stability of the separable solution for fast diffusion*, Arch. Rat. Mech. Anal. **74** (1980), 279–288.

[**BP**] M. Bertsch and L. A. Peletier, "Porous medium type equations: An overview," Mathmatical Institute, University of Leiden, Report No. 7, 1983.

[**CF1**] L. A. Caffarelli and A. Friedman, *Regularity of the free boundary for the one-dimensional flow of gas in a porous medium*, Amer. J. Math. **101** (1979), 1193–1281.

[**CF2**] L. A. Caffarelli and A. Friedman, *Regularity of the free boundary of a gas flow in an n-dimensional porous medium*, Indiana Univ. Math. J. **29** (1980), 361–391.

[**CVW**] L. A. Caffarelli, J. L. Vasquez and N. I. Wolanski, *Lipschitz continuity of solutions and interfaces of the* N*-dimensional porous medium equation*, Indiana U. Math. J. **36** (1987), 373–401.

[**CW**] L. A. Caffarelli and N. I. Wolanski, *The differentiability of the free boundary for the* N*-dimensional porous medium equation*, preprint.

[**D**] H. Darcy, "Les Fontaines Publiques de la Ville de Dijon," V. Dalmont, Paris, 1856, pp. 305–311.

[**DK1**] B. E. J. Dahlberg and C. E. Kenig, *Nonnegative solutions of the porous medium equation*, Comm. PDE **9** (1984), 409–437.

[**G**] B. H. Gilding, *Hölder continuity of solutions of parabolic equations*, J. London Math. Soc. **13** (1976), 103–106.

[Gr] J. Graveleau, personal communication.
[GM] M. E. Gurtin and R. C. MacCamy, *On the diffusion of biological populations*, Math. Biosc. **33** (1977), 35–49.
[HP] M. A. Herrero and M. Pierre, *The Cauchy problem for* $u_t = \Delta u^m$ *when* $0 < m < 1$, Trans. Amer. Math. Soc. **291** (1985), 145–158.
[HK] K. Höllig and H. O. Kreiss, C^∞*-regularity for the porous medium equation*, Math. Z. **192** (1986), 217–224.
[K] B. F. Knerr, *The porous medium equation in one dimension*, Trans. Amer. Math. Soc. **234** (1977), 381–415.
[Kr] S. N. Kruzhkov, *Results on the character of the regularity of solutions of parabolic equations and some of their applications*, Math. Notes **6** (1969), 517–523.
[LP] E. W. Larsen and G. C. Pomraning, *Asymptotic analysis of nonlinear Marshak waves*, SIAM J. Appl. Math. **39** (1980), 201–212.
[LSU] O. A. Ladyzhenskaya, V. A. Solonnikov and N. N. Ural'ceva, "Linear and Quasilinear Equations of Parabolic Type," Transl. Math. Monographs 23, Amer. Math. Soc., Providence, 1968.
[O] O. A. Oleinik, "Mathematical Problems of Boundary Layer Theory," Lecture Notes, Departemtn of Mathematics, University of Minnesota, 1969.
[OKC] O. A. Oleinik, A. S. Kalashnikov and Chzhou Yui-Lin, *The Cauchy problem and boundary problems for equations of the type of unsteady filtration*, Izv. Akad. Nauk SSSR Ser. Mat. **22** (1958), 667–704.
[V] J. L. Vasquez, *The interface of one-dimensional flows in porous media*, Trans. Amer. Math. Soc. **285** (1984), 717–737.

Ground States for the Prescribed Mean Curvature Equation: The Supercritical Case

F. V. ATKINSON, L. A. PELETIER AND J. SERRIN

1. Introduction.

We shall consider here the question of existence and nonexistence of ground states for the prescribed mean curvature equation in $\mathbb{R}^n (n > 2)$, that is, we consider solutions of the problem

$$
(1.1) \quad \left.\begin{aligned}
&\operatorname{div}\left(\frac{Du}{(1+|Du|^2)^{1/2}}\right) + f(u) = 0 \quad \text{in } \mathbb{R}^n \\
&u > 0 \qquad \text{in } \mathbb{R}^n \\
&u(x) \to 0 \quad \text{as } x \to \infty,
\end{aligned}\right\} \quad (I)
$$

where Du denotes the gradient of u. The function $f(u)$, defined for $u \geq 0$, will be assumed throughout to satisfy the following hypotheses:

(H1) $\quad f \in C^1[0,\infty)$

(H2) $\quad f(0) = 0$, and there exists a number $a \geq 0$ such that

$$f(u) > 0 \text{ for } u > a;$$

if $a > 0$ we require

$$f(u) < 0 \text{ for } 0 < u < a.$$

We say that f has *supercritical growth* provided that the following condition is satisfied:

$$\text{(H3)} \qquad uf(u) \geq \frac{2n}{n-2} F(u) \quad \text{for } u > a,$$

where

$$F(u) = \int_0^u f(t)\,dt$$

denotes the primitive of $f(u)$.

When $a = 0$, Problem I possesses a one-parameter family of radial solutions [**NS2**]. Specifically, it was shown there that corresponding to every $c > 0$ for which

$$F(c) \leq 1 \tag{1.2}$$

there exists a solution $u = u(|x|)$ of (I) such that $u(0) = c$.

On the other hand, when $a > 0$ and f satisfies a similar, but slightly stronger, growth condition, namely

$$uf'(u) \geq \frac{n+2}{n-2} f(u) \quad \text{for } u > a, \tag{H3*}$$

then no radial solutions of (I) exist (see [**AP**], and also, under slightly different hypotheses, [**NS1**]).

The question remains in the first case whether a solution can exist when $F(c) > 1$. It is convenient to start with the example

$$f(u) = u^p. \tag{1.3}$$

The hypotheses H1–3, with $a = 0$, are all satisfied if $p \geq (n+2)/(n-2)$, while (1.2) holds if the value $c = u(0)$ satisfies

$$c \leq (p+1)^{1/(p+1)}.$$

Hence radial solutions exist for this range of values of c. On the other hand, by Theorem 2 of [**S**] there can be no radial solutions when

$$c \geq (4n^2 p)^{1/(p+1)}.$$

Thus, at least at first glance, it appears there is little further to be said in the supercritical case: when $a = 0$, then for suitably small values of $u(0)$ there exist radial ground states while for larger values there need not exist any; and when $a > 0$ there exist none either (at least if H3 is strengthened to H3*).

Adopting a more sophisticated perspective, however, the problem of nonexistence for large values of $u(0)$ can be seen to lie in the *meaning* which is attached to a solution of (I). Indeed, from the standpoint of spatial geometry, equation (1.1) defines a graph $(x, u(x))$ whose mean curvature at each point is just $-f(u(x))/2n$. A more refined approach to Problem I would therefore be to ask for a *twice continuously differentiable embedded surface* S in $\mathbb{R}^n \times \mathbb{R}^+$ whose mean curvature K at each point (x, y) is given by

$$K(x,y) = -\frac{1}{2n} f(y),$$

and which, in addition, is asymptotic to the plane $y = 0$ as $x \to \infty$.

One must of course attach a continuous normal direction to the surface in order to define the sign of the mean curvature: in order to be consistent with (1.1) we agree that the vertical component of the normal is positive as $x \to \infty$.

A radial solution of this generalized problem would then be a (possibly) multivalued function $u = u(r)$ $(r = |x|)$ with a single central value $c = u(0) > 0$, which moreover is single-valued for large values of r and satisfies $u(r) \to 0$ as $r \to \infty$. We shall show in Section 5 that a surface of this type can be represented in radial coordinates by a *single-valued* relation

$$r = r(u), \quad 0 < u \leq c.$$

Moreover necessarily $f(c) > 0$ and $r(u)$ is positive and of class C^2 for $0 < u < c$. See Figure 1.

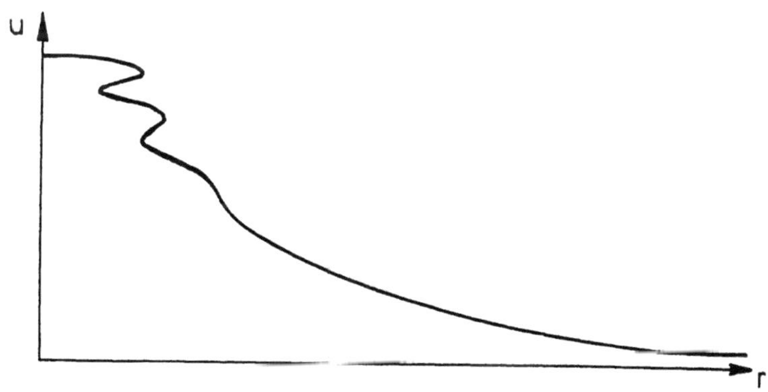

Figure 1. Solution graph; $f(u) = u^3, n = 4$.

It is easy to see that equation (1.1) is satisfied on any forward branch of $u(r)$, where the normal direction is upward, while on any reverse branch the normal must point downward so that one has

$$\text{(1.4)} \qquad \operatorname{div}\left(\frac{Du}{(1+|Du|^2)^{1/2}}\right) - f(u) = 0.$$

Finally, an easy calculation shows that the inverse function $r(u)$ satisfies the equation

$$\frac{r''}{(1+r'^2)^{3/2}} - \frac{n-1}{r}\frac{1}{(1+r'^2)^{1/2}} + f(u) = 0, \tag{1.5}$$

where the prime now denotes differentiation with respect to u. Naturally we also have the conditions

$$r(u) > 0 \quad \text{for } u \in (0,c), \tag{1.6}$$

$$r(c) = 0, \quad r'(u) \to -\infty \quad \text{as } u \to c, \tag{1.7}$$

the second condition in (1.7) ensuring regularity of the function $u(r)$ at $r = 0$, and finally

$$r(u) \to \infty \quad \text{as } u \to 0. \tag{1.8}$$

Surfaces of the sort proposed here have been studied extensively for certain special functions $f(u)$. When

$$f(u) = \text{constant}$$

they are the celebrated *Delaunay surfaces* of constant mean curvature, while when

$$f(u) = \kappa u, \quad \kappa \in \mathbb{R}^+,$$

they represent the pendant drop under (reversed) gravity. For the latter case the reader is referred to the paper by Concus and Finn [**CF**] and the recent monograph of Finn [**F**].

Numerical calculations have been carried out by Evers and Levine [**EL**] for particular functions f satisfying H1–2. These calculations stirred our original interest in the problem of generalized ground states, and also confirm the proposed form of the solution, as well as indicating an unexpected degree of regularity in the successive oscillations of the graph of $u(r)$.

As part of our work we shall investigate the qualitative behavior of radially symmetric generalized ground states, assuming that f satisfies H1–2 as well as the following weak version of H3*,

$$f'(u) > 0 \quad \text{if} \quad u > a. \tag{H4}$$

In this part of the analysis we shall assume merely that $n > 1$ (rather than $n > 2$ as in the rest of the paper).

PROPOSITION 1. *The inverse function $r(u)$ for a generalized radial ground state satisfies*

$$0 < r(u) < \frac{n}{f(u)} \quad \textit{for all } u \in (a,c).$$

For the model case (1.3) this shows in particular that

$$0 < r(u) < nu^{-p},$$

a generalized hyperbolic upper bound.

PROPOSITION 2. *If $r(u)$ has a local maximum (minimum) at a point u_0 in (a,c) then*

$$r(u_0) > \frac{n-1}{f(u_0)} \quad \left(< \frac{n-1}{f(u_0)} \right).$$

PROPOSITION 3. *The successive maximum values, as well as the successive minimum values, of $r(u)$ are increasing as u decreases from $u = c$ to $u = a$.*

PROPOSITION 4. *There exist at most a finite number of oscillations of $r(u)$ on (a,c).*

A more detailed and comprehensive study of the qualitative properties of radial solutions of equation (1.1) with initial values given by (1.7) will be the subject of a subsequent paper.

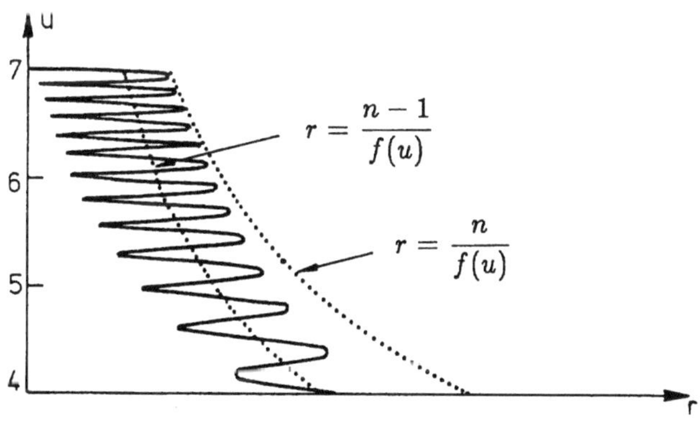

Figure 2. Oscillations for $f(u) = u^2, n = 3, c = 7$ in the ordinate range $4 \le u \le 7$.

Our further results concern the existence and non-existence of generalized radial ground states of Problem I. Theorem 1 asserts the existence of a generalized radial ground state corresponding to an arbitrarily assigned central value $c > 0$ when $f(u) > 0$ for all $u > 0$ (that is, $a = 0$).

THEOREM 1. *Let $f(u)$ satisfy hypotheses H1, H2 and H3* with $a = 0$. Then for every $c > 0$ there exists a solution $r(u)$ of (1.5) satisfying the conditions (1.6)–(1.8). Moreover the inverse function $u(r)$ satisfies (1.1) in some neighborhood of the origin and on all forward branches, and satisfies (1.4) on all reverse branches.*

In general, as noted earlier, the function $u(r)$ cannot be single valued for large values of c. When oscillations appear, the function $r(u)$ possesses the properties listed above in Propositions 1–4.

In our second theorem we assert the non-existence of radial ground states, *classical or generalized*, if $f(u)$ has one positive zero.

THEOREM 2. *Let $f(u)$ satisfy the hypotheses H1, H2, and H3* with $a > 0$. Then there exists no solution $r(u)$ of (1.5) with the properties (1.6)–(1.8).*

While Theorems 1 and 2 give a fairly detailed answer to the question of existence and nonexistence of radial ground states when the function $f(u)$ possesses *supercritical* growth, no general answer is presently available when the growth of f is *subcritical*. For example, for the particular function

$$f(u) = -ku + u^p,$$

in which

$$k > 0 \quad \text{and} \quad p < (n+2)/(n-2),$$

existence has been shown for k suitably near 0 [**PS3**], and non-existence for *single valued solutions* when k is large enough [**FLS**]. The question remains whether *generalized* ground states exist for all positive values of k.

The paper is organized as follows: In Section 2 we prove Propositions 1–4 under mild conditions on $f(u)$, in Section 3 we prove Theorem 1 and finally in Section 4 we prove Theorem 2.

2. Properties of solutions.

In this section we shall investigate some general properties of radial solutions of Problem I, and in particular prove the Propositions 1–4 stated in the Introduction. Throughout this section we shall assume that f satisfies H1 and H2.

Let S be a generalized radial ground state corresponding to f, and let its central value be $u = c$. At least for small values of r it is evident that S can be represented in the form $u = u(r)$, where u satisfies the initial value problem

$$\text{(2.1)} \qquad \left.\begin{array}{l} \left(\dfrac{r^{n-1}u'}{(1+u'^2)^{1/2}}\right)' + r^{n-1}f(u) = 0 \quad r > 0 \\ u(0) = c, \quad u'(0) = 0. \end{array}\right\} \qquad \text{(II)}$$

Naturally we have $f(c) > 0$, so $c > a$ by H2. To analyze the behavior (and indeed the existence) of radial ground states it is obviously of interest to consider Problem II in some detail.

In particular, for every $c > a$ this problem has a unique solution $u(r,c)$ in a neighborhood of $r = 0$, and

$$u(r,c) < c$$

for r small and positive. If c is small, specifically if $F(c) \leq 1$, then $u'(r,c)$ remains bounded and negative as we continue $u(r,c)$ to larger values of r [**NS2**]. Thus we can continue $u(r,c)$ as long as $u > 0$.

On the other hand, as we observed in the Introduction, if c is larger, it may not be possible to continue $u(r,c)$, at least as a single valued function, because $u'(r,c)$ might become unbounded as r tends to some finite value; see [**S**]. The next theorem provides a similar result, under somewhat different hypotheses and using somewhat different methods.

We shall denote by R the greatest number — which may be infinite — such that $u'(r,c) > -\infty$ for $0 < r < R$, and we shall write $U = u(R,c)$. Thus, when R is finite, the point (R,U) is the first vertical point on the graph of $u(r,c)$ when we trace it from the point $(0,c)$.

THEOREM 2.1. *Suppose for all sufficiently large values of u that $f(u) > 0$ and that*

$$\text{(2.2)} \qquad 0 \leq \frac{f'(u)}{f^2(u)} \leq \frac{1}{4n^2}.$$

Then for c sufficiently large there exists a first vertical point (R, U) on the graph of $u(r, c)$, with

$$c - \frac{2n}{f(c)} < U \leq c - \frac{n}{f(c)}, \tag{2.3}$$

$$\frac{n}{f(c)} \leq R < \frac{2n}{f(c)}. \tag{2.4}$$

Moreover, the inverse function $r(u)$ is of class C^2 on $[U, c)$ and satisfies $r'(U) = 0$, $r''(U) < -f(c)/2n$.

PROOF: Define the function

$$t(r) = r^{1-n} \int_0^r s^{n-1} f(u(s))ds, \quad 0 \leq r < R.$$

Then by (2.1) we can write

$$u'(r) = -\frac{t(r)}{\{1 - t^2(r)\}^{1/2}}.$$

Since $u'(r)$ stays finite on $[0, R)$ it follows that also $t(r) < 1$ on $[0, R)$. Another integration yields

$$c - u(r) = \int_0^r \frac{t(s)}{\{1 - t^2(s)\}^{1/2}} ds$$

or, if we take t as the variable of integration,

$$c - u(r) = \int_0^{t(r)} \frac{t}{(1 - t^2)^{1/2}} \frac{dt}{Q(r)}, \tag{2.5}$$

where $Q(r) = t'(r)$, provided of course that $Q > 0$ on $(0, r)$.

To begin with, for $0 \leq r < R$ and for c suitably large, we shall prove that

$$u(r) > c - \frac{2n}{f(c)}. \tag{2.6}$$

Suppose to the contrary that $u(r^*) = c - 2b$, for some first point $r^* \in (0, R)$, where $b = n/f(c)$. Then $c - 2b \leq u \leq c$, on $[0, r^*]$, and so

$$\begin{aligned} Q(r) &= f(u(r)) - (n-1)r^{-n} \int_0^r s^{n-1} f(u(s))ds \\ &\geq f(c - 2b) - \frac{n-1}{n} f(c) \\ &= \frac{f(c)}{n} - 2n \frac{f'(c - 2b\theta)}{f(c)} \end{aligned} \tag{2.7}$$

for some $\theta \in [0,1]$. Here we have used the fact that $f' \geq 0$ for large u (see (2.2)). Thus in turn $f(u)$ ultimately increases and $c - 2b \to \infty$ as $c \to \infty$. In particular (2.2) then holds for all $u \geq c - 2b$ when c is large enough. For the same reasons, we have

$$f'(c - 2b\theta) \leq \frac{1}{4n^2} f^2(c - 2b\theta) \leq \frac{1}{4n^2} f^2(c).$$

Therefore (note that equality cannot hold)

$$Q(r) > \frac{1}{2b} \quad \text{when} \quad 0 \leq r \leq r^*. \tag{2.8}$$

Now from (2.5) we get a contradiction after an easy integration:

$$\begin{aligned} 2b = c - (c - 2b) &= \int_0^{t(r^*)} \frac{t}{(1-t^2)^{1/2}} \frac{dt}{Q(r)} \\ &< \int_0^1 \frac{t}{(1-t^2)^{1/2}} dt \cdot 2b = 2b. \end{aligned}$$

With (2.6) proved, we now obtain, exactly as before, that (2.7) and (2.8) hold for $0 \leq r < R$ and for c suitably large. Thus in this interval $t(r)$ is montonically increasing and

$$t(r) = \int_0^r Q(s)ds > \frac{f(c)}{2n} r.$$

But $t(r) < 1$ on $[0, R)$, whence a vertical point (R, U) must occur with

$$R < \frac{2n}{f(c)}, \quad U > c - \frac{2n}{f(c)}.$$

To obtain the remaining estimates of (2.3) and (2.4) we need an upper bound for Q. From (2.7) we deduce from an integration by parts that

$$Q(r) = \frac{1}{n} f(u(r)) + \frac{n-1}{n} r^{-n} \int_0^r s^{-n} f'(u(s))u'(s)ds \leq \frac{1}{n} f(c).$$

for $0 < r < R$. This implies by (2.5) that

$$u(R) \leq c - \frac{n}{f(c)}.$$

Moreover we see without difficulty that

$$\frac{f(c)}{n} R \geq \int_0^R Q(r)dr = t(R) = 1,$$

whence

$$R \geq \frac{n}{f(c)}.$$

Plainly $r'(U) = 0$ and it follows from (1.5) by letting $u \to U$ that

$$r''(U) = \frac{n-1}{r(U)} - f(U) = \frac{n-1}{R} - f(U)$$

or, in view of the lower bound for R,

$$r''(U) \leq \frac{n-1}{n} f(c) - f(U) = -\frac{f(c)}{n} + f(c) - f(U). \tag{2.9}$$

Because of the growth condition (2.2) on f and the lower bound for U we have

$$f(c) - f(U) = f'(\xi)(c - U) < \frac{f^2(c)}{4n^2} \frac{2n}{f(c)} = \frac{f(c)}{2n}.$$

Substitution in (2.9) now yields

$$r''(U) < -\frac{f(c)}{2n},$$

as required.

REMARK: It is interesting to state a more precise version of Theorem 2.1 in which the condition for c is explicit. To do this requires slightly stronger hypotheses on the behavior of the function f, namely that for some $a^* \geq 0$

$$f'(u) \geq 0 \quad \text{and} \quad f''(u) \geq 0 \quad \text{for} \quad u \geq a^*.$$

In this case there exists a first vertical point (R, U) satisfying (2.3) and (2.4) provided only that c $(> a)$ satisfies the conditions

$$c \geq a^* + \frac{2n}{f(c)}, \qquad \frac{f'(c)}{f^2(c)} \leq \frac{1}{4n^2}.$$

PROOF: It is enough to check that (2.7) and (2.8) hold for such c. First we observe that $c - 2b \geq a^*$ so that

$$f(c - 2b) \leq f(u) \leq f(c)$$

for $c - 2b \leq u \leq c$. This proves (2.7). But then, since $f''(u) \geq 0$ for $u \geq a^*$, we have also

$$f'(c - 2b) \leq f'(c) \leq \left(\frac{f(c)}{2n}\right)^2$$

and (2.8) follows at once.

Example. Consider the function

$$f(u) = k,$$

where k is some positive constant. Then the solution $u(r,c)$ becomes

$$u(r,c) = c - b - (b^2 - r^2)^{1/2}$$

where $b = n/k$. Thus

$$R = \frac{n}{k} \quad \text{and} \quad U = c - \frac{n}{k}.$$

To continue beyond the point (R, U) we consider the inverse function $r(u)$. As observed in the Introduction, this is a solution of the initial value problem (see (1.5)–(1.7))

$$\left.\begin{aligned} &\left(\frac{r^{n-1}}{(1+r'^2)^{1/2}}\right)' - f(u)r^{n-1}r' = 0, \quad u < c \\ &r(u) \to 0, \quad r'(u) \to -\infty \text{ as } u \to c \\ &r(u) > 0, \quad \text{for } u < c. \end{aligned}\right\} \quad \text{(III)} \tag{2.10}$$

Local existence and uniqueness of a solution $r(u)$ is ensured by the results for (II). In the next lemma we shall obtain bounds for the solution of (III) which will enable us to continue it for all u in (a, c).

We introduce the functional

$$J(u) = \frac{r^{n-1}(u)}{(1+r'^2(u))^{1/2}} - \frac{1}{n}f(u)r^n(u). \tag{2.11}$$

Note that

$$J'(u) = -\frac{1}{n}f'(u)r^n(u) < 0 \tag{2.12}$$

if $f' > 0$.

LEMMA 2.2. *Suppose H4 holds, and let $r(u)$ be the solution of (III) as long as it exists. Then, assuming always that $u > a$, we have*

$$r(u) < \frac{n}{f(u)}.$$

PROOF: It follows from (2.12) that

$$J(u) > J(c) = 0. \tag{2.13}$$

By (2.11), this implies

$$\frac{r^{n-1}(u)}{(1+(r'(u)^2)^{1/2}} > \frac{1}{n}f(u)r^n(u) \tag{2.14}$$

and the result follows since $r(u) > 0$ for $u < c$.

THEOREM 2.3. *Suppose H4 holds. Then the solution $r(u)$ of (III) can be continued over the interval $(a,c]$.*

PROOF: By Lemma 2.2 it is evident that $r(u)$ is bounded on any compact subset of $(a,c]$, as long as it exists. Moreover, from (2.11)–(2.13) if $r(u)$ can be continued over an interval $(b,c]$ with $b > a$ then

$$\liminf r(u) > 0 \quad \text{as} \quad u \to b.$$

Finally, by (2.14) the derivative r' is also bounded,

$$|r'(u)| \leq \frac{n}{f(u)} \frac{1}{r(u)} \quad \text{on} \quad (b,c).$$

Thus $r(u)$ can be continued on any interval $(d,c]$, $d > a$, which proves the assertion.

Let $r(u)$ be a solution of (III). Suppose that there is a first point (R,U) where $r'(U) = 0$, $r''(U) < 0$, as occurs for example in the case of Theorem 2.1. For convenience in notation, from here on we write u_1 instead of U. Clearly $r' > 0$ in a left neighborhood of u_1. Thus we can define

$$u_2 = \inf\{0 < u < u_1 : r' > 0 \text{ on } (u,u_1)\}.$$

In the following lemma we shall show that if $u_2 > a$, then $r(u)$ has an isolated minimum at u_2.

LEMMA 2.4. *Suppose H4 holds. Then*

(a) *The zeros of $r'(u)$ in (a,c) are isolated.*

Let $r'(u^*) = 0$ and $u^* > a$. Then

(b) *If $r' > 0$ in a right neighborhood of u^*, we have $r''(u^*) > 0$.*
(c) *If $r' > 0$ in a left neighborhood of u^*, we have $r''(u^*) < 0$.*

PROOF: (a) If $r'(u^*) = 0$ and $r''(u^*) \neq 0$ the assertion is proved. On the other hand, if $r''(u^*) = 0$ then by equation (1.5) and H4 we have $r'''(u^*) = -f'(u^*) < 0$, and the assertion is also proved.

(b) Clearly, $r''(u^*) \geq 0$. If $r''(u^*) = 0$, then $r'''(u^*) < 0$ and hence $r' < 0$ in a right neighborhood of u^*, contradicting the assumption.

(c) Now $r''(u^*) \leq 0$. Again, if we have equality, $r'''(u^*) < 0$, which implies that $r' > 0$ in a left neighborhood of u^*, in contradiction with the assumption.

Let $\{u_{2m}\}$, $m = 1, 2, \ldots$, denote the sequence of minima of $r(u)$, numbered so that

$$u_{2(m+1)} < u_{2m}, \quad m = 1, 2, \ldots,$$

and let

$$u_{2m-1} = \sup\{u > u_{2m} : r' > 0 \text{ on } (u_{2m}, u)\}.$$

Then by Lemma 2.4, $r(u)$ has an isolated maximum at u_{2m-1} for every $m \geq 1$ and $r' < 0$ on (u_{2m-1}, u_{2m-2}) except possibly at a finite number of possible inflection points.

LEMMA 2.5. *Suppose H4 holds. Then for u_{2m} and u_{2m-1} greater than a, we have*

$$r(u_{2m}) < \frac{n-1}{f(u_{2m})}, \quad m \geq 1$$

and

$$r(u_{2m-1}) > \frac{n-1}{f(u_{2m-1})}, \quad m \geq 1.$$

PROOF: By Lemma 2.4, $r''(u_{2m}) > 0$ and $r''(u_{2m-1}) < 0$. Using this in (1.5) yields the desired result.

Write $r_k = r(u_k)$, $k = 1, 2, \ldots$. Then we have

LEMMA 2.6. *Let H4 hold. Then for $u > a$ the sequence of minima $\{r_{2m}\}$ and the sequence of maxima $\{r_{2m-1}\}$ are both increasing as u decreases from c.*

PROOF: Consider two consecutive minima r_{2m} and r_{2m+2}. For convenience we set $u_{2m} = \alpha$, $u_{2m+1} = \beta$ and $u_{2m+2} = \gamma$. We shall prove (see Figure 3) that

$$r(\alpha) < r(\gamma).$$

Since $r' \neq 0$ on (γ, β) and a.e. on (β, α), we can define the respective inverses $u^-(r)$ and $u^+(r)$ on these intervals. Of course these functions satisfy (1.4) and (1.1) respectively. Consider values of r satisfying

$$\min\{r(\alpha), r(\gamma)\} \leq r \leq r(\beta).$$

If we write

$$I(u) = (1 + r'(u)^2)^{1/2},$$

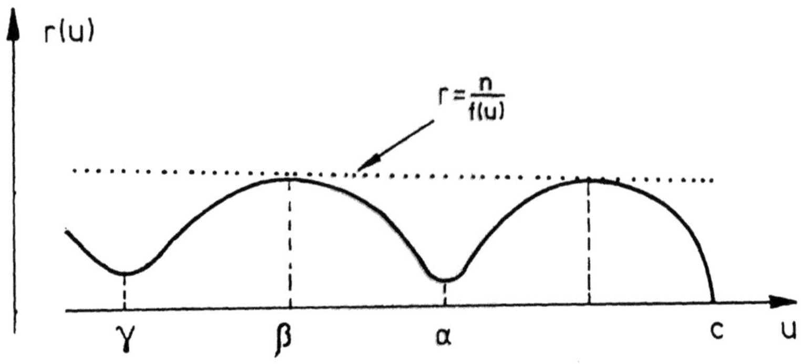

Figure 3. Graph of $r(u)$.

then integration of (2.10) over $u^-(r), u^+(r)$ yields

$$
\begin{aligned}
r^{n-1}\{I(u^+) - I(u^-)\} &= \int_{u^-}^{u^+} f(s) r^{n-1}(s) r'(s) ds \\
&= \int_{u^-}^{\beta} f(s) r^{n-1}(s) r'(s) ds + \int_{\beta}^{u^+} f(s) r^{n-1}(s) r'(s) ds \\
&= \int_{r}^{r(\beta)} \{f(u^-(\rho)) - f(u^+(\rho))\} d\rho.
\end{aligned}
\tag{2.15}
$$

Since $f(u^-(\rho)) < f(u^+(\rho))$, the right-hand side of (2.15) is negative. In turn we have $I(u^+) < I(u^-)$ and so

$$\left|r'(u^+(r))\right| > \left|r'(u^-(r))\right|.$$

It follows that if we decrease r, then $r'(u^-(r))$ must vanish before $r'(u^+(r))$, which means that $r(\gamma) > r(\alpha)$.

The proof of the assertion that the maxima also increase with decreasing u proceeds in a similar manner.

We conclude this section with a lemma which limits the number of minima and maxima.

LEMMA 2.7. *Let H4 hold, and suppose $r(u)$ is a solution of (III) on $[b, c]$, where $0 \leq b < c$. Then $r(u)$ can have at most a finite number of minima and maxima.*

PROOF: By Lemma 2.4 it is enough to consider the case $b \leq a$.

If $b < a$, then by inspection of (1.5) it is evident that any critical point of $r(u)$ in $(b, a]$ is a strict local minimum. It follows that $r(u)$ can have at most one critical point on $(b, a]$ and also, by Lemma 2.4 again, that $r(u)$ can have at most a finite number of maxima and minima on $(a, c]$.

When $b = a$, suppose to the contrary of the lemma that $\{u_k\}$ is an infinite decreasing sequence of critical points of $r(u)$ on (a, c). Then,

$$\lim_{k\to\infty} u_k = u^* \geq a.$$

By Lemma 2.4 critical points are isolated, so that $u^* = a$.

If we integrate equation (1.5) over (u, u_k), for some $k \geq 1$, we obtain

$$\int_0^{r'(u)} \frac{dp}{(1+p^2)^{3/2}} \\ = -(n-1)\int_u^{u_k} \frac{du}{r(u)(1+r'(u)^2)^{1/2}} + F(u_k) - F(u). \tag{2.16}$$

By Lemma 2.6 we see that $r(u)$ is bounded away from zero on (a, u_k) uniformly in k. Hence, carrying out the integration in (2.16) and taking absolute values,

$$\frac{|r'(u)|}{(1+r'(u)^2)^{1/2}} \leq C(u_k - a) \quad \text{on } (a, u_k),$$

where C is a constant which does not depend on k. Since we can choose k arbitrarily large, and thus u_k arbitrarily near a, this means that

$$\lim_{u\to a} r'(u) = 0.$$

It follows in turn that

$$\lim_{u\to a} r(u) = r_\infty > 0.$$

Returning to (2.16) we now find that, for any $u > a$,

$$\begin{aligned}\frac{r'(u)}{(1+r'(u)^2)^{1/2}} &= -(n-1)(1-\varepsilon)\frac{u_k - u}{r_\infty} + \int_u^{u_k} f(s)ds \\ &\leq -(n-1)(1-\varepsilon-\varepsilon')\frac{u_k - u}{r_\infty},\end{aligned}$$

where $\varepsilon, \varepsilon' \to 0$ as $u_k \to a$. Hence for all u sufficiently near a we have $r'(u) < 0$, and no oscillations occur.

COROLLARY 2.8. *Let H4 hold, and let u be a solution of (III) on $(0, c]$. Then $r(u)$ can have at most a finite number of maxima and minima. In particular, a generalized radial ground state can have at most a finite number of oscillations.*

3. An existence theorem.

In this section we assume that $a = 0$ and thus that $f(u) > 0$ for all $u > 0$. We prove Theorem 1.

Let c be any positive number, and let $r(u)$ be the solution of equation (1.5) with the properties

$$r(u) \to 0, \quad r'(u) \to -\infty \text{ as } u \to c. \tag{3.1}$$

As was shown in the previous section, this solution can be continued as a positive function to $u = 0$. Thus, to prove that it is a (generalized) ground state it is enough to show that

$$r(u) \to \infty \text{ as } u \to 0. \tag{3.2}$$

To prove (3.2) we use an argument first used in [**BLP**] to establish the existence of ground states for semilinear elliptic equations in the supercritical case; it exploits the properties of a certain functional, related to one first introduced by Pohožaev [**P**] for semilinear equations. The appropriate functional for equation (1.5) turns out to be

$$H(u) = r^n + \frac{r^{n-1}}{(1+r'^2)^{1/2}}\left\{rr' - \frac{n-2}{2}u\right\} + \frac{n-2}{2n}r^n u f(u), \tag{3.3}$$

where $r = r(u)$.

An elementary computation shows that

$$H'(u) = \frac{nr^{n-1}}{(1+r'^2)^{1/2}}\left\{r'^2 + \tfrac{1}{2} + r'(1+r'^2)^{1/2}\right\} + \frac{n-2}{2n}r^n\left\{uf'(u) - \frac{n+2}{n-2}f(u)\right\}.$$

It is easy to check that $H'(u) > 0$ if

$$uf'(u) - \frac{n+2}{n-2}f(u) \geq 0$$

as is assumed in H3*. Since $H(c) = 0$ it follows then that

$$\lim_{u \to 0} H(u) < 0. \tag{3.4}$$

By Lemma 2.7, $r(u)$ is ultimately monotonic as $u \to 0$, so

$$r_\infty = \lim_{u \to 0} r(u) \text{ exists (finite or infinite).}$$

Suppose by way of contradiction that $r_\infty < \infty$. Using (3.3) we then get

$$H(u) > -\frac{1}{2}(n-2)r^{n-1}u > -\operatorname{const} u.$$

Hence

$$\lim_{u \to 0} H(u) \geq 0.$$

This inequality however disagrees with (3.4). Thus $r_\infty = \infty$, and (3.2) is proved.

Since c was an arbitrary positive number, we have proved Theorem 1, the existence of a one-parameter family of generalized ground states $r(u)$ of (1.5)–(1.8).

4. A nonexistence theorem.

In this section we assume that $f(u)$ has a *positive* zero, namely that H2 holds with $a > 0$. We prove Theorem 2.

Suppose for contradiction that $r(u)$ is a solution of (1.5) which satisfies (1.6)–(1.8). As in the previous section we shall compute the function $H(u)$ for this solution, but now derive a contradiction.

As before, by H3*,

$$H(c) = 0 \quad \text{and} \quad H'(u) > 0 \quad \text{for} \quad a < u < c$$

and hence

$$H(a) < 0. \tag{4.1}$$

To discuss the behavior of H on $(0, a)$ we follow an argument from [**AP**]. We write

$$H(u) = K(u) + L(u),$$

where

$$K(u) = r^n + \frac{r^{n-1}}{(1+r'^2)^{1/2}}\left(rr' - \frac{n-2}{2}u\right)$$

and

$$L(u) = \frac{n-2}{2n} r^n u f(u).$$

Clearly, since $r(a)$ is finite,

$$L(a) = 0.$$

In the remainder of this section we shall show that

$$K(a) > 0, \tag{4.2}$$

and thus that

$$H(a) > 0.$$

Since this does not agree with (4.1) we have then obtained the desired contradiction.

Before turning to the proof of (4.2) we observe from the differential equation (1.5) that $r(u)$ can have at most one critical point on $(0, a)$, which can only be an isolated minimum. Thus if d is such a critical point then

$$r'(u)\begin{cases} < 0 & \text{on } (0, d) \\ > 0 & \text{on } (d, a). \end{cases}$$

If $r(u)$ has no critical point on $(0, a)$ then $r' < 0$ on $(0, a)$, for ultimately $r(u) \to \infty$ as $u \to 0$.

LEMMA 4.1. $K(0) = 0$.

PROOF: We write for convenience

$$K(u) = K_1(u) - K_2(u),$$

where

$$K_1(u) = r^n\{1 + r'(1 + r'^2)^{-1/2}\}$$

and

$$K_2(u) = \frac{n-2}{2} u r^{n-1}(1 + r'^2)^{-1/2}.$$

For u sufficiently small, the inverse $u(r)$ of $r(u)$ exists, and we can write (recall $r' < 0$)

$$K_1(u(r)) = r^n\{1 - (1 + u'(r)^2)^{-1/2}\}.$$

As in [**PS1**] one can show, since $n > 2$, that

(4.3) $$\lim_{r\to\infty} r^{n-1}u'(r) \text{ exists (finite)}$$

(see also [**FLS**]). Thus

$$K_1(u(r)) = O(r^{2-n}) \text{ as } r \to \infty$$

and so

$$\lim_{u\to 0} K_1(u) = 0.$$

Also, by (4.3)

$$\lim_{u\to 0} K_2(u) = 0$$

and therefore, as required,

$$\lim_{u\to 0} K(u) = 0.$$

LEMMA 4.2. *$K'(u) > 0$ for $0 < u < a$.*

PROOF: An elementary computation yields

$$K'(u) = \frac{nr^{n-1}}{(1+r'^2)^{1/2}}\left\{r'^2 + \tfrac{1}{2} + r'(1+r'^2)^{1/2}\right\} + r^{n-1}|f(u)|\left(r + \frac{n-2}{2}ur'\right)$$

because $f(u) < 0$ on $(0,a)$. Therefore

(4.4) $$K'(u) > r^{n-1}|f(u)|\left(r + \frac{n-2}{2}ur'\right), \quad 0 < u < a$$

as in the previous section.

Suppose $r(u)$ has a critical point d on $(0,a)$. Then $r' > 0$ on (d,a) and hence $K' > 0$ on $[d,a)$. Thus it remains to show that

$$K' > 0 \text{ on } (0,d),$$

where it will be understood that $d = a$ if $r(u)$ has no critical point on $(0,a)$.

We write

(4.5) $$r(u) + \frac{n-2}{2}ur'(u) = |r'(u)|z(r(u)),$$

where

$$z(r) = -ru'(r) - \frac{n-2}{2}u(r).$$

Then by (1.1), since $r' < 0$ and $u' < 0$ for $0 < u < d$,

$$\begin{aligned} z'(r) &= -ru''(r) - \frac{n}{2}u'(r) \\ &= u'\left\{(n-1)(1+u'^2) - \frac{n}{2}\right\} + rf(u)(1+u'^2)^{3/2} < 0. \end{aligned}$$

By (4.3) and the fact that $n > 2$ we have

$$\lim_{r\to\infty} z(r) = 0.$$

Hence

$$z(r) > 0 \quad \text{when } u(r) < d,$$

and therefore, by (4.4) and (4.5),

$$K'(u) > 0 \quad \text{when } 0 < u < d,$$

and Lemma 4.2 is proved.

Lemmas 4.1 and 4.2 together imply (4.2). This completes the proof of Theorem 2.

5. The form of generalized radial ground states.

Here we consider the form of a generalized ground state. Such a solution, as defined in the Introduction, is an embedded surface S in $\mathbb{R}^n \times \mathbb{R}^+$, whose mean curvature at each point $(x, u) \in \mathbb{R}^n \times \mathbb{R}^+$ is given by

$$K(x, u) = -\frac{1}{2n}f(u)$$

and which is asymptotic to the plane $u = 0$ as $x \to \infty$. Assuming additionally that S is radially symmetric about the axis $x = 0$, $u > 0$, and that the hypotheses H1 and H2 hold for $f(u)$, we show that S can be represented in radial coordinates by the single-valued relation

$$r = r(u), \quad 0 < u \leq c,$$

where $r = |x|$ and the point $u(0) = c$ is the central value of S. We show moreover that $f(c) > 0$ and that $r(u)$ is positive and of class C^2 for $0 < u < c$.

PROOF: Since S is a radial ground state, it follows from (1.6) that S must be cut by the line $x = 0$, $u > 0$ in only a single point, say $(0, c)$.

Suppose first that $f(c) > 0$, the case $f(c) \leq 0$ being dealt with later. It is evident from the fact that S is an embedded surface that the distinguished normal at the central point $(0, c)$ must be directed vertically upward, consistent with its direction at large values of x. The mean curvature $K(0, c) = -\frac{1}{2n} f(c)$ associated with the central point therefore makes S concave downward in the neighborhood of $(0, c)$. Hence at least for some neighborhood of $(0, c)$ the surface S can be represented in the form $r = r(u)$, $c - \varepsilon < u \leq c$.

The proof proceeds by showing that this representation must continue to hold for all u, $0 < u \leq c$. In carrying out the argument it will be convenient to think of S as a non-self-intersecting curve (trajectory) in the first quadrant of the (r, u) plane, starting at $(0, c)$ and ultimately asymptotic to $u = 0$ as $r \to \infty$.

Since S is of class C^2 with bounded curvature it follows that continuation of the representation to all values of $u > 0$ can fail to occur only if as we proceed along the trajectory there occurs a first point $(\bar{r}, \bar{u})$ such that $\bar{u} > 0$ and

$$r'(u) \to \pm\infty \quad \text{as} \quad u \to \bar{u}.$$

Let us suppose first that $f(\bar{u}) > 0$, that is $\bar{u} > a$. Since S can be represented in the form $u = u(r)$ in the neighborhood of $(\bar{r}, \bar{u})$, it is clear from the construction that

$$u'(\bar{r}) = 0, \quad u''(\bar{r}) > 0$$

(the condition $u''(\bar{r}) = 0$ is impossible by either (1.1) or (1.4)). Clearly only (1.4) matches this behavior, so the trajectory of S must have reached (at least one) vertical point and be running in the "reverse" direction when it reaches $(\bar{r}, \bar{u})$.

It follows now that S moves into the region

$$D = \{(r, u) : \bar{u} < u < c, \quad 0 < r < r(u)\}.$$

An easy analysis of the possible critical points of S in this region shows that S has no way now to escape from D and thus cannot possibly be a ground state. Hence failure of the representation cannot occur before the trajectory reaches the plane $r = a$.

We remark that this much of the argument already completes the proof when $a = 0$.

Next (assuming $a > 0$) consider the case when $\bar{u} = a$. In this case an obvious uniqueness argument in ordinary differential equations shows that S must be exactly the surface $u \equiv a$, which again is impossible.

Finally, we suppose $\bar{u} < a$. The argument given earlier now shows that the point $(\bar{r}, \bar{u})$ lies on a "forward" moving part of S, so that the curve then must enter the region

$$E = \{(r, u) : u > \bar{u}, \quad r > \tilde{r}(u)\},$$

where $\tilde{r}(u)$ is defined to be $r(u)$ when $\bar{u} \leq u \leq c$ and 0 when $u > c$.

Since S is a gound state it follows from the construction that there will be a first point after $(\bar{r}, \bar{u})$ on the trajectory where $u = \bar{u}$. Call this point $(\bar{r}', \bar{u})$. We thus have the situation envisaged in Figure 1 of [**PS1**], except that possibly between the points $(\bar{r}, \bar{u})$ and $(\bar{r}', \bar{u})$ the trajectory will have several direction reversals. In any case, an adaptation of the proof given in [**PS1**] leads to a contradiction, and completes the proof. (If there are no direction reversals, the proof is almost exactly as in [**PS1**] except that the term $\frac{1}{2}u'^2$ must be replaced by $1 - (1 + u'^2)^{-1/2}$. Otherwise, if there are direction reversals, the trajectory between $(\bar{r}, \bar{u})$ and $(\bar{r}', \bar{u})$ may be divided into a finite number of pieces, each of which has no vertical points of the trajectory in its interior. The argument of [**PS1**] can then be applied separately to each of these pieces and a contradiction is then obtained as before.)

When $f(c) \leq 0$, the case previously avoided, we proceed exactly as in the last part of the proof above to obtain a contradiction.

REMARK: In some cases the conclusion that a radial generalized ground state has a single-valued inverse can be obtained even if we assume only that S is *immersed* rather than embedded. In particular, this occurs when H1, H2 and H4 hold. To see this, consider the continuation of the trajectory of S from the initial point $(0, c)$, as in Theorem 2.3. Then, as shown there, one cannot reach any point where $u > a$ and either $r(u) = 0$ or $r'(u)$ is infinite. The case $u \leq a$, finally, can be treated exactly as above.

In consequence of this remark, the result of Theorem 2 shows that there can be no ground states, either immersed or embedded, when H1, H2 and H3* hold.

Department of Mathematics, University of Toronto.

Mathematical Institute, University of Leiden.

Department of Mathematics, University of Minnesota.

REFERENCES

[AP] F. V. Atkinson and L. A. Peletier, *On non-existence of ground states*, to appear in Quart. J. Math. (Oxford Series) (1988).

[BLP] H. Berestycki, P. L. Lions and L. A. Peletier, *An ODE approach to the existence of positive solutions for semilinear problems in* $\mathbb{R}^n$., Indiana Univ. Math. J. **30** (1981), 141–157.

[CF] P. Concus and R. Finn, *The shape of a pendant liquid drop*, Phil. Trans. Roy. Soc. London (A) **292** (1979), 307–340.

[EL] T. K. Evers, *Numerical search for ground state solutions of the capillary equation*, M.Sc. Thesis, Department of Mathematics, Iowa State University (1985). See also: H. Levine, *Numerical searches for ground state solutions of a modified capillary equation and for solutions of the charge balance equation*, in "Proceedings of the Microprogram on Nonlinear Diffusion Equations and their Equilibrium States," Springer-Verlag, New York, 1988.

[F] R. Finn, "Equilibrium capillary surfaces," Springer-Verlag, New York, 1986.

[FLS] B. Franchi, E. Lanconelli and J. Serrin, *Esistenze e unicità degli stati fondamentali per equazioni ellittiche quasilineari*, Atti Accad. Naz. Lincei (8) **79** (1985), 121–126.

[NS1] W.-M. Ni and J. Serrin, *Non-existence theorems for quasilinear partial differential equations*, Supplemento ai Rendiconti del Circolo Mat. di Palermo (II) **8** (1985), 171–185.

[NS2] W.-M. Ni and J. Serrin, *Existence and nonexistence theorems for ground states of quasilinear partial differential equations. The anomalous case*, Accad. Naz. Lincei, Convegni del Lincei **77** (1986), 231–257.

[P] S. I. Pohožaev, *Eigenfunctions of the equation* $\Delta u + \lambda f(u) = 0$, Dokl. Akad. Nauk. SSSR **165** (1965), 36–39 (in Russian); *II*, Sov. Math. **6** (1965), 1408–1411 (in English).

[PS1] L. A. Peletier and J. Serrin, *Uniqueness of positive solutions of semilinear equations in* $\mathbb{R}^n$, Arch. Rational Mech. Anal. **81** (1983), 181–197.

[PS2] L. A. Peletier and J. Serrin, *Uniqueness of nonnegative solutions of semilinear equations in* $\mathbb{R}^n$, J. Diff. Eq. **61** (1986), 380–397.

[PS3] L. A. Peletier and J. Serrin, *Ground states for the prescribed mean curvature equation*, Proc. Amer. Math. Soc. **100** (1987), 694–700.

[S] J. Serrin, *Positive solutions of a prescribed mean curvature problem*, to appear in Springer Lecture Notes in Math.

Geometric concepts and methods in nonlinear elliptic Euler–Lagrange Equations

BY ILYA J. BAKELMAN

This paper is concerned with global connections between the integrand $F(x, u, p)$ of n-multiple integrals

$$I(u) = \int_B F(x, u, Du(x))\, dx$$

and a priori estimates for solutions of the corresponding elliptic Euler–Lagrange equations, whose gradients satisfy some prescribed limitations. Such problems arise in the calculus of variations, differential geometry and continuum mechanics. Typical examples of these problems are presented in Section 3.

Let B be a bounded domain in Euclidean space E^n, ∂B a closed continuous hypersurface in E^n and $\overline{B} = B \cup \partial B$. We denote by D an open domain in E^n with closure $\overline{D} \subset B$. We shall be concerned with C^0-estimates for solutions $u(x)$ belonging to $W_2^n(B) \cap C(\overline{B})$ of the Euler–Lagrange equation, that is, whose first and second (Sobolev) generalized derivatives are summable with power n in every such domain D. We are also concerned with the minimal assumptions of smoothness on the integrand $F(x, u, p)$ which provide C^0-estimates for solutions of Euler–Lagrange equations. It turns out that these assumptions are convenient to state in terms of the convexity of suitable hypersurfaces, constructed according to properties of the integrand $F(x, u, p)$ and in terms of the composition of generalized tangential mappings of solutions and convex hypersurfaces.

The main conclusions will be obtained in the form of necessary and sufficient conditions, which either interlock or coincide. They will be expressed by geometric inequalities between fundamental invariants of equation. The estimates obtained in this paper have various applications. From them follow the well known Bernstein estimates for two-dimensional problems (see [**1**]). It should be pointed out that Bernstein's technique is specifically two-dimensional and that he used the overly strong condition that $F(p)$

has order of growth more than one as $|p| \to \infty$. The new estimates also lead to different variants of the geometric maximum principle for solutions of linear and quasilinear elliptic equations whose coefficients satisfy appropriate conditions with respect to their gradients. They also cover some results of Alexandrov [**2**] and Bakelman [**6**], [**8**], [**9**] for equations without any limitations.

Finally I want to emphasize that various techniques based on different ideas of differential geometry were applied in the papers [**1**], [**2**], [**3**], [**4**], [**5**], [**6**], [**7**], [**8**], [**9**], [**11**], [**12**], [**13**], [**14**], [**15**], [**17**], [**18**], [**19**], [**20**], [**22**], [**23**] and many others to obtain C^0-estimates for solutions of linear and quasilinear elliptic equations.

CONTENTS

§1. Geometric constructions. Two-sided C^0-estimates of functions with prescribed Dirichlet data.

1.1. Geometric constructions. Let E^n, P^n and Q^n be three n-dimensional Euclidean spaces and let $x = (x_1, \dots, x_n)$, $p = (p_1, \dots, p_n)$, $q = (q_1, \dots, q_n)$ be points in E^n, P^n, Q^n respectively. We also introduce two $(n+1)$-Euclidean spaces $E^{n+1} = E^n \times R = \{(x, z)\}$ and $P^{n+1} = P^n \times R = \{(p, w)\}$. Let $0'$, $0''$, $0'''$ be the origins of E^n, P^n and Q^n. We assume that B is a bounded open domain in E^n. Let G be an open domain in P^n and let $0'$ be an interior point of G. The only possibilities are either $\partial G \neq \emptyset$ or $\partial G = \emptyset$. In the first case G is a bounded or unbounded domain in P^n, and $G \neq P^n$; but in the second case G must coincide with the space P^n. Let $r(G) = \operatorname{dist}(0', \partial G)$.

Clearly $r(G) < +\infty$ if and only if $\partial G \neq \emptyset$. Hence $r(G) = +\infty$ only in the case when $G = P^n$.

We denote by $U(\rho)$ the open n-ball: $|p| < \rho$ in P^n. Many important applications are related to convex n-domains in G. In particular, if G is the n-ball: $|p| < a$, then $r(G) = a$ and $U(r(G)) = G$.

Now let $\phi(p)$ be a convex function in G. We introduce the set function

$$\omega(\phi, e') = |\chi_\phi(e')|_Q \quad (1.1)$$ *

where $\chi_\phi \colon G \to Q^n$ is the normal mapping of the convex function $\phi(p)$. The set function $\omega(\phi, e')$ is non-negative and absolutely additive on the family of Borel subsets e' of G. If $\chi_\phi(G)$ is an unbounded set in the space Q^n, then $\omega(\phi, e')$ can take the value $+\infty$ for some Borel subset $e' \subset G$. But

$$\omega(\phi, e') < +\infty \quad (1.2)$$

for every Borel subset $e' \subset G$ such that $\mathrm{dist}(e', \partial G) > 0$.

Now we introduce the function

$$c(\rho) = \omega(\phi, U(\rho)) \quad (1.3)$$

for $0 \leq \rho < r(G)$ and denote by $C(\phi)$ the number

$$C(\phi) = \lim_{\rho \to r(G)} c(\rho). \quad (1.4)$$

Clearly

$$C(\phi) = \omega(\phi, U(r(G)) \quad \text{and} \quad 0 \leq C(\phi) < +\infty. \quad (1.5)$$

The case $C(\phi) = 0$ can be realized if ϕ is a linear function in G, and more generally when the graph of ϕ is a convex cylinder. The case $C(\phi) = +\infty$ can be realized if meas $\chi_\phi(G) = +\infty$. The simple examples are as follows:

1) $G = P^n$; $\phi(p) = |p|^2$;
2) G: $|p| < 1$, $\phi(p) = -(1 - |p|^2)^{1/2}$.

Clearly $C(\phi)$ coincides with the total area of the normal mapping χ_ϕ if $G = U(r(G))$.

Assume that the function $\phi(p)$ satisfies the following Assumptions.

---

* We denote by $|e|_E$, $|e'|_P$, $|e''|_Q$ the Lebesgue measures for the sets $e \subset E^n$, $e' \subset P^n$, $e'' \subset Q^n$.

Assumption 1. The function $\omega(\phi, e')$ is absolutely continuous on the ring of Borel subsets of G, i.e.

$$\omega(\phi, e') = \int_{e'} \det\left(\frac{\partial^2 \phi(p)}{\partial p_i \partial p_j}\right) dp \tag{1.6}$$

Assumption 2. Let W be the subset of G such that

$$\det\left(\frac{\partial^2 \phi(p)}{\partial p_i \partial p_j}\right) > 0 \tag{1.7}$$

everywhere. Then the n-volume of the set $W \cap A(\rho_1, \rho_2)$ is strictly positive for all numbers ρ_1, ρ_2 such that $0 < \rho_1 < \rho_2 < r(G)$, where $A(\rho_1, \rho_2)$ is the annulus $0 < \rho_1 < |p| < \rho_2 < r(G)$.

If a convex function $\phi(p)$ satisfies Assumptions 1 and 2, then the function $c(\rho)$ constructed above is non-negative, continuous and strictly increasing in $[0, r(G))$; clearly $c(\rho) > 0$ for $\rho > 0$. Let $\rho = b(t)$ be the inverse for the function $t = c(\rho)$. Then $b(t)$ is also non-negative, continuous and strictly increasing in $[0, C(\phi))$ and $b(t) > 0$ for $t > 0$.

If $r(G) < +\infty$, then the inverse $\rho = b(t)$ can be extended to $[0, C(\phi)]$ as a non-negative, continuous and strictly increasing function. Clearly

$$b(C(\phi)) = r(G) < +\infty. \tag{1.8}$$

We will use (1.8) in our main estimates.

1.2. Convex and concave supports of functions $u(x) \in W_2^n(B) \cap C(\overline{B})$. Let B be a bounded domain in E^n and let ∂B be a closed continuous hypersurface in E^n. We denote by $W_2^n(B) \cap C(\overline{B})$ a set of functions, which are continuous in $\overline{B}$ and whose first and second Sobolev generalized derivatives exist and are summable with degree n in every compact subdomain D of B, i.e. $\operatorname{dist}(D, \partial B) > 0$.

Let $u(x)$ be any function from $W_2^n(B) \cap C(\overline{B})$. We set

$$m = \inf u(x), \qquad M = \sup u(x) \tag{1.9}$$

for all $x \in \partial B$. The two-sided C^0-estimates for $u(x)$ in $\overline{B}$ are non-trivial only if

$$u_0 = \inf_{\overline{B}} u(x) < m \tag{1.10}$$

and

$$u_1 = \sup_{\overline{B}} u(x) > M. \tag{1.11}$$

It is sufficient only to establish the estimate of $u(x)$ from below; the estimate from above can be obtained in the similar way.

Let $\delta > 0$ be any real number satisfying the condition

$$m - \delta > u_0.$$

We will use the notation $m_\delta = m - \delta$. Let S_{m_δ} be the part of the graph of $u(x)$ located under the hyperplane

$$\gamma_{m_\delta} : z = m_\delta \tag{1.12}$$

in the space $E^{n+1} = E^n \times R$. Let $\overline{B}_{m_\delta}$ be a set in γ_{m_δ} whose projection on E^n coincides with $\overline{B}$ and let Γ_{m_δ} and Γ be the corresponding boundaries of $\overline{H}_{m_\delta}$ and $\overline{H}$, where $\overline{H}_m$ and $\overline{H}$ are the closed convex hulls of $\overline{B}_{m_\delta}$ and $\overline{B}$.

We denote by C_{m_δ} the closed convex hull of the set $\overline{B}_{m_\delta} \cup S_{m_\delta}$. Then

$$\partial C_{m_\delta} = \overline{H}_{m_\delta} \cup S_{v_\delta}, \tag{1.13}$$

where S_{v_δ} is the graph of a convex function $v_\delta(x) \in C(\overline{H})$ such that

$$v_\delta(x) = m_\delta$$

for all $x \in \partial H$, and

$$v_\delta(x) \leq u(x) \tag{1.14}$$

for all $x \in \overline{B}$. The convex function $v_\delta(x)$ is called the *convex support of the function* $u(x)$ (according to the number $\delta > 0$). If $S_{M_{\delta'}}$ is the part of the graph $u(x)$ located over the hyperplane $\gamma_{M_{\delta'}} : z = M_{\delta'}$ ($\delta' > 0$ is any number such that $M + \delta' < u_1$), then a similar geometric construction leads to the *concave support* $w_{\delta'}(x) \in C(\overline{H})$ *of the function* $u(x)$. Clearly $w_{\delta'}(x) = M_{\delta'}$ for all $x \in \partial H$ and $w_{\delta'}(x) \geq u(x)$ for all $x \in \overline{B}$.

1.3. Two-sided C^0-estimates for functions $u(x) \in W_2^n(B) \cap C(\overline{B})$. We assume as before that B is a bounded n-domain in E^n and ∂B is a closed continuous hypersurface in E^n. We also assume that

$$u(x) \in W_2^n(B) \cap C(\overline{B}) \tag{1.15}$$

and

$$\chi_u(B) \subset G, \tag{1.16}$$

where G is a prescribed n-domain in P^n, and the origin O' of P^n is an interior point of G. The concepts of the tangential mapping $\chi_u\colon B \to P^n$ and inclusion (1.16) for functions $u(x) \in W_2^n(B) \cap C(\overline{B})$ are explained below in this subsection.

Let $u_0 = \inf_{\overline{B}} u(x)$, $u_1 = \sup_{\overline{B}} u(x)$. Clearly only the cases

$$u_0 < m \quad \text{and} \quad u_1 > M \tag{1.17}$$

are interesting for consideration. Actually we have the inequalities $m \leq u(x)$ and $u(x) \leq M$ for all other cases.

It is sufficient to investigate the estimate for $u(x)$ from below, because the estimate for $u(x)$ from above can be obtained in a similar way. Let $\delta > 0$ satisfy the inequality

$$m - \delta > u_0 \tag{1.18}$$

and let $v_\delta(x)$ be the convex support of the function $u(x)$. We denote by x_0 an interior point of B such that

$$u_0 = \inf_{\overline{B}} u(x) = u(x_0). \tag{1.19}$$

Clearly the point $M_0(x_0, u_0)$ belongs simultaneously to both the graph of the function $u(x)$ and the graph of its convex support $v_\delta(x)$. According to our notation $m_\delta = m - \delta$. Therefore from (1.18) it follows that

$$m_\delta > u_0 \quad \text{or} \quad m_\delta - u_0 > 0. \tag{1.20}$$

Now consider two convex cones K_0 and K_1 with a common vertex at the point M_0. The cone K_0 has Γ_{m_δ} as its base and the cone K_1 is a convex cone of revolution, whose base is an $(n-1)$ sphere $\sum_{m_\delta} \subset \gamma_{m_\delta}$. The center of $\sum_{m_\delta}$ is a point (x_0, m_δ) in the hyperplane γ_{m_δ} and the radius of $\sum_{m_\delta}$ is equal to diam B. We denote by T_{m_δ} an open ball in γ_{m_δ}, whose boundary is $\sum_{m_\delta}$. Let T be the projection of T_{m_δ} on the hyperplane $E^n\colon z = 0$. Then clearly

$$\chi_{K_1}(T) \subset \chi_{K_0}(\text{int } H) \subset \chi_{v_\delta}(\text{int } H), \tag{1.21}$$

where H is the closed convex hull of the set $\overline{B}_m$.

Since K_1 is the cone of revolution, then $\chi_{K_1}(T)$ is a closed n-ball in P^n. The inequality

$$(1.22) \qquad |p| \le \frac{h_\delta}{\text{diam } B}$$

describes the n-ball $\chi_{K_1}(T)$, where

$$(1.23) \qquad h_\delta = m_\delta - u_0 > 0$$

is the height of the convex cone K_1.

Now we want to obtain the estimate

$$(1.24) \qquad h_\delta \le A,$$

where $A =$ const $< +\infty$ depends only on some geometric invariants of the graph of the function $u(x)$ in the space E^{n+1}. These invariants will be constructed by means of prescribed convex function $\phi(p)$, which is defined in a given domain G in P^n and which satisfies Assumptions 1 and 2. Then from (1.18), (1.20), (1.23) and (1.24) it follows that

$$(1.25) \qquad \inf_{\overline{B}} u(x) = u_0 = m_\delta - h_\delta \ge m - \delta - A,$$

where, as we outlined above, the constant A does not depend on δ. Finally we will prove that δ can be omitted in (1.25). Note that we consider estimates for functions $u(x) \in W_2^n(B) \cap C(\overline{B})$.

Clearly the inclusion

$$(1.16) \qquad \chi_u(B) \subset G$$

is the natural Assumption, which appears in the proof of estimate (1.24). But inclusion (1.16) is convenient to use only for functions $u(x) \in W_2^n(B) \cap C^1(\overline{B})$; this inclusion is not convenient for functions $u(x) \in W_2^n(B) \cap C(\overline{B})$. We will see below that some modification of Assumption (1.16) is convenient in the proof of estimate (1.24) for functions $u(x) \in W_2^n(B) \cap C(\overline{B})$. This modified Assumption is based on the concept of global convex and global concave points of the graph of functions $u(x) \in C(\overline{B})$. A point $M(x, u(x))$ of the graph of $u(x)$, where $x \in B$ is said to be *global convex (global concave)*, if there exists at least one supporting hyperplane α passing through M and lying under (over) the graph of $u(x)$. If such a supporting hyperplane α has the equation $z = p_1^0 x_1 + \cdots + p_n^0 x_n + b^0$, then the point

$$p_0 \stackrel{\text{def}}{=} \chi(\alpha) = (p_1^0, p_2^0, \ldots, p_n^0) \in P^n$$

is called the normal image of α. We denote by

$$\nu_u^+(B) = \bigcup_{\alpha^+} \chi(\alpha^+), \tag{1.26}$$

$$\nu_u^-(B) = \bigcup_{\alpha^-} \chi(\alpha^-), \tag{1.27}$$

$$\nu_u(B) = \nu_u^+(B) \cup \nu_u^-(B), \tag{1.28}$$

where α^+ (or α^-) is a supporting hyperplane passing through a global convex (or global concave) point of the graph of $u(x)$. All such supporting hyperplanes α^+ (or α^-) at all global convex (or global concave) points are considered in (1.26) (or (1.27)).

Assumption 3. Let a convex function $\phi(p)$ be defined in a domain G in P^n. Then we consider only functions $u(x) \in W_2^n(B) \cap C(\overline{B})$ for which the inclusion

$$\nu_u(B) \subset G \tag{1.29}$$

holds.

Let

$$D_{m_\delta} = S_{v_\delta} \cap S_{m_\delta} \tag{1.30}$$

(see p. 8 for notations in (1.30)), and let D be the projection of D_{m_δ} on the hyperplane E^n. Clearly D_δ is a closed subset of B, and $\operatorname{dist}(D_\delta, \partial B) > 0$. From the definition of a convex hull it follows that every supporting hyperplane α of the convex hypersurface S_{v_δ} is also a supporting hyperplane of the hypersurface $S_{m_\delta}: z = u(x) \in W_2^n(B) \cap C(\overline{B})$ at some global convex point. Clearly $\alpha \cap D_{m_\delta} \neq \emptyset$, if $\alpha \cap S_{v_\delta}$ contains a point $(x, v_\delta(x))$, where $x \in \text{ int } H$. Now we assume that the function $u(x)$ satisfies Assumption 3. Then

$$\chi_{v_\delta}(\text{int } H) \subset \nu_u^+(B) \subset \nu_u(B) \subset G. \tag{1.31}$$

Thus from (1.21) and (1.31) it follows that

$$\chi_{K_1}(T) \subset \chi_{v_\delta}(\text{int } H) \subset \nu_u^+(B) \subset \nu_u(B) \subset G. \tag{1.32}$$

Clearly

$$\chi_{v_\delta}(D_\delta) = \chi_{v_\delta}(B) = \chi_{v_\delta}(\text{int } H). \tag{1.33}$$

Thus we obtain the equalities

$$\begin{aligned} |(\chi_\phi \circ \chi_{v_\delta})(D_\delta)|_Q &= |(\chi_\phi \circ \chi_{v_\delta})(B)|_Q = \\ &= |(\chi_\phi \circ \chi_{v_\delta})(\text{int } H)|_Q, \end{aligned} \tag{1.33a}$$

where $(\chi_\phi \circ \chi_{v_\delta})(e) = \chi_\phi(\chi_v(e))$ for all Borel subsets e of the set int H. Now we can see that for every convex function $\phi(p)$ satisfying Assumptions 1 and 2 and for every function $u(x) \in W_2^n(B) \cap C(\overline{B})$ satisfying Assumption 3, there exists the number

$$\omega_+(\phi, u, \delta) = |(\chi_\phi \circ \chi_{v_\delta})(\text{int } H)|_Q. \tag{1.34}$$

Clearly

$$\omega_+(\phi, u, \delta) = |(\chi_\phi \circ \chi_{v_\delta})(B)|_Q = |(\chi_\phi \circ \chi_{v_\delta})(D_\delta)|_Q \tag{1.34a}$$

if $\delta \in (0, \delta_u)$, where $\delta_u = m - u_0 > 0$, $m = \inf_B u(x)$ and $u_0 = \inf_{\overline{B}} u(x)$. Clearly $\omega_+(\phi, u, \delta)$ is the area of the normal mapping $\chi_\phi\colon G \to Q^n$ computed on the set $\chi_{v_\delta}(\text{int } H)$. From (1.33) it follows that int H can be replaced either by D_δ or by B in the last statement.

LEMMA 1. *Let $\partial G \neq \emptyset$ (i.e. $r(G) < +\infty$), $\phi(p)$ satisfy Assumptions 1 and 2, and the function $u(x) \in W_2^n(B) \cap C(\overline{B})$ satisfy Assumption 3. Let also $\delta_u = m - u_0 > 0$. Finally let the inequality*

$$\omega_+(\phi, u, \delta) \leq C(\phi) \tag{1.35}$$

hold for any $\delta \in (0, \delta_u)$, where $u_0 = \inf_{\overline{B}} u(x)$. Then the inequalities

$$0 < h_\delta \leq b(\omega_+(\phi, u, \delta)) \text{ diam } B \tag{1.36}$$

hold. Moreover, if $\omega_+(\phi, u, \delta) = C(\phi)$, then $b(\omega_+(\phi, u, \delta)) = \delta(G)$ in (1.36).

PROOF: Since

$$0 < \delta < \delta_u = m - u_0 \tag{1.37}$$

then from (1.37) we obtain

$$h_\delta = m_\delta - u_0 = (m-\delta) - u_0 = \delta_u - \delta > 0.$$

The positive number h as we have seen above is the height of the convex cone of revolution K_1. According to our constructions

$$|(\chi_\phi \circ \chi_{K_1})(T)|_Q \leq |(\chi_\phi \circ \chi_{v_\delta})(\text{int } H)|_Q = \omega_+(\phi, u, \delta). \tag{1.38}$$

Now from (1.35) and the properties of the function $\rho = b(t)$, we obtain the inequality

$$b(\omega_+(\phi, u, \delta)) \leq b(C(\phi)) = r(G) < +\infty. \tag{1.39}$$

where the equality corresponds to the case

$$\omega_+(\phi, u, \delta) = C(\phi). \tag{1.40}$$

Since $\chi_{K_1}(T)$ is the n-ball

$$|p| \leq \frac{h_\delta}{\text{diam } B} \tag{1.41}$$

in the space P^n, then from (1.38–39) and (1.41) it follows that

$$\chi_{K_1}(T) \subset U(r(G)) \cap \chi_{v_\delta}(\text{int } H)$$

and

$$\frac{h_\delta}{\text{diam } B} \leq b(\omega_+(\phi, u, \delta)) \leq r(G) < +\infty \tag{1.42}$$

consecutively. Now the desired estimates (1.36) follow directly from inequalities (1.37a) and (1.42). The proof of Lemma 1 is completed.

LEMMA 2. *Let $\partial G = \emptyset$ (i.e. $r(G) = +\infty$), let the convex function $\phi(p)$ satisfy the conditions of Lemma 1 and let $u(x) \in W_2^n(G) \cap C(\overline{B})$. Let also $\delta_u = m - u_0 > 0$. Finally let the inequality*

$$\omega_+(\phi, u, \delta) < C(\phi) \tag{1.43}$$

hold for any $\delta \in (0, \delta_u)$. Then the inequalities

$$0 < h_\delta \leq b(\omega_+(\phi, u, \delta)) \text{ diam } B \tag{1.44}$$

hold for any $x \in \overline{B}$.

Since $r(G) = +\infty$ and $U(r(G)) = P^n$, then clearly

$$\chi_{K_1}(T) \subset \chi_v(int\ H) \subset P^n = U(r(G)).$$

Therefore Lemma 2 is a simplification of Lemma 1 both in the statement and in the proof.

Now we consider the properties of $\omega_+(\phi, u, \delta)$ which depend on the number $\delta \in (0, \delta_u)$, where $\delta_u = m - u_0 > 0$. If $0 < \delta_2 \leq \delta_1 < \delta_u$, then for every supporting hyperplane of the graph of function $v_{\delta_1}(x)$, $x \in$ int H there exists a parallel supporting hyperplane of the graph of function $v_{\delta_2}(x)$, $x \in$ int H. This statement follows directly from the definition of a supporting hyperplane to the set and the constructions of the functions $v_{\delta_1}(x)$ and $v_{\delta_2}(x)$ for $x \in$ int H. Thus

$$\omega_+(\phi, u, \delta_1) \leq \omega_+(\phi, u, \delta_2).$$

Hence the limit (finite or infinite) $\omega_+(\phi, u) = \lim_{\delta \to 0^+} \omega_+(\phi, u, \delta)$ exists and is positive. The case $\omega_+(\phi, u) = +\infty$ is not excluded from our considerations.

If $\delta_u = m - u_0 = 0$, then $u(x) \geq m$ for all $x \in \overline{B}$. Thus $\omega_+(\phi, u) = 0$ in this case. Conversely, if $\omega_+(\phi, u) = 0$, then clearly $\delta_u = m - u_0 = 0$. The number $\omega_+(\phi, u)$ is called the *ϕ-total area of convex support* for a function $u(x) \in W_2^n(B) \cap C(\overline{B})$, subject to Assumption 3. The *ϕ-total area* $\omega_-(\Phi, u)$ *of concave support* for the same function $u(x)$ can be constructed in a similar way.

THEOREM 1. *Let $\partial G \neq \emptyset$ (i.e. $r(G) < +\infty$) and a convex function $\phi(p)$, defined in $G \subset P^n$, satisfy Assumptions 1 and 2. Let the function $u(x) \in W_2^n(B) \cap C(\overline{B})$ satisfy Assumption 3. Then inequalities*

$$m - b(\omega_+(\phi, u))\ diam\ B \leq u(x) \leq M + b(\omega_-(\phi, u))\ diam\ B \tag{1.45}$$

hold for all $x \in \overline{B}$, if

$$\omega_\pm(\Phi, u) \leq C(\phi), \tag{1.46}$$

where, as usual, $m = \inf_{\partial B} u(x)$, $M = \sup_{\partial B} u(x)$.

PROOF: The cases $\omega_+(\phi, u) = 0$ or $\omega_-(\phi, u) = 0$ lead to the respective estimates $u(x) \geq m$ or $u(x) \leq M$ for all $x \in \overline{B}$. Therefore only the cases

$$u_0 = \inf_{\overline{B}} u(x) < m \quad \text{and} \quad u_1 = \sup_{\overline{B}} u(x) > M$$

are interesting. Let $\delta > 0$ be an arbitrary number, subject to the condition $\delta < m - u_0$. Clearly

$$\omega_+(\phi, u, \delta) \leq \omega_+(\phi, u) \leq C(\phi).$$

Since

$$m_\delta - h_\delta \leq u_0 \leq u(x) \tag{1.47}$$

for all $x \in \overline{B}$, then from (1.47) and Lemma 1 it follows that

$$m - \delta - b(\omega_+(\phi, u, \delta)) \text{ diam } B \leq u_0 \leq u(x) \tag{1.48}$$

for all $x \in \overline{B}$. The function $\rho = b(t)$ is continuous and strictly increasing in $[0, C(\phi)]$ and

$$\lim_{\delta \to 0^+} \omega_+(\phi, u, \delta) = \omega_+(\phi, u).$$

Thus (1.48) becomes

$$m - b(\omega_+(\phi, u)) \text{ diam } B \leq u(x) \tag{1.49}$$

for all $x \in \overline{B}$, if $\delta \to 0^+$ in inequalities (1.48).

If we apply the same considerations to the function $-u(x)$, then we obtain the inequalities

$$-u(x) \geq -M - b(\omega_-(\phi, u)) \text{ diam } B \tag{1.50}$$

for all $x \in \overline{B}$. From (1.49) and (1.50) it follows that inequalities (1.45) hold in $\overline{B}$. The proof of Theorem 2 is completed.

REMARK: Since $r(G) < +\infty$, then $\omega_+(\phi, u)$ or $\omega_-(\phi, u)$ can be replaced by $C(\phi)$ in (1.45), if $\omega_+(\phi, u) = C(\phi)$ or $\omega_-(\phi, u) = C(\phi)$.

THEOREM 2. *Let $\partial G = \emptyset$ (i.e. $r(G) = +\infty$) and let a convex function $\phi(p)$ defined in $G = P^n$ satisfy Assumptions 1 and 2. Let $u(x) \in W_2^n(B) \cap C(\overline{B})$. Then inequalities*

$$m - b(\omega_+(\phi, u)) \text{ diam } B \leq u(x) \leq M + b(\omega_-(\phi, u)) \text{ diam } B, \tag{1.51}$$

hold for all $x \in \overline{B}$, if

$$\omega_\pm(\phi, u) < C(\phi). \tag{1.52}$$

The proof of this Theorem can be obtained by Lemma 2 in the same way as the proof of Theorem 1 by Lemma 1.

REMARK: Since $b(C(\phi)) = +\infty$, then inequalities (1.52) must be strict. In the opposite case, inequalities (1.51) are not interesting.

Now consider the computation of the number $C(\phi) = |\chi_\phi(U(r(G)))|_Q$, which is one of the main invariants in the statements of Theorems 1 and 2. The total area $\sigma(\phi)$ of the normal mapping χ_ϕ is expressed by the formula

$$\sigma(\phi) = |\chi_\phi(G)|_Q = \int_G \det\left(\frac{\partial^2\phi(p)}{\partial p_i \partial p_j}\right) dp. \tag{1.53}$$

Clearly $C(\phi) \leq \sigma(\phi)$. Below we assume that G is either the n-ball: $|p| < a$ or the entire space P^n. Then $U(r(G)) = G$ and therefore $C(\phi) = \sigma(\phi)$. Now we additionally assume that the graph of $\phi(p)$ in the space P^{n+1} is a complete convex hypersurface. Let K_ϕ be the asymptotic cone of this complete convex hypersurface $w = \phi(p)$. Since $\chi_\phi(G) = \chi_{K_\phi}(G)$, then $\sigma(\phi) = |\chi_{K_\phi}(G)|_Q$. If G is any convex bounded domain, then K_ϕ is a ray orthogonal to P^n. Hence $\chi_{K_\phi}(G) = Q^n$ and $\sigma(\phi) = +\infty$. Thus if G is the n-ball: $|p| < a$ ($a =$ const $< +\infty$) and if the graph of $\phi(p)$ is a complete convex hypersurface, then

$$C(\phi) = \sigma(\phi) = +\infty. \tag{1.54}$$

Hence the crucial inequalities (1.46) become

$$\omega_\pm(\phi, u) \leq +\infty. \tag{1.55}$$

(see Theorem 1). Now let K_ϕ be projected one-to-one onto P^n. Then K_ϕ is a non-degenerate convex cone. Therefore $\chi_{K_\phi}(G)$ is a bounded closed set in Q^n. Clearly $\sigma(\phi) = |\chi_{K_\phi}(G)|_Q < +\infty$ and $G = P^n$. Thus the crucial inequalities (1.52) (see Theorem 2) become

$$\omega_\pm(\phi, u) < \sigma(\phi), \tag{1.56}$$

because $C(\phi) = \sigma(\phi)$ in this case.

§2. Applications to the Dirichlet problem for the Euler–Lagrange equation.

$$(2.0)\qquad \sum_{i,k=1}^{n} F_{p_i p_k}(Du(x))u_{ik} = nf_u(x, u(x))$$

As we mentioned above, the functional, corresponding Euler–Lagrange equation (2.0), is

$$(2.1)\qquad J(u) = \int_B [F(Du) + nf(x,u)]dx$$

We assume that B is a bounded domain in E^n and ∂B is a closed continuous hypersurface in E^n, $u(x) \in W_2^n(B) \cap C(\overline{B})$ is a solution of equation (2.0) satisfying prescribed continuous Dirichlet data on ∂B and also satisfying Assumption 3:

$$(2.2)\qquad \nu_u(B) \subset G$$

(see Section 1), where G is prescribed bounded domain in P^n, for which $r(G) = \operatorname{dist}(0', \partial G) > 0$. Remember that $G = P^n$ if and only if $r(G) = +\infty$. We also assume that $F(p)$ is a convex function, defined in G, and that $F(p)$ satisfies Assumptions 1 and 2 (see Section 1). Now we formulate Assumptions 4 and 5 concerning the properties of the function $f(x,u)$.

Assumption 4. A function $f(x,u) \in C(\overline{B} \times R)$ is convex with respect to u for every fixed $x \in \overline{B}$.

Therefore the derivative $f_u(x,u)$ is an increasing function of u for every fixed $x \in \overline{B}$. We denote by $f_u^+(x,u)$ and $f_u^-(x,u)$ the positive and negative parts of $f_u(x,u)$.

Assumption 5. The functions $f_u^+(x,k)$ and $f_u^-(x,k)$ are locally summable with degree n in B, where $-\infty < k < +\infty$ is any constant.

THEOREM 3. *Let $u(x) \in W_2^n(B) \cap C(\overline{B})$ be a solution of the Euler–Lagrange equation for the functional $J(u)$ with prescribed continuous Dirichlet data on ∂B. If Assumptions 1–5 are satisfied, then the inequalities*

$$(2.3)\qquad \omega_+(F,u) \le \int_B [f_u^+(x,m)]^n dx$$

and

$$(2.4) \qquad \omega_-(F,u) \le \int_B [f_u^-(x,M)]^n dx$$

hold, where $m = \inf_{\partial B} u(x)$, $M = \sup_{\partial B} u(x)$ *and the numbers* $\omega_\pm(F,u)$ *were defined in subsection 1.3.*

The integrals in (2.3) and (2.4) can optionally take the value $+\infty$.

PROOF: It is sufficient to prove inequality (2.3), because inequality (2.4) can be proved in the same way. If $\omega_+(F,u) = 0$, then (2.3) is trivial. Let

$$(2.5) \qquad \omega_+(F,u) > 0,$$

then from the definition of $\omega_+(F,u)$ it follows that

$$(2.6) \qquad m > u_0 = \inf_{\overline{B}} u(x).$$

Let δ be any number from the open interval $(0,\delta_u)$, where $\delta_u = m - u_0 > 0$ and let $v_\delta(x)$ be the corresponding convex support of the function $u(x)$ (see subsection 1.2). Then from (1.33) and (1.34) we obtained

$$(2.7) \qquad \omega_+(F,u,\delta) - |\chi_F(\chi_{v_\delta}(D_\delta))|_Q$$

where D_δ is the set, on which $u(x) = v_\delta(x)$. We remind the reader that D is a closed subset of B and

$$(2.8) \qquad \operatorname{dist}(D_\delta, \partial B) > 0.$$

It is well known that $\chi_{v_\delta}(D_\delta)$ is a closed subset of P^n.* From the definition of the convex function $v_\delta(x)$ it follows that every supporting hyperplane α of the hypersurface $z = v_\delta(x)$ is also a supporting hyperplane of the hypersurface $z = u(x)$. Moreover there exists a point $M_0(x_0, v(x_0)) \in \alpha$ such that $x_0 \in D_\delta$. Thus either

$$(2.9) \qquad \chi_{v_\delta}(D_\delta) \subset \nu_u^+(B) \subset G$$

if $r(G) < +\infty$,** or

$$(2.10) \qquad \chi_{v_\delta}(D_\delta) \subset U(a) \subset \nu_u^+(B) \subset G = P^n,$$

*An elementary proof of this fact can be found in [**12**], Chapter 1, §7.

**Recall that $r(G) = \operatorname{dist}(0', \partial G)$ and G is an open domain in P^n.

if $r(G) = +\infty$, where $U(a)$ is the n-ball $|p| \leq a$ and the number a depends only on the number δ. Thus $\chi_{v_\delta}(D_\delta)$ is a compact subset of G. Hence

$$|\chi_{v_\delta}(D_\delta)|_P < +\infty. \tag{2.11}$$

Since $u(x) \in W_2^n(B) \cap C(\overline{B})$, then from (2.8) and (2.10) it follows that

$$\det(u_{ij}) \geq 0 \quad \text{and} \quad |\chi_{v_\delta}(D_\delta)|_P = \int_{D_\delta} \det(u_{ij})\,dx. \tag{2.12}$$

We also use the information that every point $(x, u(x))$ of the graph of $u(x)$ is convex for $x \in D_\delta$, which follows directly from the definition of the set D_δ.

The same considerations lead to the proof that the set $\chi_F(\chi_{v_\delta}(D_\delta))$ is compact in the space Q^n. Hence

$$|\chi_F(\chi_{v_\delta}(D_\delta))|_Q < +\infty. \tag{2.13}$$

According to Assumption 1 for the function $F(p)$ we obtain the formula

$$|\chi_F(\chi_{v_\delta}(D_\delta))|_Q = \int_{\chi_{v_\delta}(D_\delta)} \det(F_{p_i p_j}(p))\,dp \tag{2.14}$$

Now from the theory of multiple integrals (see [**21**], Chapter IV, §10) and the facts presented above it follows that

$$0 < |\chi_F(\chi_{v_\delta}(D_\delta))|_Q = \int_{D_\delta} \det(F_{p_i p_j}(Du(x))\det(u_{ij}(x))\,dx. \tag{2.15}$$

Thus $|D_\delta|_E > 0$. The quadratic form $\sum_{i,k=1}^n u_{ik}(x)\xi_i\xi_k$ is non-negative almost everywhere in D_δ, because all points $(x, u(x))$, $x \in D_\delta$ of the graph of $u(x)$ are convex and because $u(x)$ has first and second Sobolev generalized derivatives almost everywhere in B. Since the function $F(p)$ is convex, then

$$\sum_{i,j=1}^n F_{p_i p_j}(p)\xi_i\xi_j \tag{2.16}$$

is defined almost everywhere in G. Since $u(x) \in W_2^n(B) \cap C(\overline{B})$, then the composite non-negative quadratic form

$$\sum_{i,j=1}^n F_{p_i p_j}(D(u(x)))\xi_i\xi_j \tag{2.17}$$

is defined almost everywhere in B.

Thus the final part of the proof of Theorem 3 is reduced to the upper estimate of the integral in the right side of equality (2.15). Now we denote by D'_δ the subset of D_δ consisting of all points $x \in D_\delta$, where all generalized derivatives of the first and second orders of the function $u(x)$ take finite values, and where quadratic forms

$$\sum_{i,j=1}^{n} u_{ij}(x)\xi_i\xi_j \tag{2.18a}$$

and

$$\sum_{i,j=1}^{n} F_{p_ip_j}(D(u(x)))\xi_i\xi_j \tag{2.18b}$$

are defined and non-negative. Clearly

$$|D'_\delta|_E = |D_\delta|_E > 0. \tag{2.19}$$

It is also clear that

$$\det(u_{ij}(x)) \geq 0 \quad \text{and} \quad \det(F_{p_ip_j}(Du(x))) \geq 0$$

for all $x \in D'_\delta$. Thus inequality (2.15) becomes

$$0 < \omega_+(F,u,\delta) = \int_{D'_\delta} \det(F_{p_ip_j}(D(u(x))))\det(u_{ij}(x))dx. \tag{2.20}$$

Now we prove the inequality

(2.21)

$$[\det(F_{p_ip_j}(Du(x)))]^{1/n} \cdot [\det(u_{ij}(x))]^{1/n} \leq \frac{1}{n}\sum_{i,j=1}^{n} F_{p_ip_j}(Du(x))u_{ij}(x)$$

for all $x \in D'_\delta$.

Both determinants and the sum in inequality (2.21) are invariants of orthogonal transformations of the Euclidean space $R^n = \{\xi = (\xi_1, \xi_2, \dots, \xi_n)\}$. If we fix any point $x \in D'_\delta$ and bring quadratic forms (2.18a and b) to canonical form, then (2.21) becomes the well known Cauchy inequality between the arithmetic and geometric means. Thus the proof of inequality (2.11) is completed.

Since $u(x)$ is a solution of the Dirichlet problem

$$\sum_{i,j=1}^{n} F_{p_ip_j}(Du(x))u_{ij}(x) = nf_u(x,u), \quad u|_{\partial B} = h(x) \in C(\partial B),$$

then

$$0 \le \frac{1}{n}\sum_{i,j=1}^{n} F_{p_ip_j}(Du(x))u_{ij}(x) = f_u(x,u(x)) \le f_u^+(x,u(x)) \tag{2.22}$$

for all $x \in D'_\delta$. From (2.21), (2.22) and Assumptions 4 and 5 we obtain

$$0 < \omega_+(F,u,\delta) \le \int_{D'_\delta} [f_u^+(x,m)]^n dx,$$

where $m = \inf_{\partial B}\{h(x)\}$ as usual. Since

$$\int_{D'_\delta} [f_u^+(x,m)]^n dx \le \int_B [f_u^+(x,m)]^n dx,$$

we obtain the inequalities

$$0 < \omega_+(F,u,\delta) \le \int_B [f_u^+(x,m)]^n dx. \tag{2.23}$$

Since $\omega_+(F,u,\delta)$ is a non-increasing function of $\delta > 0$, then from (2.23) and the definition of the number $\omega_+(F,u)$ it follows that

$$0 < \omega_+(F,u) = \lim_{\delta\to 0+} \omega_+(F,u,\delta) \le \int_B [f_u(x,m)]^n dx.$$

The inequality (2.3) is proved. The inequality (2.4) can be established in the same way. The proof of Theorem 3 is now complete.

THEOREM 4. *(The main Theorem of estimates of solutions for Euler–Lagrange equation (2.17), the case $\partial G \neq \emptyset$.) Let $\partial G \neq \emptyset$ (i.e. $r(G) < \infty$) and let $u(x) \in W_2^n(B) \cap C(\overline{B})$ be a solution of the Euler–Lagrange equation (2.17). If Assumptions 1–5 are valid and if the inequalities*

$$\Omega_m^+ \le C(F) \tag{2.24}$$

and

$$\Omega_M^- \le C(F) \tag{2.25}$$

hold, then the estimates

$$(2.26) \qquad m - b(\Omega_m^+)\,diam\ B \le u(x) \le M + b(\Omega_M^-)\,diam\ B$$

hold for all $x \in \overline{B}$, *where*

$$(2.27) \qquad C(F) = |\chi_F(U(r(G)))|_Q = \int_{U(r(G))} \det(F_{p_ip_j}(p))dp,$$

$$(2.28) \qquad \Omega_m^+ = \|f_u^+(x,m)\|_{L^n}^n,$$

$$(2.29) \qquad \Omega_M^- = \|f_u^-(x,M)\|_{L^n}^n,$$

$m = \inf_{\partial B} u(x)$, $M = \sup_{\partial B} u(x)$. *The function* $\rho = b(t)$ *is introduced above (see Section 1.1).*

The proof of Theorem 4 follows directly from Theorems 1 and 3. Inequalities (2.24) and (2.25) are sharp. The corresponding examples will be considered in §3. The number n in Assumption 3: $u(x) \in W_2^n(B) \cap C(\overline{B})$ is not interchangeable with $n' < n$.

THEOREM 5. *(Main Theorem of estimates for Euler–Lagrange equation (2.17); the case* $\partial G = \emptyset$.*) Let* $\partial G = \emptyset$ *(i.e.* $r(G) = \infty$ *or* $G = P^n$*) and let* $u(x) \in W_2^n(B) \cap C(\overline{B})$ *be a solution of the Euler–Lagrange equation (2.17). If Assumptions 1–5 are valid and the strict inequalities*

$$(2.30) \qquad \Omega_m^+ < C(F)$$

$$(2.31) \qquad \Omega_M^- < C(F)$$

hold, then estimates (2.26) hold for all $x \subset \overline{B}$.

The notation is explained above (see Theorem 4). The proof of Theorem 5 follows directly from Theorems 2 and 3. Inequalities (2.30) and (2.31) are sharp. The corresponding examples will be considered in §3.

It is possible to obtain an essential development of Theorems 4 and 5 for a few wide classes of nonlinear elliptic Euler–Lagrange equations, which correspond to multiple integrals

$$(2.32) \qquad I(u) = \int_B F(x, u(x), Du(x))dx.$$

This development is also related to solutions $u(x) \in W_2^n(B) \cap C(\overline{B})$ of these equations. We also assume that the function $F(x,u,p)$ is convex with respect to $p \in G$ for any fixed $x \in B$, $u \in R$ and $F(x,u,p) \in W_2^n(B \times R \times G)$. Since the space of this paper is limited we shall not present the main statements of these results. The proof of these results is based on combination of existence theorems for generalized solutions of Monge–Ampere equations (see [**4**], [**10**] and some new theorems) and the technique developed in this paper. We also refer the reader to our paper [**3**], where a part of these results were formulated for C^2-solutions of nonlinear Euler–Lagrange elliptic equations with C^2-function $F(x,u,p)$. The proofs of estimates for C^2-solutions were sketched in [**11**].

§3. Applications to calculus of variations, differential geometry and continuum mechanics.

3.1. Application to calculus of variations. In the well known paper [**13**] by S. Bernstein the following theorem was proved.

THEOREM 6 (S.N. BERNSTEIN). *Let $u(x,y) \in C^2(\overline{B})$ be a solution of the Euler–Lagrange equation for the following two-dimensional functional:*

$$J(u) = \int_B \{F(u_x, u_y) + f(x,y,u)\} dxdy. \tag{3.1}$$

Let the following conditions be fulfilled:

1) B is a domain in a two-dimensional Euclidean plane $E^2 = \{(x,y)\}$ and ∂B is a closed curve in E^2;

2) $F(p,q) \in C^2(P^2)$, where $P^2 = \{(p,q)\}$ is the second two-dimensional Euclidean space;

3) Let

$$|F(p,q)| \geq N(p^2+q^2)^{a/2} \tag{3.2}$$

for $p^2+q^2 \geq r^2$, where $N =$ const > 0, $r =$ const > 1 and $a =$ const > 1;

4)

$$F_{pp}F_{qq} - F_{pq}^2 \geq F_0 = \text{const} > 0 \tag{3.3}$$

for all $(p,q) \in P^2$.

5) $f(x, y, u)$ is a convex C^2-function in $\overline{B} \times R$ with respect to u.
Then for $u(x, y)$ the estimate

$$|u(x,y)| \le U_0, \quad (x,y) \in \overline{B}, \tag{3.4}$$

can be obtained, where the constant U_0 depends only on the properties of the functions $F(p,q)$, $f(x,y,u)$ and their derivatives up to second order, constants of conditions 3 and 4, and the numbers $m = \inf_{\partial B} u(x,y)$, $M = \sup_{\partial B} u(x,y)$.

First of all Bernstein estimated the integral

$$\iint_B \{|u_{xx}| + |u_{yy}| + |u_{xy}|\} dxdy \tag{3.5}$$

in the terms of data mentioned in his theorem and then he obtained the desired estimate (3.4). His technique is essentially two-dimensional.

The conditions and the proof of Theorem 5 are based on ideas and a technique different from the considerations of Bernstein. This permits us to omit the overly strong Bernstein's condition 3 (see inequality (3.2)) and also to consider generalized solutions $u(x) \in W_2^n(B) \cap C(\overline{B})$ of Euler–Lagrange equations instead of classical ones in the Bernstein's Theorem. Moreover, convex functions $F(p,q)$ and $f(x,y,u)$ can be sufficiently non-smooth, and all considerations can be made for functions depending on n variables, where $n > 2$.

Thus the Bernstein Theorem can be significantly developed in many directions.

Now we present the statement and the proof of our theorem.

THEOREM 7. *Let $F(p)$ be a convex function defined over the entire space P^n and satisfying Assumption 1. We also assume that the inequality*

$$\det(F_{p_ip_j}(p)) \ge F_0 > 0 \tag{3.6}$$

holds almost everywhere in P^n, where F_0 is any constant. Let $f(x,u)$ satisfy Assumption 4 and let the numbers*

$$\Omega_k^{\pm} = \|f_u^{\pm}(x,k)\|_{L^n(B)} < +\infty \tag{3.7}$$

---

*Clearly this restriction on $F(p)$ is somewhat stronger than the restriction in Assumption 2 (see subsection 1.1).

for any constant $k \in (-\infty, +\infty)$.**

Then the inequalities

$$m - \left[\frac{\Omega_m^+}{\mu_n F_0}\right]^{1/n} \cdot \text{diam } B \le u(x) \le M + \left[\frac{\Omega_M^-}{\mu_n F_0}\right]^{1/n} \cdot \text{diam } B \tag{3.8}$$

hold for all $x \in \overline{B}$ *for any solution* $u(x)$ *of the Dirichlet problem*

$$\sum_{i,j=1}^{n} F_{p_i p_j}(Du(x))u_{ij}(x) = nf_u(x, u(x)), \tag{3.9}$$

$$u|_{\partial B} = h(x) \in C(\partial B), \tag{3.10}$$

which belongs to $W_2^n(B) \cap C(\overline{B})$, *where as usual* $m = \inf_{\partial B} h(x)$, $M = \sup_{\partial B} h(x)$.

PROOF: We consider two convex functions $\phi_1(p) = \frac{1}{2}\sum_{i=1}^{n} p_i^2$ and $\phi_2(p) = F(p)$ in the entire space P^n. According to the definition of the functions $c_1(\rho)$, $c_2(\rho)$ (see subsection 1.1, formula (1.3)) we obtain

$$c_1(\rho) = \int_{U(\rho)} \det(\phi_{1,ij}(p))dp = \int_{U(\rho)} 1dp = \mu_n \rho^n, \tag{3.11}$$

$$c_2(\rho) = \int_{U(\rho)} \det(F_{ij}(p))dp, \tag{3.12}$$

where as usual $U(\rho)$ is the n-ball $|p| < \rho$ in P^n and μ_n is the volume of $U(1)$. Inequality (3.6) yields to the inequality

$$c_1(\rho) \le \frac{1}{F_0} c_2(\rho), \tag{3.13}$$

which holds for all $\rho \in [0, +\infty)$. Clearly $c_1(\rho)$ and $c_2(\rho)$ are continuous and strictly increasing in $[0, +\infty)$. Since $C(\phi_1) = \lim_{\rho \to +\infty} c_1(\rho) = +\infty$, then it follows from (3.13) that $C(\phi_2) = \lim_{\rho \to \infty} c_2(\rho) = +\infty$. Therefore for any $\rho \in [0, +\infty)$ there exists only one number $\rho^* \ge \rho$ such that

$$c_1(\rho^*) = \frac{1}{F_0} c_2(\rho). \tag{3.14}$$

**The restriction (3.7) on $f_u^{\pm}(x, u)$ is somewhat stronger than the restriction in Assumption 5.

Let $\rho = b_1(t)$ and $\rho = b_2(t)$ be inverses of $c_1(\rho)$ and $c_2(\rho)$. Then they are defined in $[0, +\infty)$, and are strictly increasing and continuous in $[0, +\infty)$. Thus

$$\rho^* = b_1\left(\frac{1}{F_0} c_2(\rho)\right) \tag{3.15}$$

where $0 \leq \rho < \rho^* < +\infty$ are numbers considered in (3.14). Since the non-negative numbers Ω_m^+ and Ω_M^- are finite, then

$$\Omega_m^+ < C(\phi_2) = +\infty \tag{3.16}$$

and

$$\Omega_M^- < C(\phi_2) = +\infty. \tag{3.17}$$

Thus we can use Theorem 5 and obtain the estimates

$$m - b_2(\Omega_m^+)\text{diam } B \leq u(x) \leq M - b_2(\Omega_M^-)\text{diam } B \tag{3.18}$$

for all $x \in \overline{B}$. According to inequality (3.16), the non-negative number $\rho = b_2(\Omega_m^+)$ is finite. Hence

$$\Omega_m^+ = c_2(\rho). \tag{3.19}$$

From (3.15) and (3.19) it follows that there exists only one number ρ^* such that

$$\rho \leq \rho^* < +\infty \tag{3.20}$$

and

$$\rho^* = b_1\left(\frac{1}{F_0} c_2(\rho)\right) = b_1\left(\frac{\Omega_m^+}{F_0}\right). \tag{3.21}$$

Inequalities (3.18), (3.20) and identity (3.21) now yield to the inequality

$$m - b_1\left(\frac{\Omega_m^+}{F_0}\right) \text{diam } B \leq u(x). \tag{3.22}$$

Similar considerations lead to the inequality

$$u(x) \leq M + b_1\left(\frac{\Omega_M^-}{F_0}\right) \text{diam } B. \tag{3.23}$$

Since $t = c_1(\rho) = \mu_n \rho^n$, then

$$b_1(t) = \left(\frac{t}{\mu_n}\right)^{1/n}. \tag{3.24}$$

Thus inequalities (3.22–23) and the formula (3.24) yield to the desired inequalities (3.8). The proof of Theorem 7 is complete.

3.2. Applications to differential geometry. In this subsection we consider two-sided C^0-estimates of solutions of the Dirichlet problem for mean curvature equation in Euclidean and Minkowski $n+1$-dimensional spaces.

a) Hypersurfaces with prescribed mean curvature in Euclidean space E^{n+1}.

Let a hypersurface S with prescribed mean curvature H be a graph of a function $u(x) \in W_2^n(B) \cap C(\overline{B})$. We assume that $H(x) \in L^n(B)$ and $u|_{\partial B} = h(x) \in C(\partial B)$. Clearly $u(x)$ is a solution of the Euler–Lagrange equation for the functional

$$I(u) = \int_B \left[\sqrt{1+(Du(x))^2} + nH(x)u(x)\right] dx, \tag{3.25}$$

satisfy the Dirichlet boundary condition $u(x) = h(x)$ for all $x \in \partial B$.

The convex function $F(p) = \sqrt{1+|p|^2} \in C^\infty(P^n)$. Hence $G = P^n$ and $r(G) = +\infty$. The asymptotic cone K_F with the vertex at O' has the equation $w = 1 + |p|$, $p \in P^n$. Therefore $\chi_{K_F}(P^n)$ is the unit n-ball $|q| \le 1$ in the space Q^n. According to (1.56)

$$C(F) = |\chi_{K_F}(P^n)|_Q = \mu_n, \tag{3.26}$$

where μ_n is the volume of the n-unit Euclidean ball.

Clearly $f_u^+(x,u) = H^+(x)$ and $f_u^-(x,u) = H^-(x)$, where $H^+(x) \ge 0$ and $H^-(x) \ge 0$ are the positive and negative parts of $H(x)$. If all conditions of Theorem 5 are fulfilled then the crucial inequalities (2.30–31) become

$$\|H^+(x)\|_{L^n(B)} < \mu_n \quad \text{and} \quad \|H^-(x)\|_{L^n(B)} < \mu_n \tag{3.27}$$

They provide the estimates (2.26) for the function $u(x)$, where the function

$$b(t) = \left[\frac{\mu_n^{2/n}}{\mu_n^{2/n} - t^{2/n}}\right]^{1/2}$$

is defined for $t \in [0, \mu_n)$.

The inequalities (3.27) are sharp. Actually the Dirichlet problem

$$\sum_{i=1}^{n} \frac{\partial}{\partial x_i} \left\{ \frac{u_i}{[1 + (Du)^2]^{1/2}} \right\} = nH_0$$

$$u|_{\partial B} = 0$$

where $H_0 =$ const > 0 and B is the n-ball $|x| < \frac{1}{H_0}$ is the corresponding example of this assertion.

The existence theorems of the Dirichlet problem for mean curvature equation by the interlocked necessary and sufficient conditions were established by Serrin [**22**], [**23**] and myself [**6**], [**12**]. Two-sided C^0-estimates of solutions are very important in these investigations. The estimates considered in this subsection were established in [**6**].

b) Spacelike hypersurfaces with prescribed mean curvature in the Minkowski space M^{n+1}.

The space $R^{n+1} = \{(x,t)\} = \{(x_1, x_2, \ldots, x_n, t)\}$ with metric

$$ds^2 = \sum_{i=1}^{n} dx_i^2 - dz^2 \tag{3.28}$$

is called the Minkowski space and is denoted by M^{n+1}. Let S be a hypersurface such that ds^2 is restricted to a positive form on S. Such an S is called spacelike. If S is the graph of a function $z = u(x)$, then S is spacelike if and only if $|Du(x)| < 1$ for any $x \in B$, where B is the domain of the function $u(x)$. The spacelike hypersurfaces were studied by Calabi [**16**] and Cheng and Yau [**17**] in connection with the Bernstein conjecture in M^{n+1}. The spacelike solutions $u(x) \in W_2^n(B) \cap C(\overline{B})$ of the Euler–Lagrange equation for the

$$M(u) = \int_B [-[1 - (Du)^2]^{1/2} + nH(x)u]\, dx \tag{3.29}$$

have prescribed mean curvature $H(x)$ in M^{n+1}. Clearly $H(x)$ is locally summable with degree n in the open domain B. According to our general considerations we conclude that the convex function $F(p) = -(1 - |p|^2)^{1/2}$ is defined in the open ball $|p| < 1$ in P^n. Thus G is the ball $|p| < 1$ and $r(G) = 1$. Clearly

$$\chi_F(G) = Q^n. \tag{3.30}$$

Hence $C(F) = +\infty$. The crucial inequalities (2.24) and (2.25) in Theorem 4 become

$$\int_B H_\pm^n(x)dx \leq +\infty \tag{3.31}$$

for the Dirichlet problem for mean curvature equation in the space M^{n+1}. Since all other conditions of Theorem 4 are fulfilled, then we can apply estimates (2.26) if and only if the functions $H^+(x)$ and $H^-(x)$ are locally summable with degree n in B.

Thus there is an essential difference between the solutions of the Dirichlet problem for the mean curvature equation in Euclidean and Minkowski spaces.

3.3. Applications to continuum mechanics.

a) The problem of torsion of hardening rods.

Let P be a prismatic rod represented by the cylinder with the base $\overline{B} = B \cup \partial B$ and generators parallel to z-axis, where B is a bounded domain in the x, y-plane. Let the base of P be clamped and let the rod P twist under the action of a moment M. We denote by w the torsion per unit length of the rod. Let $u(x, y)$ be the stress function of the rod P then $T = (u_x^2 + u_y^2)^{1/2}$ is the intensity of the tangential stress tensor and

$$\frac{\partial}{\partial x}[g(T^2)u_x] + \frac{\partial}{\partial y}[g(T^2)u_y] = -2w \tag{3.32}$$

is *the equation of torsion of hardening rods.* The function $g(T^2)$ is called *the modulus of plasticity of the rod P.* It describes the dependence between the intensity Γ of the shear strain tensor and the intensity T of the tangential stress tensor by the formula $\Gamma = g(T^2)T$. The experimental law

$$\frac{d\Gamma}{dT} > 0 \tag{3.33}$$

is the necessary and sufficient condition of ellipticity for equation (3.32). The problem of hardening rods can be reduced to the Dirichlet problem for equation (3.32) with zero boundary data.

Now consider the Dirichlet problem

$$\sum_{i=1}^{n}[g(|Du|^2)u_{x_i}]_{x_i} = nH(x,u), \quad u|_{\partial B} = 0 \tag{3.34}$$

for equations which somewhat generalize equation (3.32) to n dimensions. Clearly (3.34) is the Euler–Lagrange equation for the functional

(3.35)
$$J(u) = \int_B \frac{1}{2}\left\{\int_0^{|Du(x)|^2} g(s)ds\right\} dx + n\int_B \left\{\left\{\int_0^{u(x)} [H(x,s)]dx\right\} dx.\right.$$

It is appropriate to consider $g(|p|^2)$ either in the n-balls $U(a)\colon |p| < a$ or in the entire space P^n, where $0 < a =$ const $< +\infty$. This makes it possible to consider $g(T^2)$ either as a function of a single variable T^2 or as a composite function of a single variable T. Now we introduce the function $\Gamma(T) = g(T^2)T$, which is defined in the same domain as the function $g(T^2)$, and formulate assumptions for functions $g(T^2)$, $\Gamma(T)$ and $H(x,u)$, allowing Theorems 4 and 5 to be applied for two-sided C^0-estimates of solutions of equations (3.34).

Below we assume that a positive constant a also takes the value $+\infty$. Thus both finite intervals $[0,a)$ and the ray $[0,+\infty)$ can be represented by $[0,a)$.

Let $s = T^2$ and $g(s) = g(T^2)$. Clearly $g(s)$ is defined on $[0,a^2)$.

Assumption 6. A function $g(s)$ is positive and absolutely continuous on $[0,a^2)$, i.e. $\frac{dg(s)}{ds}$ exists everywhere on $[0,a^2)$ and

(3.36)
$$g(s) = g(0) + \int_0^s \frac{dg(s)}{ds} ds$$

for all $s \in [0,a^2)$.

Assumption 7. Let $g(s)$ satisfy Assumption 6, then

(3.37)
$$\frac{d\Gamma}{dT} > 0$$

on the set of positive measure in any interval (a',a''), $0 < a' < a'' < a$, where $\Gamma(T) = g(T^2)\cdot T$ and $s = T^2$.

Assumption 8. A function $H(x,u)$ is increasing with respect to u for every fixed $x \in B$ and the functions $H^+(x,k)$ and $H^-(x,k)$ are locally summable with degree n in B for all values of the constant $k \in (-\infty,+\infty)$.

Below we consider functions $g(s)$ and $H(x,u)$ which satisfy Assumptions 6, 7, 8. From (3.35) it follows that the graph S_F of the function

(3.38)
$$F(p) = \frac{1}{2}\int_0^{|p|^2} \{g(s)\}ds$$

is a hypersurface of revolution. The meridian of this hypersurface has equation

$$w = F(T)$$

where $p = (T, 0, 0, \dots, 0) \in P^n$ and $0 \leq T < a$. Clearly

$$\frac{dw}{dT} = g(T^2)T = \Gamma(T)$$

for all $T \in [0, a)$. From Assumption 6 it follows that $\frac{dw}{dT}$ is an absolutely continuous function of T. Since $\Gamma(0) = 0$, then

$$\frac{dw}{dT} = \int_0^T \frac{d\Gamma(\xi)}{dT} d\xi \tag{3.39}$$

for all $T \in [0, a)$. Now from (3.39) and Assumption 7 it follows that the function $w = F(T)$ is convex and the derivative $\frac{dF(T)}{dT}$ is strictly increasing absolutely continuous on $[0, a)$. Hence the area of the normal image $\omega(S_F, e')$ of the convex hypersurface of revolution S_F is absolutely continuous and

$$(S_F, e') = \int_{e'} \det(F_{p_i p_j}(p))dp. \tag{3.40}$$

The formula

$$\det(F_{p_i p_j}(p)) = g^{n-1}(|p|^2)\frac{d\Gamma(|p|)}{d|p|} \tag{3.41}$$

holds almost everywhere in the ball $U(a)$: $|p| < a$. Now from Assumptions 6 and 7 and equalities (3.39–41) it follows that

$$\omega(S_F, U(\rho)) = \mu_n \int_0^\rho g^{n-1}(T)\frac{d\Gamma(T)}{dT}T^{n-1}dT =$$

$$= \mu_n \int_0^\rho \Gamma^{n-1}(T)d\Gamma(T) = \mu_n\Gamma^n(\rho),$$

where μ_n is the volume of the n-unit ball in E^n and $U(\rho)$ is the n-ball $|p| < \rho$ in P^n for $0 \leq \rho < a$.

Thus from Assumptions 6 and 7 it follows that the function $F(p)$ introduced by (3.38) satisfies Assumptions 1 and 2 (see subsection 1.1). Clearly Assumption 8 is equivalent to Assumptions 4 and 5 (see Section 2). Thus the following theorems can be obtained from Theorems 4 and 5.

THEOREM 8. *Let G be the n-ball $|p| < a$, where $a < +\infty$ and let $u(x) \in W_2^n(B) \cap C(\overline{B})$ be a solution of the Dirichlet problem (3.34). If Assumptions 3, 6, 7, 8 are valid and if the inequalities*

$$\Omega_0^{\pm} \leq \mu_n \Gamma^n(a) \tag{3.42}$$

hold, then the estimates

$$-b(\Omega_0^+)\text{diam } B \leq u(x) \leq b(\Omega_0^-)\text{diam } B \tag{3.43}$$

hold for all $x \in \overline{B}$, where

$$\Omega_0^{\pm} = \int_B [H^{\pm}(x,0)]^n dx, \tag{3.44}$$

and $\rho = b(t)$ is the inverse to the strictly increasing absolutely continuous function $\mu_n \Gamma^n(\rho)$, $0 \leq \rho \leq a$.

THEOREM 9. *Let $G = P^n$ and let $u(x) \in W_2^n(B) \cap C(\overline{B})$ be a solution of the Dirichlet problem (3.34). If Assumptions 6, 7, 8 are valid and if the strict inequalities*

$$\Omega_0^{\pm} < \mu_n \Gamma^n(+\infty) \tag{3.45}$$

hold, then estimates (3.43) hold for all $x \in \overline{B}$.

All notations are explained in Theorem 8.

b) Equations relating to gas dynamics.

It is well known that the stationary irrational flow of an ideal compressible fluid can be described by the following equation of continuity

$$\operatorname{div}(\sigma \cdot Du) = 0, \tag{3.46}$$

where the fluid density σ satisfies a density-speed relation $\sigma = \sigma(|Du|)$. For a perfect gas this relation is

$$\sigma = (1 - \frac{\gamma - 1}{2}|Du|^2)^{\frac{1}{\gamma-1}}, \tag{3.47}$$

where the constant γ is the ratio of specific heats of the gas and $\gamma > 1$.

In this subsection we consider non-homogeneous equations

$$\operatorname{div}(\sigma \cdot Du) = nH(x), \tag{3.48}$$

where σ is defined by (3.47). Clearly (3.48) is the Euler–Lagrange equation for the functional

$$S(u) = \int_B \frac{1}{2}\left\{\int_0^{T^2} \sigma(s)dx\right\} dx + \int_B \{nH(x)u\}dx, \tag{3.49}$$

where as usual, $T = |\text{grad } u(x)|$. Since $\gamma > 1$, then the function

$$\sigma(|p|) = (1 - \frac{\gamma - 1}{2}|p|^2)^{\frac{1}{\gamma-1}}$$

and its derivatives are defined only in the ball: $|p| < \left(\frac{2}{\gamma-1}\right)^{1/2}$.

According to the previous subsection we should consider the function

$$F(T) = \frac{1}{2}\int_0^{T^2} \left(1 - \frac{\gamma-1}{2}s^2\right)^{\frac{1}{\gamma-1}} ds$$

for $0 \le T < \left(\frac{2}{\gamma-1}\right)^{1/2}$. The function $F(T)$ is convex in $\left[0, \left(\frac{2}{\gamma+1}\right)^{1/2}\right)$ and it is concave in $\left(\left(\frac{2}{\gamma+1}\right)^{1/2}, \left(\frac{2}{\gamma-1}\right)^{1/2}\right)$. Thus the function

$$F(p_1, p_2, \ldots, p_n) = \frac{1}{2}\int_0^{|p|^2} (1 - \frac{\gamma-1}{2}s^2)^{\frac{1}{\gamma-1}} ds$$

is C^∞-strictly convex function only in the ball

$$G_\gamma\colon |p| < \left(\frac{2}{\gamma+1}\right)^{1/2}. \tag{3.50}$$

Hence equation (3.49) is *elliptic* and the flow is *subsonic*, when $p = Du \in G_\gamma$.

THEOREM 10. *Let $G = G_\gamma$ and let $u(x) \in W_2^n(B) \cap C(\overline{B})$ be a solution of the Dirichlet problem (3.34), satisfying Assumption 3 (see subsection 1.3). If the inequalities*

$$\Omega^\pm < \mu_n \left(\frac{2}{\gamma+1}\right)^{\frac{n(\gamma+1)}{2(\gamma-1)}} \tag{3.51}$$

hold, then the estimates (3.43) hold.

We use the notation

$$\Omega^\pm = \int_B [H^\pm(x)]^n dx,$$

where $H^{\pm}(x)$ are the respective positive and negative parts of the function $H(x)$ and $\rho = b(t)$ is the inverse of the function $c(\rho) = \mu_n(1-\frac{\gamma-1}{2}\rho^2)^{\frac{n}{\gamma-1}}\cdot\rho^n$, $0 \leq \rho \leq \left(\frac{2}{\gamma+1}\right)^{1/2}$.

Theorem 10 follows directly from Theorem 8. If we consider the boundary data $u|_{\partial B} = h(x) \in C(\partial B)$, then we replace inequalities (3.43) by inequalities

$$m - b(\Omega^+)\text{diam } B \leq u(x) \leq M + b(\Omega^-)\text{diam } B \tag{3.52}$$

for all $x \in \overline{B}$, where as usual $m = \inf_{\partial B} h(x)$, $M = \sup_{\partial B} h(x)$.

School of Mathematics, Institute for Advanced Study, Princeton, NJ 08540 and Department of Mathematics, Texas A&M University, College Station, TX 77843

This work was supported in part by NSF grant DMC–8420850.

References

1. Alexandrov, A.D., *Intrinsic geometry of convex surfaces*, GETTL (1948), 1–368.
2. Alexandrov, A.D., *Uniqueness conditions and estimates of solutions of the Dirichlet problem*, Vestnik LGU, No. 13 (1963), 5–29.
3. Bakelman, I., *Convex functions methods in the Dirichlet problem for elliptic Euler–Lagrange equations*, Proc. of the International Conference on Variational Methods for free Surface Interface, Springer–Verlag, (1987), 127–137.
4. Bakelman, I., *Generalized elliptic solutions of the Dirichlet problem for n-dimensional Monge–Ampere equations*, Proc. of Symposia in Pure Math., Vol. 45, Part 1, (1986), Nonlinear Functional Analysis and its Applications, F. Browder – editor, AMS, 73–102.
5. Bakelman, I., *Geometric methods of solutions of elliptic equations*, Nauka, Moscow (1965), 1–340.
6. Bakelman, I., *Geometric problems in quasilinear elliptic equations*, Uspekhi Math. Nauk, 25, No. 3, (1970), 49–112; Engl. Translation: Russian Math. Surveys (1970), 25, No. 3, 45–109.
7. Bakelman, I., *Notes concerning torsion of hardening rods and its generalizations*, IMA Preprint Series, University of Minnesota, No. 208 (1986), 1–38.
8. Bakelman, I., *On the theory of quasilinear elliptic equations*, Sibirsk. Math. Zh., No. 2 (1961), 179–186.
9. Bakelman, I., *R-curvature, estimates and stability of solutions for the general elliptic equations*, Journal of Diff. Equations, Vol. 43, No. 1 (1982), 106–133.
10. Bakelman, I., *The boundary value problems for nonlinear elliptic equations and the maximum principle for Euler–Lagrange equations*, Archive for Rational Mechanics and Analysis, Vol. 93, No. 3 (1986), 271–300.
11. Bakelman, I., *The boundary value problems for non-linear elliptic equations II.*, IMA Preprint Series, University of Minnesota, No. 209 (1986), 1–23.
12. Bakelman, I.; Verner, A. and Kantor, B., *Introduction into Global Differential Geometry*, Nauka, Moscow (1973), 1–440.
13. Bernstein, S.N., *Equations of Calculus of Variations*, Collection of papers, Vol. 3, Akad. Nauk of USSR (1960), 191–241.
14. Brezis, H., *Some Variational Problems with Lack of Compactness*, Proc. of Symposia in Pure Math., Vol. 45, Part 1, (1986), Nonlinear Functional Analysis and its Applications, F. Browder – editor, AMS, 165–201.
15. Caffarelli, L.; Nirenberg, L. and Spruck, J., *The Dirichlet problem for nonlinear second order elliptic equations*, I., Comm. Pure Appl. Math., 37, (1984), 369–402; II. (jointly with J.J. Kohn), Comm. Pure and Appl. Math., 38, (1985), 209–252; III., Acta Math., Vol. 155, 261–301.
16. Calabi, E., *Examples of Bernstein problem for some nonlinear equations*, Proc. Symposia Global Analysis, University of California, Berkeley, (1968).
17. Cheng, S.Y. and Yau, S.T., *Maximal spacelike hypersurfaces in the Lorentz–Minkowski spaces*, Annals of Math. **104** (1976), 495–516.
18. Gidas, B.; Ni, W.-M. and Nirenberg, L., *Symmetry of positive solutions of nonlinear elliptic equations in R^n*, Mathematical Analysis and Applications, Part A, Adv. in Math. Suppl. Studies, Vol. 7A, Academic Press, New York (1981).
19. Ni, W.-M. and Serrin, J., *Existence and non-existence theorems for ground states of quasilinear partial differential equations*, Rend. Acad. Naz. Lincei (to appear).
20. Peletier, L. and Serrin, J., *Uniqueness of Positive Solutions of Semilinear Equations in R^n*, Arcive of Rat. Mechanics and Analysis, Vol. 81, No. 2, (1983), 181–197.
21. Schwartz, L., *Analyse Mathématique I*, Hermann, Paris, (1967), 1–824.

22. Serrin, J., *Boundary curvature and the solvability of Dirichlet problem*, Actes Congress, Inst. Math., Nice 2, (1970).
23. Serrin, J., *The Dirichlet problem for quasilinear elliptic equations with many independent variables*, Trans. Royal Phil. Soc. Series, No. 153, (1969), 417–496.

Nonlinear Parabolic Equations with Sinks and Sources

CATHERINE BANDLE AND MARIA ASSUNTA POZIO

1. Introduction.

1.1 Let $D \subset R^N$ be a bounded domain whose boundary is in C^1. We shall put

$$Q_T := D \times (0,T) \qquad \text{and} \qquad \Gamma_T := \partial D \times (0,T).$$

Suppose that $a(x) \in C^\alpha(D), \alpha \in (0,1]$, is an arbitrary function of *variable* sign and that $u_0(x) \geq 0$ is continuous in $\bar{D}$.

This paper deals with parabolic problems of the following kind:

$$(1.1) \qquad \begin{cases} u_t - \Delta u^m = a(x)u^p & \text{in } Q_T \\ u(x,0) = u_0(x) \geq 0 & \text{in } D \\ u \geq 0 & \text{in } Q_T \end{cases}$$

u satisfying either *Dirichlet* boundary conditions

$$(1.2') \qquad u = \psi \geq 0 \qquad \text{on } \Gamma_T$$

where ψ is an assigned, continuous time independent function, or the *Neumann* boundary condition

$$(1.2'') \qquad \frac{\partial u^m}{\partial n} = 0 \qquad \text{on } \Gamma_T$$

where $\frac{\partial}{\partial n}$ denotes the outer normal derivative.

Here, we shall only consider the cases

$$\text{(I)} \qquad m > p \geq 1$$

and

$$\text{(II)} \qquad m = 1 > p.$$

Problems of type (I) arise in population dynamics [**8**] and those of type (II) are used to describe reaction-diffusion processes of order p [**1**].

For convenience we shall use the notation: $(\mathrm{I})_D$ or $(\mathrm{II})_D$ referring to the Dirichlet problem (1.1), (1.2′) of type (I) or (II). Correspondingly, we set $(\mathrm{I})_N$ and $(\mathrm{II})_N$ for the Neumann problems (1.1), (1.2″).

Observe that for $m > 1$, (1.1) is degenerate. Next, we discuss the equations for the stationary states which are common to both problems (I) and (II).

1.2 If we set $v = u^m$ and $q = p/m$, the equation for the stationary states becomes

$$\Delta v + a v^q = 0 \quad \text{in } D, \qquad 0 < q < 1 \tag{1.3}$$

with either Dirichlet boundary conditions

$$v = \psi^m \quad \text{on } \partial D \tag{1.4'}$$

or Neumann boundary conditions

$$\frac{\partial v}{\partial n} = 0 \qquad \text{on } \partial D. \tag{1.4''}$$

Let us first focus on the Dirichlet problem (1.3), (1.4′). It always possesses a solution [**3**]. Moreover, if a is of constant sign there is at most one non-trivial solution. However, if a changes sign, the situation is completely different; many solutions may appear.

It turns out [**3**], that there is a natural way to classify them. For this purpose consider

$$D^{\pm} := \{x \in D : a(x) \gtrless 0\}$$

and denote by

$$D_k^+, \qquad k \in \mathcal{M} := \{1, 2, 3, \ldots r\}$$

the connected components of D^+. Suppose that

(A-1) $\quad \partial D_k^+$ *satisfies an inner sphere condition with repect to* D^+.

DEFINITION 1.1: For any $I \subseteq \mathcal{M}$ define $D_I^+ = \bigcup_{k \in I} D_k^+$,

$$\mathcal{S}_I := \{v(x) : v \text{ solution of (1.3), (1.4') with } v(x) > 0 \text{ in } D_I^+\},$$
$$\mathcal{N}_I := \{v \in \mathcal{S}_I : v \equiv 0 \text{ in } D^+ - D_I^+\}.$$

Concerning the solutions of (1.3), (1.4′) we have [**3**]

THEOREM 1.1.

(i) *Every solution belongs to some* $\mathcal{N}_I$.

(ii) $\mathcal{N}_I$ *contains at most one element, if* I *is finite.*

(iii) *For any* $I \subseteq \mathcal{M}$, $\mathcal{S}_I \neq \phi$.

(iv) *Every* $\mathcal{S}_I$ *possesses a minimal and a maximal solution such that* $v_I(x) \leq \tilde{v}(x) \leq V_I(x)$ *for every other solution* $\tilde{v}(x)$.

From this theorem we deduce immediately the

COROLLARY 1.2.

(i) *If* $D^+ \neq \phi$ *the Dirichlet problem has always a non-trivial solution.*

(ii) *There exists exactly one solution* V *which is positive in* D^+, *provided* $\mathcal{M}$ *is finite. It is a maximal solution for all* $\mathcal{S}_I$.

(iii) *All other solutions must vanish in some* D_k^+, *provided* $\mathcal{M}$ *is finite.*

The last property is due to the appearance of a "dead core", a phenomenon which occurs only when the power of the nonlinearity q is smaller than one [**10**].

The number of solutions depends on the number $r \leq \infty$ of connected components of D^+, on their distance and on the size of the negative part $a^-(x)$ of $a(x)$. Roughly speaking, if $r > 1$, it increases as $a^-(x)$ increases, and it is bounded above by 2^r if $r < \infty$. A study on the number of solutions has been carried out in [**3**]. It turns out that there is an essential difference between problems with a finite or an infinite number of components D_k^+. In fact, it can be shown that if $\mathcal{M}$ is finite and $|a^-|_\infty$ is sufficiently small, then Problem (1.3), (1.4′) has at most one non-trivial solution. This isn't true in general when $\mathcal{M}$ is infinite; a counterexample has been constructed in [**3**].

In contrast to the Dirichlet problem, additional conditions are needed for the existence of non-trivial solutions to the Neumann problem (1.3), (1.4″). It is clear that no non-trivial solution exists if $a(x)$ is of constant sign. The classification of the solutions can be done as for the Dirichlet problem. Let us introduce the classes $\mathcal{S}_I$ and $\mathcal{N}_I$ with respect to Problem (1.3), (1.4″). Then the same arguments as for Theorem (1.1)(i) and (ii) yield [**4**].

THEOREM 1.3.

(i) *All solutions of the Neumann problem belong to some* $\mathcal{N}_I$.

(ii) $\mathcal{N}_I$ *has at most one element for any finite* $I \subseteq \mathcal{M}$.

S_I does not necessarily contain an element as it was the case for the Dirichlet problem. The following results have been established in [**4**].

THEOREM 1.3′.

(i) *A necessary condition for* S_I *to be non-empty is*

$$\int_{D^- \cup D_I^+} a(x)dx < 0.$$

(ii) *There exists a solution* v *which is positive in* D^+ *(i.e.* $v \in S_M$*) if and only if the condition*

$$\int_D a(x)dx < 0 \tag{A-2}$$

is satisfied.

Notice that if (A-2) holds, there is an element in each S_I. This condition, however, is not necessary for the existence of an element in a fixed S_I.

1.3 For the discussion of the time-dependent problems we distinguish between Problem (I) and Problem (II). It follows from the work of Aronson, Crandall and Peletier [**2**] and de Mottoni, Schiaffino and Tesei [**6**] that $(\mathrm{I})_D$ has a unique local solution in a weak sense. According to [**9**] this solution is continuous and satisfies

$$\begin{aligned}&\int_D \eta(x,T)u(x,T)dx - \int_D \eta(x,0)u_0(x)dx - \int_{Q_T} \eta_t u\, dx\, dt \\ &- \int_{Q_T} u^m \Delta\eta dx\, dt + \int_0^T \left(\oint_{\partial D} \psi^m \frac{\partial \eta}{\partial n} ds\right) dt = \int_{Q_T} a u^p \eta dx\, dt \\ &\qquad \forall \eta \in C^\infty(\bar{Q}_T), \quad \eta = 0 \text{ on } \partial D \times [0,T].\end{aligned} \tag{1.5_D}$$

The same arguments extend to the Neumann problem $(\mathrm{II})_N$. Thus for T sufficiently small there exists a unique solution satisfying

$$\begin{aligned}&\int_D \eta(x,T)u(x,T)dx - \int_D \eta(x,0)u_0(x)dx - \int_{Q_T} \eta_t u\, dx\, dt \\ &- \int_{Q_T} u^m \Delta\eta dx\, dt = \int_{Q_T} a u^p \eta dx\, dt \quad \forall \eta \in C^\infty(\bar{Q}_T) \\ &\qquad \text{such that } \frac{\partial \eta}{\partial n} = 0 \text{ on } \partial D \times [0,T].\end{aligned} \tag{1.5_N}$$

Concerning the behavior of the solutions as t increases we have [**3**]

THEOREM 1.4. *Assume (A-1) and suppose that $\mathcal{M}$ is finite. All solutions of $(I)_D$ converge to a stationary state as $t \to \infty$.*

For infinite $\mathcal{M}$ see [**3**]. Results in this direction can be obtained for the Neumann problem too, though unbounded solutions exist if (A-2) doesn't hold. Indeed, in Section 2 the following theorem is proved:

THEOREM 1.5. *Assume (A-1) and suppose that $\mathcal{M}$ is finite. For the Neumann Problem $(I)_N$ the following statements are equivalent:*

(i) *All solutions exist globally in time and are bounded.*
(ii) *All solutions converge to a stationary state.*
(iii) *Condition (A-2) holds.*

The discussion of Problem (II) differs from that of Problem (I). It follows from the classical theory that there is always a classical solution for sufficiently small t. Since the source term fails to be differentiable at zero, the maximum principle doesn't apply. Hence uniqueness cannot be expected. In Section 3 we derive the following non-uniqueness result.

THEOREM 1.6. *Assume (A-1).*

(i) *Regardless of $u_0(x)$, Problem (II) has always a solution $u(x,t)$ which is positive in $D^+ \times (0,T)$. There is only one solution with this property, provided that $\mathcal{M}$ is finite.*
(ii) *If $u_0 \equiv 0$, and $\psi \equiv 0$, (II) has infinitely many solutions.*
(iii) *If $I \subsetneq \mathcal{M}$, $\mathcal{N}_I \neq \phi$ and if $u_0(x) \leq v_s(x)$ for $v_s(x) \in \mathcal{N}_I$, Problem (II) has infinitely many solutions. Here $\mathcal{N}_I$ is defined similarly as before.*

The methods used in Section 3 have grown out of those developed for handling Problem (I). They are based on upper and lower solutions and on an extended version of the maximum principle. Most results can be generalized to nonlinearities of the type $a(x)f(u)$ where f behaves like u^p near the origin and is sublinear at ∞, see [**3**] and [**4**]. At the end we classify certain solutions in the same spirit as for the time-independent problem.

ACKNOWLEDGEMENT: The results of this paper were obtained during the Microprogram at the Mathematical Sciences Research Institute in Berkeley. We would like to thank the organizers, Professors W.-M. Ni, L.A. Peletier, and J. Serrin for having given us the opportunity to attend this workshop.

2. Asymptotic Behavior of the Solutions of the Neumann Problem (I)$_N$.

2.1 The asymptotic behavior of the solutions to (I)$_N$ can be completely described as for Problem (I)$_D$ (see Theorem 1.4). The aim of this section is to prove Theorem 1.5. Let us first add some remarks.

(1) If (A-2) doesn't hold, there exist initial values $u_0(x)$ such that $u(\cdot, T)$ becomes unbounded in its interval of existence $[0,T), T \leq \infty$.

(2) An example where solutions blow up in finite time can easily be obtained if $a(x) \geq \varepsilon > 0$ and $p > 1$. In this case for any $\alpha > 0$,

$$\tilde{u}(x,t) = \tilde{u}(t) := \alpha / \big(1 - (p-1)\varepsilon\alpha^{p-1}t\big)^{1/(p-1)}$$

is a lower solution in $D \times [0,T), T := 1/((p-1)\varepsilon\alpha^{p-1})$ and blows up as $t \to T - 0$. Hence the solution with initial data $u_0 \geq \tilde{u}(0)$ blows up in finite time (see Proposition 2.1).

(3) If (A-2) doesn't hold, bounded solutions may exist too. This is the case, for instance, if there are non-trivial stationary solutions belonging to some $\mathcal{N}_I, I \neq \mathcal{M}$ (see the example in [**4**]).

(4) If (A-2) doesn't hold and if u_0 is a lower solution such that $u_0 \not\equiv 0$ in D_k^+ for any $k \in \mathcal{M}$, then the solution with initial values u_0 is unbounded (see Lemma 2.2). As far as we know it is still an open question whether such solutions blow up in finite time.

2.2 In order to prove Theorem 1.5 we need the following comparison result. The ideas of the proof are the same as in [**6**] for the Dirichlet problem (I)$_D$, and it will therefore be omitted.

PROPOSITION 2.1. *Let $\underline{u}$ and $\bar{u}$ be lower and upper stationary solutions of (I)$_N$ of class $C(\bar{D})$ and denote by $u(x,t:\underline{u})$ and $u(x,t:\bar{u})$ the solutions of (I)$_N$ with $u_0 = \underline{u}$ or $u_0 = \bar{u}$, respectively. Then we have*

(i) *$u(x,t:\underline{u})$ is a non-decreasing and $u(x,t:\bar{u})$ a non-increasing function of t.*

(ii) *If $\underline{u} \leq \bar{u}$, then $u(x,t:\underline{u}) \to u_*$ and $u(x,t:\bar{u}) \to u^*$ in $L^r(D)$ for any $r \in [1,\infty)$ as $t \to \infty$. Here, u_* and u^* are solutions of the stationary problem such that $u_* \leq u \leq u^*$ for any other time-independent solution $u, \underline{u} \leq u \leq \bar{u}$.*

Moreover, we shall make use of the following auxiliary results proved in [**3**] and [**4**].

LEMMA 2.2.

(i) *For any ball B such that $\bar{B} \subset D^+$ there exists a family of lower stationary solutions $\{u_\rho\}$ of $(I)_N$ with $\rho \in (0, \rho_0(B)]$ such that supp $u_\rho = \bar{B}$ and $|u_\rho|_\infty = \rho$.*

(ii) *For any $I \subseteq \mathcal{M}$, any set of balls $\{B_k \subset D_k^+, k \in I\}$ and any set $\{\rho_k \in R^+ : \rho_k \leq \rho_0(B_k), k \in I\}$ let u_k be the lower solution considered in (i) with $B = B_k$, $\rho = \rho_k$, $k \in I$. Then*

$$\underline{u}(x) := \begin{cases} u_k(x), & x \in D_k^+, k \in I \\ 0, & x \in \bar{D} - D_I^+ \end{cases}$$

is a lower stationary solution to $(I)_N$.

LEMMA 2.3. *If (A-2) holds, then for any given $M > 0$ there exists an upper solution $\bar{u}$ such that $\bar{u} \geq M$ in D.*

2.3 Let us proceed to the proof of Theorem 1.5. We first show that (i) implies (iii). By Lemma 2.2 (ii) there exists a lower solution $\underline{u}(x)$ to $(\mathrm{I})_N$ such that $\underline{u} \not\equiv 0$ in D_k^+ for any $k \in \mathcal{M}$. We don't know whether an upper stationary state exists, however by our assumption $u(x, t : \underline{u})$ is bounded. From Proposition 2.1 it can be deduced that u converges to a stationary state $u_* \geq \underline{u}$ as $t \to \infty$. Since $\underline{u} \not\equiv 0$ in D_k^+ for any $k \in \mathcal{M}$, u_* belongs to $\mathcal{S}_{\mathcal{M}}$. Theorem 1.3′ (ii) thus applies and proves (iii).

Suppose that (iii) holds. By Lemma 2.3 there is an upper stationary solution $\bar{u} \geq |u_0|_\infty \geq u_0 \geq 0$ for any positive function $u_0 \in C(\bar{D})$. In view of Proposition 2.1 (i) we have

$$0 \leq u(x, t : u_0) \leq u(x, t : \bar{u}) \leq \bar{u} \text{ in } D \times R^+.$$

This implies (i).

Next, we prove that (iii) implies (ii). Let $u_0 \in C(\bar{D})$. By the previous arguments, $u(x, t : u_0)$ exists in $D \times R^+$. Let $J \subseteq \mathcal{M}$ be the set such that $u(x, t : u_0) \equiv 0$ for any $(x, t) \in D_k^+ \times R^+$ and $k \in J$. Consider first the case where $J = \phi$. Then for all $k \in \mathcal{M}$ there is a pair $(x_k, t_k) \in D_k^+ \times R^+$ such that $u(x_k, t_k : u_0) > 0$. Due to the continuity of u, Lemma 2.2 (i) and Proposition 2.1 imply that $u(x_k, t : u_0) > 0$ for all $k \in \mathcal{M}$, $t \geq t_0 := \max\{t_k, k \in \mathcal{M}\}$. By Lemma 2.2 (ii) a lower solution $\underline{u} \leq u(x, t_0 : u_0)$ exists such that $\underline{u}(x_k) > 0$ for any $k \in \mathcal{M}$. By virtue of our assumption and Lemma 2.3 we can also find an upper solution $\bar{u} \geq u(x, t_0 : u_0)$. Hence by Proposition 2.1, $u(x, t : \underline{u})$ and $u(x, t : \bar{u})$ converge to the stationary states

u_* and u^*, $u_* \le u^*$, as $t \to \infty$. Since $u^* \ge u_* \ge \underline{u}$, u^* and u_* belong to $S_{\mathcal{M}}$. Since $\mathcal{M}$ is finite, Theorem 1.3 (ii) implies that $u^* = u_*$. Thus the fact that $u(x,t:\underline{u}) \le u(x,t+t_0:u_0) \le u(x,t:\bar{u})$ yields (ii). If $J \neq \phi$, we consider Problem $(\tilde{\text{I}})_N$ where $a(x)$ is replaced by

$$\tilde{a}(x) = \begin{cases} a(x) & \text{for } x \in \bar{D} - D_J^+ \\ 0 & \text{for } x \in D_J^+ \end{cases}$$

Then $u(x,t:u_0)$ solves $(\tilde{\text{I}})_N$ in $D \times R^+$, and $J = \phi$ with respect to Problem $(\tilde{\text{I}})_N$. Hence the first part of this proof implies that $u(x,t:u_0)$ converges to the unique stationary solution u^* of $(\tilde{\text{I}})_N$ in $S_{\tilde{\mathcal{M}}}$. Since $u(x,t:u_0) \equiv 0$ in $D_k^+ \times R^+$ for any $k \in J$, its limit function u^* satisfies $u^*(x) \equiv 0$ in D_J^+. Hence u^* solves also $(\text{I})_N$ and $u^* \in \mathcal{N}_{\mathcal{M}-J}$ which completes the proof. The implication (ii) $\to$ (i) is trivial and the theorem is thus established.

REMARK: In the case of infinite $\mathcal{M}$, (i) and (iii) are equivalent as well.

3. Multiplicity Results for Problem (II).

3.1 It should be pointed out that the results of this section hold only if $D^+ \neq \phi$. Let us start with a simple observation obtained from the parabolic maximum principle.

PROPOSITION 3.1. *Let $u(x,t)$ be a solution of Problem (II) with either Dirichlet or Neumann boundary conditions.*

(i) *If $u(x_k,t_0) = 0$ for some $x_k \in D_k^+$, then $u(x,t) \equiv 0$ in $D_k^+ \times [0,t_0)$.*
(ii) *If $u_0(x_0) \neq 0$ for some $x_0 \in D_k^+$, then $u(x,t) > 0$ in $D_k^+ \times R^+$.*

The trivial solution solves (II) when $u_0 \equiv 0$ and $\psi \equiv 0$ or $\frac{\partial u}{\partial n} = 0$. The question arises whether there are also non-trivial solutions.

THEOREM 3.2. *For any $t_0 > 0$ and $k \in \mathcal{M}$ there exists a solution to $(II)_D$ with $\psi \equiv 0$ and $(II)_N$, with $u_0(x) \equiv 0$ such that $u(x,t) \equiv 0$ in $D_k^+ \times [0,t_0]$ and $u(x,t) > 0$ in $D_k^+ \times \{t > t_0\}$.*

PROOF: This assertion will be proved by the method of upper and lower solutions; in particular the version derived in [7] applies here. Take as upper solution the function $\bar{u}(t)$ determined by

$$\frac{d\bar{u}}{dt} = a_{\max}\bar{u}^p, \qquad \bar{u}(0) = c_0$$

that is

$$\bar{u}(t) = \{(1-p)a_{\max}t + c_0\}^{1/(1-p)}.$$

For the construction of the lower solution, take any open ball B, $\bar{B} \subset D_k^+$ and denote by φ the positive first eigenfunction of

$$\Delta\varphi + \lambda\varphi = 0 \text{ in } B, \quad \varphi = 0 \text{ on } \partial B.$$

Put

$$\underline{u}(x,t) = \begin{cases} 0 & \text{in } \{D - B\} \times R^+ \\ c(t-t_0)_+^r \varphi & \text{in } B \times R^+ \end{cases}$$

For $t \le t_0$ we have $\underline{u} \equiv 0$ and for $t > t_0$

$$\underline{u}_t - \Delta\underline{u} - a\underline{u}^p = \begin{cases} 0 & \text{in } \{D - B\} \times (t_0, \infty) \\ \mathcal{H} & \text{in } B \times (t_0, \infty) \end{cases} \tag{3.1}$$

where

$$\begin{aligned} \mathcal{H} &= c^p(t-t_0)^{rp}\varphi^p\{r(t-t_0)^{r-1-rp}c^{1-p}\varphi^{1-p} + \\ &\quad + \lambda(t-t_0)^{r-rp}c^{1-p}\varphi^{1-p} - a\} \\ &\le c^p(t-t_0)^{rp}\varphi^p\{r(t-t_0)^{r-1-rp}c^{1-p}\varphi_{\max}^{1-p} \\ &\quad + \lambda(t-t_0)^{r-rp}c^{1-p}\varphi_{\max}^{1-p} - a_{\min}\} \end{aligned}$$

If we choose $r > 1/(1-p)$ and $T > t_0$ we can always find a number C^* such that $\mathcal{H} \le 0$ in Q_T for all $c \le C^*$. It follows easily that $\underline{u}$ is a weak lower solution in Q_T. We can also determine c_0 in $\bar{u}$ such that $\underline{u} \le \bar{u}$ in Q_T. By [**7**] this establishes the existence of a solution $\underline{u} \le u \le \bar{u}$ in Q_T. The global existence is proved by continuation methods, and the fact that $u > 0$ in $D_k^+ \times (t_0, \infty)$ follows from Proposition 3.1 (ii).

REMARK: It should be emphasized that $\bar{u}$ and $\underline{u}$ are upper and lower solutions for both the Dirichlet and Neumann problems.

COROLLARY 3.3. *Let $u(x,t)$ be a solution of Problem (II) vanishing in $D_k^+ \times R^+$. Then for any $t_0 > 0$ there exists another solution $\tilde{u}$ of the same problem such that $\tilde{u} \equiv 0$ in $D_k^+ \times [0, t_0]$ and $\tilde{u} > 0$ in $D_k^+ \times \{t > t_0\}$.*

PROOF: Let $\hat{u}$ be the solution constructed in Theorem 3.2. Then $\underline{u} = \max\{u, \hat{u}\}$ is a lower solution and the same arguments as before yield the assertion.

Hence the solution $u(x,t)$ considered in Corollary 3.3 generates an infinite number of solutions. In general the existence of such a $u(x,t)$ cannot be guaranteed; e.g. if $u_0(x) \not\equiv 0$ in D_k^+ for all $k \in \mathcal{M}$, then by Proposition 3.1 (ii), all solutions must be positive in $D^+ \times R^+$. We shall prove that in this case the solution is unique. This will be a consequence of the more general results given below.

3.2 This part parallels some ideas of Section 1.1 concerning the classification of the stationary states.

Consider Problem (II) with either Dirichlet or Neumann boundary conditions.

DEFINITION 3.4. *$\mathcal{T}_I = \{u(x,t)$: solutions of (II) such that $u > 0$ in $D_I^+ \times R^+ + \{0\}$ and $u \equiv 0$ in $\{D^+ - D_I^+\} \times R^+ + \{0\}\}$.*

Clearly $\mathcal{T}_I$ depends on $u_0(x)$ and on the boundary conditions. It may be empty, except for $\mathcal{T}_\mathcal{M}$ which by Corollary 3.3 always contains an element.

Before we state our uniqueness result let us observe that the function

$$U_\varepsilon(x,t) = (u+\varepsilon)^{1-p}/(1-p) \tag{3.2}$$

u being a solution of (II), satisfies the differential equation

$$\frac{\partial U_\varepsilon}{\partial t} - \Delta U_\varepsilon - \frac{p}{(u+\varepsilon)^{1-p}}\left|\nabla U_\varepsilon\right|^2 = a\left(\frac{u}{u+\varepsilon}\right)^p \quad \text{in } Q_T. \tag{3.3}$$

THEOREM 3.5. *Assume (A-1) and let $\mathcal{M}$ be finite. Then for any $I \subseteq \mathcal{M}$, $\mathcal{T}_I$ has at most one element. In particular, there is at most one solution which is positive in $D^+ \times R^+$.*

PROOF: The proofs of Theorems 1.1 (ii) and 1.3(ii) extend to the present case without any difficulty. We sketch here the main steps and refer to [**3**] and [**4**] for more details.

Suppose that there exist two solutions u_1 and u_2 in $\mathcal{T}_I$. Then there is a region $Q' \subseteq Q_T$ where $u_1 > u_2$. Thus the difference $\delta = U_1 - U_2$ with $U_i := u_i^{1-p}/(1-p)$ is positive in Q' and assumes its maximum at some point $(x_0, t_0) \in \bar{Q}'$. We distinguish two cases.

$$\text{(i)}\, U_2(x_0,t_0) > 0.$$

We then deduce from (3.3), putting $\varepsilon = 0$, that

$$\begin{aligned} 0 &= \frac{\partial \delta}{\partial t} - \Delta\delta - \frac{p}{u_1^{1-p}}\left|\nabla U_1\right|^2 + \frac{p}{u_2^{1-p}}\left|\nabla U_2\right|^2 \\ &\geq \frac{\partial \delta}{\partial t} - \Delta\delta - \frac{p}{u_2^{1-p}}(\nabla U_1 + \nabla U_2, \nabla\delta) \end{aligned}$$

A straightforward application of the parabolic maximum principle leads to a contradiction. Hence we must have

$$\text{(ii)} U_2(x_0, t_0) = 0.$$

In view of our assumptions (x_0, t_0) cannot lie in $D^+ \times [0, T]$. We then look at

$$\delta_\varepsilon(x,t) := U_1(x,t) - U_{2,\varepsilon}(x,t)$$

$U_{2,\varepsilon}$ being defined in (3.2) with u replaced by u_2. ε is chosen such that δ_ε satisfies

$$\frac{\partial \delta_\varepsilon}{\partial t} - \Delta\delta_\varepsilon - \frac{p}{(u_2+\varepsilon)^{1-p}}(\nabla U_1 + \nabla U_{2,\varepsilon}, \nabla\delta_\varepsilon) \leq 0.$$

in a small neighborhood of (x_0, t_0), in the topology relative to Q_{t_0}, and such that δ_ε takes there its maximum at an interior point. This is impossible and case (ii) is thus excluded.

REMARK: The second assertion follows also from a result in [5] provided $a(x) \geq 0$ in D.

3.3 We now ask for a criterion which guarantees that $\mathcal{T}_I$ is non-empty. Results in this direction have been stated in Theorem 1.6, and we shall present here their proof.

PROOF OF THEOREM 1.6:

(i) For all $k \in \mathcal{M}$ denote by $u_k(x,t)$ the solution constructed in Theorem 3.2 with $t_0 = 0$. Then $u(x,t) = \max_k\{u_k(x,t)\}$ is a lower solution which doesn't vanish in any component of $D^+ \times R^+$. As an upper solution we take u as in the proof of Theorem 3.2 with $c_0 = \max\{|u_0|_\infty, |\psi|_\infty\}$. The assertion is now a consequence of Proposition 3.1 (ii).

(iii) We restrict ourselves to $I = \{k\}$, the general case is then immediate. Let $B, \bar{B} \subset D_k^+$, be a domain such that $u_0 > 0$ in $\bar{B}$. Denote as usual by φ the positive eigenfunction of

$$\Delta\varphi + \lambda\varphi = 0 \text{ in } B, \quad \varphi = 0 \text{ on } \partial B$$

Then

$$-\Delta\varphi - a(x)\varphi^p = \left[\lambda - a(x)\varphi^{p-1}\right]\varphi \leq \left[\lambda - a_0\varphi_{\max}^{p-1}\right]\varphi$$

where $a_0 = \min_{\bar{B}}\{a(x)\}$. If we choose

$$\varphi_{\max} \leq \min\left\{(a_0/\lambda)^{1/(1-p)}, \min_{\bar{B}} u_0(x)\right\}$$

then

$$\underline{u}(x,t) = \begin{cases} \varphi & \text{in } B \times R^+ \\ 0 & \text{elsewhere} \end{cases}$$

is a lower solution. Furthermore, $\bar{u} = u_s(x) \in \mathcal{N}_{\{k\}}$ is an upper solution satisfying $\underline{u} \leq u_0 \leq \bar{u}$ in $\bar{D} \times \{0\}$ and $\underline{u} \leq \bar{u}$ in $D \times R^+$. This establishes the existence of a solution in $\mathcal{T}_{\{k\}}$. (ii) follows by Theorem 3.2.

Mathematisches Institut, Universität Basel, Rheinsprung 21,
CH-4051 Basel, SWITZERLAND
Dipartimento di Matematica, IIa Università di Roma,
Via Orazio Raimondo, I-00173 Roma ITALY

Research at MSRI sponsored in part by NSF Grant DMS-812079-05.

REFERENCES

1. Aris, R., "The Mathematical Theory of Diffusion and Reaction in Permeable Catalysts, Vol. I and II," Clarendon Press, Oxford, 1975.
2. Aronson, D.G., Crandall, M., and Peletier, L.A., *Stabilization of solutions of a degenerate diffusion problem*, Nonlinear Analysis TMA **6** (1982), 1001–1022.
3. Bandle, C., Pozio, M.A., and Tesei, A., *The asymptotic behavior of the solutions of a class of degenerate parabolic problems*, to appear in T.R.A.M.S..
4. Bandle, C., Pozio, M.A., and Tesei, A., *Existence and uniqueness of solutions of nonlinear Neumann problems*, Quaderno IAC **1** (1987). CNR Roma, Italy.
5. Chen, Z, and Luo, X., *Comparison and uniqueness of positive solutions for the mixed problem for semi-linear parabolic equations*, Commun. Part. Diff. Equations **11** (1986), 1285–1295.
6. de Mottoni, P., Schiaffino, A., and Tesei, A., *Attractivity properties of nonnegative solutions for a class of nonlinear parabolic problems*, Ann. Mat. Pura Appl. **136** (1984), 35–48.
7. Deuel, J., *Nichtlineare parabolische Randwertprobleme mit Unter- und Oberlösungen*, Diss. Nr. 5750, ETH Zürich.
8. Namba, T., *Density-dependent dispersal and spatial distribution of a population*, J. Theor. Biol. **86** (1980), 351–363.
9. Sacks, P., *The initial and boundary value problem for a class of degenerate parabolic equations*, Comm. Part. Diff. Equ. **8** (1983), 693-733.
10. Stakgold, I., *Estimates for some free boundary problems*, W.N. Everitt and B.D. Sleeman, eds., Springer Lect. Notes in Math. **846** (1981).

Source-type Solutions of Fourth Order Degenerate Parabolic Equations

FRANCISCO BERNIS

1. Introduction.

Let $m > 1$ be a real number. We are going to prove that the degenerate parabolic equation

$$\frac{\partial W}{\partial t} + \frac{\partial^4}{\partial y^4}(|W|^m \operatorname{sgn} W) = 0 \tag{1.1}$$

has a source-type solution in $\mathbb{R}\times\mathbb{R}_+$, i.e. a solution with the Dirac measure δ as initial condition. (Here $W = W(y,t)$ is a continuous real-valued function (for $t > 0$) and $\mathbb{R}_+$ stands for the open half-line $(0,\infty)$. All differential equations will be understood in the sense of distributions.) For each $t > 0$, this solution W has bounded support and infinitely many isolated zeros accumulating near the boundary of the support. Furthermore, for $t > 0$ we find that $W \in C^{0,1/m}$ and $W_t \in L^1$, but W_t is unbounded near each isolated zero of $W(\cdot\,,t)$. These properties also hold for solutions having δ', δ'' and δ''' as initial conditions.

The scaling properties of δ and Equation (1.1) suggest that one look for a similarity solution of the form:

$$W(y,t) = t^{-\beta} w(y/t^\beta), \quad \beta = 1/(m+3). \tag{1.2}$$

Then the initial value problem for $W = W(y,t)$ is reduced to a nonlinear eigenvalue problem on $\mathbb{R}$ for $w = w(x)$. Notice that (1.2) implies that, for all $t > 0$,

$$\int_{\mathbb{R}} W(y,t)\varphi(y)\,dy = \int_{\mathbb{R}} w(x)\varphi(xt^\beta)\,dx, \tag{1.3}$$

where φ is a test function. The fact that δ is even and the invariance of Equation (1.1) under the transformation $(y,t) \to (-y,t)$ suggest that w is even. This will lead to a problem on the half-line $\bar{\mathbb{R}}_+$. These reductions are performed in Section 2. Our first theorem is as follows.

THEOREM 1.1. *There exists a unique solution W of Equation (1.1) in $\mathbb{R} \times \mathbb{R}_+$ such that:*

(I) $W = W(y,t)$ *has the form (1.2) with* $w \in C(\mathbb{R})$,

(II) $w = w(x)$ *is even,*

(III) w *has bounded support, and*

(IV)

$$W(\cdot\,,t) \to \delta \text{ in } \mathcal{D}'(\mathbb{R}) \text{ as } t \to 0^+ \tag{1.4}$$

where $\mathcal{D}'(\mathbb{R})$ *is the space of distributions on* $\mathbb{R}$.

This theorem will be proved in Sections 2-6. Uniqueness also holds in a general class of weak solutions, without assuming similarity structure (Theorem 5.1).

The solution W of Theorem 1.1 is in a sense analogous with the Barenblatt [1] and Pattle [11] source-type solution of the porous media equation

$$\frac{\partial F}{\partial t} = \frac{\partial^2}{\partial y^2}(|F|^m \operatorname{sgn} F). \tag{1.5}$$

The Barenblatt-Pattle solution reads

$$F(y,t) = t^{-b}(C - By^2t^{-2b})_+^{1/(m-1)} \tag{1.6}$$

where $(x)_+ = \max\{x, 0\}$, $b = 1/(m+1)$, $B = (m-1)/2(m+1)$ and C is determined by

$$\int_{\mathbb{R}} F(y,1)\,dy = 1.$$

Unlike W, the function (1.6) is everywhere nonnegative and has singularities only at the boundary of the support. No explicit formula is known for w.

Since w has bounded support, it follows from (1.2) that the support of W is limited by the two curves $y = \pm At^\beta$, with A constant. These curves are called *interfaces* or *free boundaries*. It is also clear that the support *propagates at finite speed* (while for linear parabolic equations the speed of propagation is infinite).

Papers [5] and [9] prove existence results including Equation (1.1), though δ and its derivatives are not allowed as initial data. In [6] it is proved that the property of finite speed of propagation holds for the classes of solutions considered in [5] and [9]. The solutions of Theorems 1.1 and 10.1 belong to these classes for $t \geq \varepsilon > 0$, but not in any neighborhood of $t = 0$.

In the existence and uniqueness proofs of Sections 3 and 5 we use (directly or indirectly) the quasilinear parabolic equation

$$V_t + (|V_{yy}|^m \operatorname{sgn} V_{yy})_{yy} = 0 \tag{1.7}$$

related formally to (1.1) via $W = V_{yy}$. The idea of using Equation (1.7) is related to the H^{-1} approach to the porous media equation of Brézis [**7**], except that H^{-1} is to be replaced by H^{-2} in the present situation. Although not explicitly developed, this relation is more clearly seen in Section 5. The method of Sections 3 and 5 does not apply if the initial condition is δ'' or δ'''. Notice that $\delta, \delta' \in H^{-2}(\mathbb{R})$, while $\delta'', \delta''' \notin H^{-2}(\mathbb{R})$.

The plan of the paper is as follows:

Section 2: The initial value problem is reduced to a problem for w on $\mathbb{R}_+$ (Theorem 2.1) and a boundary value problem for v is posed, where v is a second primitive of w.

Section 3: An existence method for elliptic quasilinear problems is applied to the v-problem.

Section 4: The support is bounded.

Section 5: Uniqueness.

Section 6: The proof of Theorem 1.1 is completed.

Section 7: The function $|w(x)|$ behaves (in an average sense) as $(A - x)^{3/(m-1)}$ at the extreme, A, of the support. The proof of this "nondegeneracy" property is closely related to that of boundedness of the support.

Section 8: Zeros and moments. Theorem 8.3 shows a delicate balance between positive and negative parts.

Section 9: Optimal Hölder and L^p regularity is obtained.

Section 10: Problems with δ', δ'', and δ''' as initial conditions.

Section 11: An existence proof via ODE initial value problems is sketched. This method has a broader field of applications.

Note: This work was partially carried out while visiting at Purdue University, and part of the work described is joint with J.B. McLeod.

2. A nonlinear eigenvalue problem on $\mathbb{R}$ and a problem on the half-line.

It follows from direct computations that (1.2) is a solution on $\mathbb{R} \times \mathbb{R}_+$ of (1.1) if and only if $w = w(x)$ is a solution on $\mathbb{R}$ of

$$(|w|^m \operatorname{sgn} w)^{IV} = \beta(xw' + w), \quad \beta = 1/(m+3). \tag{2.1}$$

It is convenient to set

$$u = |w|^m \operatorname{sgn} w, \text{ thus } w = |u|^{1/m} \operatorname{sgn} u. \tag{2.2}$$

Then (2.1) is equivalent to

$$u^{IV} = \beta x(|u|^{1/m} \operatorname{sgn} u)' + \beta |u|^{1/m} \operatorname{sgn} u. \tag{2.3}$$

Since we want $u \not\equiv 0$, we are led to a nonlinear eigenvalue problem on the whole line. Using the fact that u is even we obtain a problem on the closed half-line $\bar{\mathbb{R}}_+$. Specifically, we have

THEOREM 2.1. (equivalent to Theorem 1.1). *Let u and w be related by (2.2). There exists a unique $w \in C(\bar{\mathbb{R}}_+)$ such that:*

(I) *u is a solution on $\mathbb{R}_+$ of (2.3),*
(II) $u'(0) = u'''(0) = 0$,
(III) *u has bounded support, and*
(IV)

$$\int_0^\infty w(x)\, dx = 1/2. \tag{2.4}$$

The fact that Theorems 1.1 and 2.1 are equivalent is straightforward once we observe that (2.4) is obtained from (1.3) and (1.4). Notice that $u \in C^3(\bar{\mathbb{R}}_+)$ and Equation (2.3) is invariant under the transformation $x \to -x$. (We recall that derivatives are in the sense of distributions.)

The proof is based on the following idea. Assume for a moment that Theorem 2.1 is true. Integrating (2.3) we obtain

$$u''' = \beta x |u|^{1/m} \operatorname{sgn} u, \quad \beta = 1/(m+3). \tag{2.5}$$

Let v be the second primitive of w, defined by

$$v(x) = \int_x^\infty (s - x) w(s) ds. \tag{2.6}$$

Thus (recall (2.2))

$$w = v'', \quad u = |v''|^m \operatorname{sgn} v''. \tag{2.7}$$

The differential equation for v is obtained by means of integrating (2.5) (or integrating (2.1) twice), and reads

$$(|v''|^m \operatorname{sgn} v'')'' = \beta(xv' - v), \quad \beta = 1/(m+3). \tag{2.8}$$

(Notice the minus sign in front of the v.) Now (2.4) becomes $v'(0) = -1/2$. The point is that (2.8) has a powerful *a priori* estimate (obtained by multiplying (2.8) by v and integrating), while (2.3) has no analogous estimate. The idea is to develop existence theory for Equation (2.8), see Section 3, and then recover Equation (2.3) from (2.8). Since (2.7) and (2.8) imply (2.5), the condition $u'''(0) = 0$ will be automatically satisfied. We also use Equation (2.8) to prove that the support is bounded (Section 4), although this argument also can be adapted to Equation (2.3). Uniqueness will be proved in Section 5 in a more general way.

3. Existence theorem for a related boundary value problem.

Let $(0, a)$ be a bounded interval with $a > 1$. We consider the Sobolev space

$$W^{2,p}(0,a) = \{v \in L^p(0,a) : v' \text{ and } v'' \in L^p(0,a)\}.$$

THEOREM 3.1. *Let $p = m+1$. There exists a solution $v \in W^{2,p}(0,a)$ of (2.8) such that $v'(0) = -1/2$, $u'(0) = 0$, $v(a) = v'(a) = 0$, where u is defined by (2.7). Furthermore,*

$$\int_0^a (|v''(x)|^p + v(x)^2)\, dx \text{ is bounded independently of } a. \tag{3.1}$$

(We remark that from (2.7) and (2.8) it follows that u'' is continuous and thus that $u'(0) = 0$ makes sense.)

The following proof is based on a Galerkin method (approximation from finite-dimensional subspaces) combined with Brouwer's fixed point theorem and monotonicity with respect to the top order derivatives. This is a standard existence method for quasilinear elliptic problems, whose details can be found, e.g. in the book of Lions [**10**, Chapter 2.2].

We take the following space of test functions:

$$W = \{\varphi \in W^{2,p}(0,a) : \varphi'(0) = \varphi(a) = \varphi'(a) = 0\} \tag{3.2}$$

Let $f \in C^2(\bar{\mathbb{R}}_+)$ be a fixed function (thus f is independent of a) such that

$$f'(0) = -\tfrac{1}{2}, \quad \text{support } f \subset [0,1] \tag{3.3}$$

We set the notation

$$B(v,\varphi) = \int_0^a (|v''|^m \operatorname{sgn} v''\varphi'' - \beta x v'\varphi + \beta v\varphi)dx. \tag{3.4}$$

The *weak formulation of the problem* is the following: find $v \in f + W$ such that

$$B(v,\varphi) = 0, \quad \text{for all } \varphi \in W. \tag{3.5}$$

The fact that $\varphi(0)$ is arbitrary in (3.2) implies (by virtue of an integration by parts) that v satisfies the boundary condition $u'(0) = 0$. The boundary conditions $v'(0) = -1/2$, $v(a) = v'(a) = 0$ are implied by the fact that $v - f \in W$ and (3.3). (Recall that we take $a > 1$.)

We proceed to prove that there exists a solution of the weak formulation of the problem. Integrating the lower order terms by parts, we see that, for *any* $v \in f + W$,

$$\begin{aligned} &B(v, v-f) \\ &\quad = \int_0^a |v''|^p + \frac{3}{2}\beta\int_0^a v^2 - \int_0^a |v''|^{p-1} \operatorname{sgn} v'' f'' - \beta\int_0^a (xf' + 2f)v, \end{aligned}$$

and by the Hölder and Young inequalities

$$B(v, v-f) \geq -C_1 + C_2\int_0^a (|v''|^p + v^2), \quad \forall v \in f + W, \tag{3.6}$$

where the constants C_1 and C_2 depend only on f and m (and thus are independent of a).

Let $\{b_1, \ldots, b_n, \ldots\} \subset W$ be a set of $C^2([0,a])$ linearly independent functions generating a dense subspace of W. From (3.6) and Brouwer's fixed point theorem (in the form given by [**10**,p. 53]) it follows that for each n there exists $v_n \in f + [b_1, \ldots, b_n]$ such that

$$B(v_n, \varphi) = 0 \quad \text{for all } \varphi \in [b_1, \ldots, b_n],$$

where $[b_1, \ldots, b_n]$ is the subspace generated by $b_1, \ldots, b_n$. The estimate (3.6) and the monotonicity of the operator $v \to (|v''|^m \operatorname{sgn} v'')''$ allow us to pass to the limit as the dimension $n \to \infty$ (see e.g. [**10**, pp. 171–173]), obtaining a function $v \in f+W$ satisfying (3.5). This concludes the existence proof. Finally, (3.1) follows from (3.6), since $B(v, v - f) = 0$ when v is the solution of the problem.

REMARK 3.1: The above proof of Theorem 3.1 applies to any $m > 0$ (not only to $m > 1$). It also applies to $a = \infty$ if we replace the space $W^{2,p}$ by the space $\{\varphi \in L^2 : \varphi'' \in L^p\}$ and suppose the support of b_n to be bounded for all n.

REMARK 3.2: The solution of Theorem 3.1 is unique. If v and $\bar{v}$ are two solutions, subtract Equations (2.8) for v and $\bar{v}$, multiply by $v - \bar{v}$ and integrate. Uniqueness then follows at once.

4. Boundedness of the support.

THEOREM 4.1. *Let v be the function of Theorem 3.1. Then support $v \subset [0, A]$ and $A < a$ for a large enough.*

(Thus the zero extension of v (for a large enough) is a solution of (2.8) on $\mathbb{R}_+$ and has bounded support.)

PROOF OF THEOREM 4.1: We use an energy method combined with weighted interpolation inequalities. Let $z > 0$. Multiply (2.8) by $(x - z)^2 v(x)$ and integrate with respect to x between z and a. We obtain (write $xv'(x) = (x - z)v'(x) + zv'(x)$):

(4.1)
$$\int_z^a (x-z)^2 |v''|^{m+1} dx + \frac{5}{2}\beta \int_z^a (x-z)^2 v^2\, dx + \beta z \int_z^a (x-z) v^2\, dx =$$
$$= -4 \int_z^a (x-z) v' |v''|^m \operatorname{sgn} v'' dx - 2 \int_z^a v |v''|^m \operatorname{sgn} v'' dx.$$

(The power $(x - z)^2$ is introduced to avoid boundary terms at $x = z$ when integrating by parts.) We introduce the notations

(4.2) $\qquad p = 1 + m, \quad q = 1 + 1/m$, thus $1/p + 1/q = 1$

(4.3)
$$E_s(z) = \int_z^a (x-z)^s |v''(x)|^p dx, \quad F(z) = \int_z^a (x-z) |v'(x)|^p dx$$

(4.4)
$$G_s(z) = \int_z^a (x-z)^s |v(x)|^p dx, \quad H(z) = \int_z^a (x-z) v(x)^2 dx.$$

Consider (4.1). Dropping the second term of the left-hand side and applying Hölder's inequality, we obtain

(4.5) $$E_2(z) + \beta z H(z) \le 4E_1(z)^{1/q} F(z)^{1/p} + 2E_0(z)^{1/q} G_0(z)^{1/p}.$$

REMARK 4.1: Dimensional analysis helps to understand the underlying idea of the following delicate computations. Recall Equation (2.5) and the fact that $v'' = w = |u|^{1/m} \operatorname{sgn} u$. The global dimensionality exponent of $w(x)$ with respect to x is $4/(m-1)$, while the local exponent near $x = x_0 \neq 0$ is $3/(m-1)$. *All the steps of the proof then preserve these global and local dimensionalities.* We shall also obtain precise information on the behavior at the extreme of the support (see Section 7). The proof of Theorem 4.1 can also be carried out using the second term of (4.1) (instead of the third term), but then we lose the local behavior and the precise exponent $3/(m-1)$ of Section 7.

Proceeding with the proof, let us now apply the following inequalities:

$$\begin{aligned}
&(4.6) \quad G_0(z) \leq C\, F(z)^{1/p}\, G_1(z)^{1/q} \\
&(4.7) \quad F(z) \leq C\, E_1(z)^{1/2}\, G_1(z)^{1/2} \\
&(4.8) \quad G_1(z) \leq C\, E_1(z)^{\lambda}\, H(z)^{(1-\lambda)p/2}, \quad \lambda = (m-1)/(3m+1) \\
&(4.9) \quad E_1(z) \leq E_2(z)^{1/2}\, E_0(z)^{1/2}
\end{aligned}$$

where *the constants C depend only on m* (i.e. only on p). Inequalities (4.7)–(4.8) are taken from [**4**, Lemma 8] and (4.6) from [**4**, Lemma 11], while (4.9) is implied by the Schwarz inequality. We consider the zero extension of v to $\mathbb{R}_+$ in order to apply Lemmas 8 and 11 of [**4**]. This is possible because $v(a) = v'(a) = 0$ and v'' is the highest order derivative involved in (4.6)–(4.8). In (4.8) it is essential that $m > 1$ (i.e. $p > 2$). Inserting (4.6)–(4.9) in (4.5) and computing all the exponents, we obtain

$$\begin{aligned}
(4.10) \quad E_2(z) + \beta z H(z) \leq C(m) E_2(z)^{a_1} \big(zH(z)\big)^{a_2} z^{-a_2} E_0(z)^{a_3} + \\
+ C(m) E_2(z)^{b_1} \big(zH(z)\big)^{b_2} z^{-b_2} E_0(z)^{b_3},
\end{aligned}$$

$$\begin{aligned}
&a_1 = a_3 = 3m/2(3m+1), \quad a_2 = (m+1)/2(3m+1) \\
&b_1 = m/2(m+1)(3m+1), \quad b_2 = (2m+1)/2(3m+1) \\
&b_3 = 3m(2m+1)/2(m+1)(3m+1).
\end{aligned}$$

We next apply a variation of Young's inequality. If $P, Q, R \geq 0$, $h_1, h_2 > 0$, $h_1 + h_2 < 1$, then for all $\varepsilon > 0$ there exists $C = C(\varepsilon, h_1, h_2)$ such that

$$(4.11) \qquad P^{h_1} Q^{h_2} R \leq \varepsilon(P+Q) + C R^{1/(1-h_1-h_2)}.$$

Observe that

$$\begin{aligned}
1 - a_1 - a_2 &= (2m+1)/2(3m+1), \\
1 - b_1 - b_2 &= (2m+1)^2/2(m+1)(3m+1)
\end{aligned}$$

$$(4.12)\qquad a_2/(1-a_1-a_2) = b_2/(1-b_1-b_2) = (m+1)/(2m+1)$$

$$(4.13)\qquad a_3/(1-a_1-a_2) = b_3/(1-b_1-b_2) = 3m/(2m+1)$$

The underlying reason to use Equalities (4.12)–(4.13) is the preservation of global and local dimensionalities in the sense of Remark 4.1. We apply (4.11) to (4.10), take ε small enough (depending only on m) and absorb $E_2(z) + zH(z)$ in the left-hand side. Taking into account (4.12)–(4.13), we obtain

$$E_2(z) + zH(z) \leq C(m) z^{-(m+1)/(2m+1)} E_0(z)^{3m/(2m+1)},$$

and dropping $zH(z)$ and applying (4.9)

$$(4.14)\qquad E_1(z) \leq C(m) z^{-(m+1)/2(2m+1)} E_0(z)^{(5m+1)/2(2m+1)}.$$

This is a first order differential inequality, since $E_1'(z) = -E_0(z)$. Integrating (4.14) explicitly we see that if $0 < x < y$ and $E_1(y) \neq 0$, then

$$(4.15)\qquad y^{1+(m+1)/(5m+1)} - x^{1+(m+1)/(5m+1)} \leq C(m) \left(E_1(x)^{(m-1)/(5m+1)} - E_1(y)^{(m-1)/(5m+1)} \right)$$

where $C(m) > 0$. Since $m > 1$, the exponent of E_0 in (4.14) is greater than 1. This is essential here. Letting $x \to 0$, (4.15) implies that support $v \subset [0, A]$ and $A \leq C(m) E_1(0)^{(m-1)/(6m+2)}$. Since $E_1(0) \leq A E_0(0)$, we obtain

$$(4.16)\qquad A \leq C(m) E_0(0)^{(m-1)/(5m+3)}.$$

This completes the proof of Theorem 4.1, since by (3.1) the value $E_0(0)$ is bounded independently of a.

5. A general uniqueness theorem.

In this section we consider very weak solutions of compact support of Equation (1.1), *without assuming similarity structure*. Let $0 < K, T < \infty$, and let F be a distribution of $\mathcal{D}'(\mathbb{R})$.

THEOREM 5.1. *There exists at most one*

$$W \in L^{m+1}(\mathbb{R} \times (0,T)) \tag{5.1}$$

such that W *satisfies Equation (1.1) in* $\mathcal{D}'(\mathbb{R} \times (0,T))$ *with*

$$\textit{support } W \subset (-K,K) \times [0,T] \tag{5.2}$$

and

$$W(\cdot\,,t) \to F \textit{ in } \mathcal{D}'(\mathbb{R}) \text{ as } t \to 0^+. \tag{5.3}$$

The idea of the proof is to pass from Equation (1.1) to Equation (1.7) ("integrating" twice with respect to y). In applying this idea we shall need Lemma 5.2 (below), which relies on the fact that the first moments of W are independent of t (Lemma 5.1).

Let $p = m + 1$ and $1/p + 1/q = 1$. From (5.1) and (1.1) it follows that

$$W_t \in L^q(0,T; W^{-4,q}(\mathbb{R})) \text{ and } W \in C([0,T]; W^{-4,q}(\mathbb{R})). \tag{5.4}$$

Thus $W(\cdot\,,t)$ makes sense for *all* $t \in [0,T]$. This explains the meaning of (5.3). Let W_1 and W_2 be two functions satisfying the conditions of the theorem. We set

$$W_3 = W_2 - W_1, \quad U = |W_2|^m \operatorname{sgn} W_2 - |W_1|^m \operatorname{sgn} W_1. \tag{5.5}$$

Then W_3 satisfies

$$(W_3)_t + U_{yyyy} = 0, \quad \text{in } \mathcal{D}'(\mathbb{R} \times (0,T)) \tag{5.6}$$

$$W_3(\cdot\,,t) \to 0 \text{ in } \mathcal{D}'(\mathbb{R}) \text{ as } t \to 0^+. \tag{5.7}$$

LEMMA 5.1. *For almost all* $t \in (0,T)$, *we have*

$$\int_{-K}^{K} W_3(y,t)\,dy = \int_{-K}^{K} yW_3(y,t)\,dt = 0.$$

PROOF: We first remark that $W_3(\cdot\,,t) \in L^1(\mathbb{R})$ for almost all $t \in (0,T)$, because of (5.1) and (5.2). Thus, if $\xi \in C^\infty(\mathbb{R})$ we have

$$(W_3(\cdot\,,t), \xi) = \int_{-K}^{K} W_3(y,t)\xi(y)dy$$

for almost all $t \in (0,T)$. (The support of ξ may be unbounded because of (5.2)). Now we take $\theta(t)\xi(y)$ as a test function for Equation (5.6), where $\theta \in C_0^\infty((0,T))$ and either $\xi(y) = 1$ or $\xi(y) = y$. Since $\frac{d^4\xi}{dy^4} \equiv 0$, we obtain

$$-\int_0^T \theta'(t)dt \int_{-K}^K W_3(y,t)\xi(y)dy = 0, \forall \theta \in C_0^\infty((0,T)).$$

But this means that

$$\frac{d}{dt}(W_3(\cdot\,,t),\xi) = 0 \quad \text{in } \mathcal{D}'((0,T)).$$

This and (5.7) imply

$$(W_3(\cdot\,,t),\xi) = 0 \quad \text{for all } t \in [0,T]$$

and the lemma follows.

LEMMA 5.2. *Let $E(y) = (1/2)|y|$. We define $V_3(\cdot\,,t) = E * W_3(\cdot\,,t)$, (convolution in y for each t). Then*

$$\text{support } V_3 \subset (-K,K) \times [0,T].$$

PROOF: From (5.2) and Lemma 5.1 we obtain that for almost all $t \in (0,T)$

$$V_3(y,t) = \int_y^K (x-y)W_3(x,t)dx = \int_{-K}^y (y-x)W_3(x,t)dx.$$

This implies the lemma.

REMARK 5.1: In general, $E*W_1$ and $E*W_2$ do not have compact support. In fact, they are equal to $C_1y + C_2$ for $|y| > K$.

LEMMA 5.3. *V_3 satisfies*

$$(V_3)_t + U_{yy} = 0 \quad \textit{in } \mathcal{D}'(\mathbb{R} \times (0,T)). \tag{5.8}$$

PROOF: This follows from (5.2), (5.6) and Lemma 5.2.

LEMMA 5.4.

$$V_3 \in L^p(0,T;W^{2,p}(\mathbb{R})) \textit{ and } (V_3)_t \in L^q(0,T;W^{-2,q}(\mathbb{R})).$$

PROOF: Lemma 5.4 follows from (5.1), (5.8) and Lemma 5.2.

By a standard result (see e.g. Lions [**10**, p. 321]), Lemma 5.4 implies the following:

LEMMA 5.5. *$V_3 \in C([0,T], L^2(\mathbb{R}))$ and*

$$\int_0^T ((V_3)_t, V_3)\,dt = \frac{1}{2}\int_{\mathbb{R}} \big(V_3(y,T)\big)^2 dy,$$

where $(\cdot\,,\cdot)$ stands for the duality between $W^{-2,q}(\mathbb{R})$ and $W^{2,p}(\mathbb{R})$. Notice that $V_3(\cdot\,,t) \to 0$ strongly in $L^2(\mathbb{R})$ as $t \to 0^+$, because of (5.7) and the continuity in $L^2(\mathbb{R})$.

PROOF OF THEOREM 5.1: From Equation (5.8) and Lemma 5.4, it follows that

$$\int_0^T ((V_3)_t, V_3)\,dt + \int_0^T (U_{yy}, V_3)\,dt = 0. \tag{5.9}$$

Lemma 5.4 also implies that

$$\int_0^T (U_{yy}, V_3)\,dt = \int_0^T \int_{\mathbb{R}} U (V_3)_{yy}\,dy\,dt. \tag{5.10}$$

From (5.9), Lemma 5.5, (5.10) and $(V_3)_{yy} = W_3$ we obtain

$$\frac{1}{2}\int_{\mathbb{R}} \big(V_3(y,T)\big)^2 dy + \int_0^T \int_{\mathbb{R}} U W_3\,dy\,dt = 0.$$

Recalling (5.5), we conclude that $W_3 \equiv 0$. This completes the proof.

REMARK 5.2: In Theorem 5.1 $F \in H^{-2}(\mathbb{R})$ if W exists. As noted in the Introduction, δ and $\delta' \in H^{-2}(\mathbb{R})$, but δ'' and $\delta''' \notin H^{-2}(\mathbb{R})$. The solutions of Equation (1.1) having δ'' or δ''' as initial condition do not satisfy (5.1), because of their behavior near $t = 0$ (See Section 10).

REMARK 5.3: Theorem 5.1 still holds if (5.2) is replaced by

$$\int_0^T \int_{\mathbb{R}} |y|^{2p} |W(y,t)|^p\,dy\,dt < \infty, \quad (p = m+1).$$

Then Lemma 5.4 follows from Hardy inequality (see e.g. [**6**, Appendix I]). In this form the theorem applies to any $m > 0$.

6. Proof of Theorems 2.1 and 1.1.

First, observe that Theorems 3.1, 4.1 and 5.1 imply easily Theorem 2.1

(thus also Theorem 1.1). Let v be the function in Theorem 3.1. For a large enough, the zero extension of v is a solution on $\mathbb{R}_+$ of Equation (2.8) and has bounded support (Theorem 4.1). Let u be defined by (2.7). Then u has bounded support and is a solution on $\mathbb{R}_+$ of Equations (2.5) and (2.3). The condition $u'(0) = 0$ is included in Theorem 3.1, (2.4) follows from $v'(0) = -1/2$ and $u'''(0) = 0$ from (2.5).

Finally, uniqueness follows by checking that the function $W = W(y,t)$ of Theorem 1.1 satisfies the conditions of Theorem 5.1. Notice that (5.1) follows from (1.2).

7. Behavior at the extreme of the support.

THEOREM 7.1. *Let w be the function defined in Theorem 2.1 and let $A =$* supremum support w. *Then for all $x \in [0, A)$*

$$(A-x)^{-1} \int_x^A |w(z)|\, dz \geq C(m) A^{1/(m-1)} (A-x)^{3/(m-1)},$$

where $C(m) > 0$ is a constant depending only on m.

REMARK 7.1: This theorem gives a "nondegeneracy" property at the free boundary (in an averaged sense). The exponent of $A - x$ is optimal (by Theorem 7.2, see below). This property cannot hold in the pointwise sense, since w has infinitely many zeros near A (see Section 8). Theorem 7.1 implies the following:

$$\sup_{x \leq z \leq A} |w(z)| \geq C(m) A^{1/(m-1)} (A-x)^{3/(m-1)} \quad \text{if } 0 \leq x \leq A.$$

THEOREM 7.2. *Let w and A be as in Theorem 7.1 Then*

$$|w(x)| \leq (\beta A/2)^{1/(m-1)} (A-x)^{3/(m-1)} \quad \textit{if } 0 \leq x \leq A.$$

PROOF: We use Equation (2.5). Since $u \in C^3(\mathbb{R}_+)$,

$$u(A) = u'(A) = u''(A) = u'''(A) = 0. \tag{7.1}$$

From Taylor's formula, (2.2) and (2.5), it follows that

$$|w(x)|^m \leq (\beta A/2) \int_x^A (z-x)^2 |w(z)|\,dz. \tag{7.2}$$

Starting from $|w(x)| \leq K_0$ and iterating, we obtain $|w(x)| \leq K_n(A-x)^{b_n}$. Letting $n \to \infty$ the theorem follows. (We set $z - x \leq A - x$ in (7.2). Better estimates of the constant can be given, e.g. in terms of the Γ function.)

The proof of Theorem 7.1 relies on Section 4. Recall notations (4.2)–(4.3) and $v'' = w$. Clearly, we may replace a by ∞ (or by A) in the upper limit of the integrals.

LEMMA 7.1. *For all $x \in [0, A]$*

$$E_1(x) \geq C(m) A^{(m+1)/(m-1)} (A-x)^{(5m+1)/(m-1)}, \quad C(m) > 0.$$

PROOF OF LEMMA 7.1: Consider the relation (4.15). Since A is the extreme of the support, $E_1(y) \neq 0$ if $0 \leq y < A$. Letting $y \to A^-$ the lemma is implied by (4.15) and the following numerical inequality: "Let $b > 0$ and $0 \leq x \leq A$. Then

$$A^{1+b} - x^{1+b} \geq A^b (A - x)."$$

PROOF OF THEOREM 7.1: We observe that

$$E_1(x) = \int_x^A (z-x)|w(z)|^{m+1} dz \leq (A-x)\big(\sup_{x \leq z \leq A} |w(z)|^m\big) \int_x^A |w(z)|\,dz.$$

Now Theorem 7.1 follows from Lemma 7.1 and Theorem 7.2.

8. Zeros and Moments.

In this section, u, w and v denote the functions defined by Theorem 2.1 and (2.6), while $A =$ supremum support w.

THEOREM 8.1. *The zeros of w in $(0, A)$ form an infinite increasing sequence $\{a_n\}, n \geq 1$, such that $a_n \to A$ and $u'(a_n) \neq 0$ for all n.*

PROOF: First, let us prove the u has at least one zero in (B, A) for any B such that $0 \leq B < A$. If not, either $u > 0$ or $u < 0$ on (B, A). Assume that $u > 0$ on (B, A). (The case $u < 0$ is similar). From $u''(A) = 0$ and Equation (2.5) it follows that $u'' < 0$ on (B, A), thus (from $u(A) = u'(A) = 0$) $u < 0$ on (B, A): a contradiction. Now Theorem 8.1 follows easily if we show that $u'(a_n) \neq 0$ for all n. Multiplying Equation (2.8) by v and integrating we obtain that if $x \geq 0$, then

$$\frac{\beta x}{2} v(x)^2 + \int_x^A |v''|^{m+1}\, dz + \frac{3\beta}{2} \int_x^A v^2\, dz = u'(x)v(x) - u(x)v'(x). \tag{8.1}$$

Thus $u'(a_n)v(a_n) > 0$. □

Next, we multiply Equation (2.5) by u and integrate to obtain that for all $x \geq 0$

$$\beta \int_x^A z|u(z)|^{1+1/m} dz = \frac{1}{2} u'(x)^2 - u(x)u''(x). \tag{8.2}$$

This will be used later.

THEOREM 8.2.

(I) $u(0) > 0$ *and* $u''(0) < 0$.

(II) *For each $n \geq 1$, there exists in (a_n, a_{n+1}) one and only one zero b_n of u' and one and only one zero c_n of u''. Furthermore, $a_n < b_n < c_n < a_{n+1}$.*

PROOF:

(I) $u(0) > 0$ follows from (8.1) and $v'(0) = -\frac{1}{2}$ (see Theorem 3.1); and $u''(0) < 0$ from (8.2).

(II) Let $u > 0$ on (a_n, a_{n+1}). (The other case $u < 0$ is similar.) From Theorem 8.1, we have $u'(a_n) > 0$ and $u'(a_{n+1}) < 0$. Since $u''' > 0$ on (a_n, a_{n+1}), u' has exactly one zero b_n. Also, u'' has at most one zero. From (8.2) $u''(b_n) < 0$. The theorem follows if we prove that $u''(a_{n+1}) > 0$. Suppose not; it would be $u(a_{n+1}) = 0$, $u'(a_{n+1}) < 0$ and $u''(a_{n+1}) \leq 0$, thus from Equation (2.5) if follows that $u(x) < 0$ for all $x > a_{n+1}$: a contradiction.

The following theorem on moments shows a delicate balance between positive and negative parts.

THEOREM 8.3. *Let* $M_s = \int_0^\infty x^s w(x)\,dx$. *Then* $M_0 = 1/2$, $\beta M_1 = -u''(0) > 0$, $M_2 = 0$ *and* $\beta M_3 = -2u(0) < 0$.

PROOF: The fact that $M_0 = 1/2$ is included in Theorem 2.1. The statements on M_1, M_2 and M_3 follow by multiplying Equation (2.5) by $1, x$ and x^2 (respectively) and integrating.

The maxima of $|u|$ in the arches have the following remarkable and tricky property.

THEOREM 8.4. $u(0) > |u(b_1)| > \cdots > |u(b_n)| > \ldots$. *In particular,*

$$u(0) = \max_{x \geq 0} |u(x)|.$$

PROOF: Set $h(x) = u'(x)/x$, thus $u''' = xh'' + 2h'$. Multiplying Equation (2.5) by h and integrating we obtain, for $x > 0$

$$\int_x^A z(h'(z))^2 dz = \frac{\beta m}{m+1}|u(x)|^{1+1/m} - \frac{1}{2}h(x)^2 - xh(x)h'(x).$$

The theorem follows from the observation that the right-hand side is $(\beta m/(m+1))|u(b_n)|^{1+1/m}$, for $x = b_n$, and is $(\beta m/(m+1))|u(0)|^{1+1/m} - (1/2)u''(0)^2$ as $x \to 0$.

9. Regularity.

In this section u and w again denote the functions defined by Theorem 2.1. Most of the section is devoted to proving that w' is integrable near A (the extreme of the support). Other results of the section follow easily from Theorem 8.1. In fact, since $w = |u|^{1/m} \operatorname{sgn} u$,

$$w' = (1/m)|u|^{-(m-1)/m} u' \quad \text{where } u \neq 0. \tag{9.1}$$

From $u'(a_n) \neq 0$ it follows that near $x = a_n$, $|w(x)|$ behaves as $|x - a_n|^{1/m}$ and $|w'(x)|$ behaves as $|x - a_n|^{-(m-1)/m}$. This implies the following:

THEOREM 9.1. *$w \in C^{0,1/m}(\bar{\mathbb{R}}_+)$. Moreover, near each isolated zero of w we have $w' \notin L^{m/(m-1)}$ and $w \notin C^{0,\varepsilon+1/m}$ for any $\varepsilon > 0$.*

REMARK 9.1: Let $W = W(y,t)$ be the function in Theorem 1.1. Theorem 9.1 implies that $W \in C^{0,1/m}(\mathbb{R} \times (\varepsilon, T))$ if $0 < \varepsilon < T < \infty$ and $1/m$ is the optimal Hölder exponent. Also, $U \in C^{3,1/m}(\mathbb{R} \times (\varepsilon, T))$, where $U = |W|^m \operatorname{sgn} W$.

REMARK 9.2: For the porous media equation the function $H = (m/(m-1))|W|^{m-1} \operatorname{sgn} W$ is the pressure, and H_y and H_{yy} are bounded in the support of W. But for our W the derivatives H_y and H_t are unbounded near each zero of W, while H_{yy} is not integrable.

THEOREM 9.2. *$w' \in L^p(\mathbb{R}_+)$ if and only if $p < m/(m-1)$. In particular, $w' \in L^1(\mathbb{R}_+)$ for any $m > 1$.*

REMARK 9.3: Returning to Theorem 1.1, we have

$$t \int_{\mathbb{R}} |W_t(y,t)|\, dy = \text{ constant for all } t > 0.$$

We recall (Theorem 8.2) that $|u(b_n)|$ is the maximum of $|u|$ in the interval (a_n, a_{n+1}).

THEOREM 9.3. *There exist constants C, λ with $\lambda < 1$ such that $|u(b_n)| \leq C\lambda^n$ for all $n \geq 1$.*

PROOF OF THEOREM 9.2: The "only if" part is included in Theorem 9.1. From the argument preceding Theorem 9.1 it is clear that w and $|u|^\alpha$ (for any $\alpha > 0$) are absolutely continuous near each isolated zero of w. We proceed to the behavior near $x = A$. Assuming Theorem 9.3 for the moment, we see that $|u|^\alpha$ (for any $\alpha > 0$) is absolutely continuous on $\bar{\mathbb{R}}_+$, since

$$\sum_{n \geq 1} |u(b_n)|^\alpha < \infty. \tag{9.2}$$

Let $1 \leq p < m/(m-1)$. From (9.1) and $|u'|^p \leq C|u'|$ we obtain

$$|w'|^p \leq C\big|(|u|^\alpha)'\big| \quad \text{with } \alpha = 1 - p(m-1)/m > 0.$$

This completes the proof of Theorem 9.2, provided that Theorem 9.3 holds.

PROOF OF THEOREM 9.3: Let $h_n = a_{n+1} - a_n$. In this proof C stands for a positive constant independent of n. The following lemmas will be needed.

LEMMA 9.1. *$|u(b_n)|^{(m-1)/m} \leq Ah_n^3$ for all $n \geq 1$.*

PROOF: From Theorem 8.2 u' and u'' each have a zero in (a_n, a_{n+1}). Thus

$$|u(b_n)| \leq |u'|_\infty h_n \leq |u''|_\infty h_n^2 \leq |u'''|_\infty h_n^3, \tag{9.3}$$

where $|\cdot|_\infty$ stands for the supremum in (a_n, a_{n+1}). Equation (2.5) and (9.3) imply Lemma 9.1.

LEMMA 9.2. *Let*

$$F_n(x) = (x - a_n)(a_{n+1} - x)/h_n,$$
$$Q_n(x) = |u'(a_{n+1})|F_n(x),$$
$$P_n(x) = |u'(a_n)|F_n(x).$$

Then, for all $n \geq 1$,

$$Q_n(x) \leq |u(x)| \leq P_n(x) \quad \text{if } a_n \leq x \leq a_{n+1}.$$

PROOF: Let $u > 0$ in (a_n, a_{n+1}), thus $u''' > 0$. (The case $u < 0$ is similar.) Let $f(x) = u(x) - P_n(x)$. Then $f \leq 0$ follows from $f''' \geq 0$ and $f(a_n) = f(a_{n+1}) = f'(a_n) = 0$. (Notice that $u'(a_n) = P_n'(a_n)$). In fact, f' is convex and has a zero, c, in the interior of the interval. Since $f'(a_n) = 0$, we have

$$f'(x) \leq 0 \text{ if } a_n \leq x \leq c \text{ and } f'(x) \geq 0 \text{ if } c \leq x \leq a_{n+1}.$$

This and $f(a_n) = f(a_{n+1}) = 0$ imply that $f \leq 0$ on (a_n, a_{n+1}).

Similarly, if we set $g(x) = u(x) - Q_n(x)$, then $g \geq 0$ follows from $g''' \geq 0$ and the fact that $g(a_n) = g(a_{n+1}) = g'(a_{n+1}) = 0$.

LEMMA 9.3. *$|u(b_n)| \leq |u'(a_n)|h_n/4$ for all $n \geq 1$.*

PROOF: Set $x = a_n + h_n/2$ in $P_n(x)$.

LEMMA 9.4. *$|u'(a_n)| \leq C\mu^n$ with $\mu < 1$, for all $n \geq 1$.*

PROOF: From (8.2) it follows that

$$\frac{1}{2}u'(a_n)^2 - \frac{1}{2}u'(a_{n+1})^2 = \beta \int_{a_n}^{a_{n+1}} z|u(z)|^{1+1/m} dz. \tag{9.4}$$

Observe that $z \geq a_1$,

$$|u(z)|^{1+1/m} \geq |u(b_n)|^{-(m-1)/m}|u(z)|^2$$

(because $m > 1$) and $|u(z)| \geq Q_n(z)$ (from Lemma 9.2). From these observations and the computation of the integral of Q_n^2 on (a_n, a_{n+1}), it follows that

(9.5)

$$\int_{a_n}^{a_{n+1}} z|u(z)|^{1+1/m} dz \geq C|u'(a_{n+1})|^2 |u(b_n)|^{-(m-1)/m} h_n^3 \geq C|u'(a_{n+1})|^2,$$

where the last inequality follows from Lemma 9.1. Now (9.4)–(9.5) imply Lemma 9.4.

Finally, Theorem 9.3 follows from Lemmas 9.3–9.4 and $h_n < A$. □

10. Problems with δ', δ'' and δ''' as initial conditions.

Considering similarity solutions of the form

$$W(y,t) = t^{-k\beta_k} w(y/t^{\beta_k}), \quad \beta_k = (4 + k(m-1))^{-1}, \tag{10.1}$$

we are able to prove the following theorem (see [**3**]). (Notice that for $k = 1$ we have (1.2).)

THEOREM 10.1. *Let $k = 2, 3$ or 4. Then there exists a solution W of Equation (1.1) in $\mathbb{R} \times \mathbb{R}_+$ such that:*

(I) *$W = W(y,t)$ has the form (10.1) with $w \in C(\mathbb{R})$ and $w \not\equiv 0$,*
(II) *$w = w(x)$ is odd for $k = 2$ or 4 and even for $k = 3$,*
(III) *w has compact support, and*
(IV)

$$W(\cdot, t) \to C_k \delta^{(k-1)} \text{ in } \mathcal{D}'(\mathbb{R}) \text{ as } t \to 0^+. \tag{10.2}$$

The methods of Sections 4 and 7 apply to all the three cases. For $k = 2$ (the δ' case) the existence method of Section 3 and the uniqueness theorem of Section 5 apply (thus $C_2 \neq 0$), while for $k = 3$ or 4 (the δ'' and δ''' cases) they do not apply (and it is quite delicate to know whether $C_k \neq 0$). The reason is the following: consider (10.1) with w continuous and support w

bounded. Then (5.1) is satisfied if and only if $k < \frac{5}{2}$. Also, an estimate similar to (3.6) breaks down for $k \geq \frac{5}{2}$. In Section 11 we will sketch an existence method which applies to all the three cases (as well as to the δ case) and may be useful in some other situations. For $k = 3$ and $k = 4$ we know uniqueness results within a class of similarity solutions (see [3]). Relation (10.2) follows from (10.1) and the fact that the appropriate moments of W (on $\mathbb{R}$) are zero for all $t > 0$.

Setting $u = |w|^m \operatorname{sgn} w$ as before, the equation for u reads:

$$u^{IV} = \beta_k x(|u|^{1/m} \operatorname{sgn} u)' + k\beta_k |u|^{1/m} \operatorname{sgn} u. \tag{10.3}$$

If we multiply (10.3) by x^{k-1} (with $k = 2$, 3 or 4) and integrate we obtain third order equations which play the role of (2.5). Assuming that $u''(0) = 0$ for $k = 2$, $u'(0) = 0$ for $k = 3$ and $u(0) = 0$ for $k = 4$, we may rewrite these third order equations in the following way:

$$(u''/x)' = \beta_2 |u|^{1/m} \operatorname{sgn} u, \quad (\delta' \text{ case}) \tag{10.4}$$

$$(u'/x)'' = \beta_3 |u|^{1/m} \operatorname{sgn} u, \quad (\delta'' \text{ case}) \tag{10.5}$$

$$(u/x)''' = \beta_4 |u|^{1/m} \operatorname{sgn} u, \quad (\delta''' \text{ case.}) \tag{10.6}$$

Theorems 8.1, 9.1 and 9.2 hold for all the three cases.

REMARK 10.1: Let $k = 3$ or $k = 4$ in Theorem 10.1. Then the first moments of W are zero and W has a second primitive (with respect to y), V, *of bounded support.* Furthermore, V satisfies Equation (1.7) and $V(\cdot, t) \to C_k \delta^{(k-3)}$ as $t \to 0^+$. The inverse is also true, in the following sense. Let $j = 0$ or $j = 1$. If V satisfies Equation (1.7) and $V(\cdot, t) \to B_j \delta^{(j)}$ as $t \to 0^+$, then V_{yy} satisfies Equation (1.1) and $V_{yy}(\cdot, t) \to B_j \delta^{(j+2)}$ as $t \to 0^+$.

REMARK 10.2: The solution of the porous media equation (1.5) with $-\delta'$ as initial condition reads (see [2] and [8])

$$F(y,t) = t^{-1/m} f(y/t^{1/2m}),$$
$$f(x) = |x|^{1/m} \operatorname{sgn} x (C - B|x|^{1+1/m})_+^{1/(m-1)},$$

where $B = (m-1)/2(m+1)$ and C is determined by

$$\int_{\mathbb{R}} x f(x)\, dx = 1.$$

No explicit formula is known for the functions w in Theorem 10.1.

REMARK 10.3: What happens for noninteger k? Consider (10.1) with w continuous and having exponential decay at $\pm\infty$. Then Equation (1.1) with $m = 1$ (linear equation) has a solution $W \not\equiv 0$ (on $\mathbb{R} \times \mathbb{R}_+$) of the form (10.1) if and only if k is integer ≥ 1. Returning to the nonlinear equation ($m > 1$), we know that such a solution does not exist if $k \leq \frac{5}{2}$ with $k \neq 1$ and $k \neq 2$.

11. Existence proof via ODE initial value problems.

We take the δ'' case. To avoid the singularity at $x = 0$, we replace Equation (10.5) by the system

$$u' = \beta_3 x v, \quad v'' = |u|^{1/m} \operatorname{sgn} u. \tag{11.1}$$

In this way $u'(0) = 0$ automatically. Since we look for an even solution, we impose $v'(0) = 0$ (which is equivalent to $u'''(0) = 0$). The existence proof proceeds as follows (see [**3**] for details).

Let (u_λ, v_λ) be a solution of (11.1) with initial conditions

$$u_\lambda(0) = 1, \quad v_\lambda(0) = -\lambda, \quad v_\lambda'(0) = 0. \tag{11.2}$$

Since $m > 1$, (u_λ, v_λ) is defined for all $x \geq 0$. (The lack of uniqueness where $u = 0$ can be circumvented.) Consider the set of λ, S_1, such that for some finite x

$$u_\lambda, v_\lambda, v_\lambda' > 0, \tag{11.3}$$

and the set of λ, S_2, such that for some finite x

$$u_\lambda, v_\lambda, v_\lambda' < 0. \tag{11.4}$$

Both S_1 and S_2 are open (continuity with respect to initial data, once the uniqueness problem has been circumvented) and non-empty (S_1 contains λ negative, S_2 contains λ large). Also, they are mutually exclusive. Hence there is some λ for which neither (11.3) nor (11.4) ever happens. Let (u, v) be the corresponding solution. Then u has infinitely many zeros. (If not, either (11.3) or (11.4) would happen for x large enough.)

To obtain that the support is bounded, we first prove that the integrals involved in Section 4 (with $a = \infty$) are meaningful and finite. (In this step it is useful to know that u has infinitely many zeros.) Then the method of Section 4 applies. (The coefficients of the left-hand side of (4.1) are still positive.)

REMARK 11.1: The above existence proof also applies if $0 < m < 1$. In this case we obtain a solution (u, v), defined on a right maximal interval $(0, b)$, such that neither (11.3) nor (11.4) ever happens on $(0, b)$. Let us prove that $b = +\infty$, i.e. that the solution is defined for all $x \geq 0$. Multiplying $v'' = |u|^{1/m} \operatorname{sgn} u$ by $xv = u'/\beta_3$ (see (11.1)) and integrating we obtain for all $x \in (0, b)$

$$(11.5)\quad \frac{1}{2}v(x)^2 + \frac{m}{(m+1)\beta_3}|u(x)|^{1+1/m} + \int_0^x zv'(z)^2\,dz = C + xv(x)v'(x) = C + u'(x)v'(x)/\beta_3,$$

wherer C is a constant. Assume (for contradiction) that b is finite. From (11.1) it follows that all the three functions u, v' and v must be unbounded near b. If u' has infinitely many zeros near b, then (11.5) implies that u is bounded near b (let $x \to b^-$ along the zeros of u'). Thus either $u' > 0$ or $u' < 0$ near b. Assume $u' > 0$ (the case $u' < 0$ is similar). Then $\lim u(x)$ as $x \to b^-$ exists, and thus is $+\infty$. Now from (11.1) it follows that the limits of $v'(x)$ and $v(x)$ as $x \to b^-$ also exist and are $+\infty$. Thus

$$u(x), v(x), v'(x) > 0 \quad \text{for } x \text{ near } b.$$

This contradiction concludes the proof.

Of course, the support of u is unbounded if $0 < m < 1$. See [**3**] for the behavior of u at infinity.

Dept. of Mathematics, Polytechnic University of Barcelona
Apdo 30002, 08034 Barcelona, SPAIN.

Research at MSRI sponsored in part by NSF Grant DMS-812079-05.

REFERENCES

1. G.I. Barenblatt, *On some unsteady motions of a liquid and a gas in a porous medium*, Prikl. Mat. Mekh. **16** (1952), 67–78.
2. G.I. Barenblatt and Ya. B. Zel'dovich, *On the dipole-type solution in problems of unsteady gas filtration in the polytropic regime*, Prikl. Mat. Mekh. **21** (1957), 718–720.
3. F. Bernis and J.B. McLeod, *Solutions of the equation $u_t + (|u|^m \operatorname{sgn} u)_{xxxx} = 0$, $0 < m < \infty$, with $\delta, \delta', \delta''$ or δ''' as initial conditions*, to appear.
4. F. Bernis, *Compactness of the support for some nonlinear elliptic problems of arbitrary order in dimension n*, Comm. Partial Differential Equations **9** (1984), 271–312.
5. F. Bernis, *Existence results for doubly nonlinear higher order parabolic equations on unbounded domains*, to appear in Math. Ann.
6. F. Bernis, *Qualitative properties for some nonlinear higher order degenerate parabolic equations*, IMA Preprint 184, University of Minnesota (1985).
7. H. Brézis, *Monotonicity methods in Hilbert spaces and some applications to nonlinear partial differential equations*, in "Contributions to Nonlinear Functional Analysis, E.H. Zarantonello, ed.," Academic Press, New York, 1971, pp. 101–156.
8. B.H. Gilding and L.A. Peletier, *On a class of similarity solutions of the porous media equation*, J. Math. Anal. Appl. **55** (1976), 351–364.
9. O. Grange and F. Mignot, *Sur la résolution d'une équation et d'une inéquation paraboliques non linéaires*, J. Func. Anal. **11** (1972), 77–92.
10. J.L. Lions, "Quelques méthodes de résolution des problèmes aux limites non linéaires," Dunod, Paris, 1969.
11. R.E. Pattle, *Diffusion from an instantaneous point source with concentration-dependent coefficient*, Quart. J. Mech. Appl. Math. **12** (1959), 407–409.

Nonuniqueness and Irregularity Results for a Nonlinear Degenerate Parabolic Equation

M. BERTSCH, R. DAL PASSO, AND M. UGHI

Consider the problem

$$\text{(I)} \qquad \begin{cases} u_t = u\Delta u - \gamma|\nabla u|^2 & \text{in } \mathbb{R}^N \times \mathbb{R}^+ \\ u(x,0) = u_0(x) & x \in \mathbb{R}^N, \end{cases}$$

where γ is a real constant and the initial function u_0 satisfies:

$$\text{(H}_1\text{)} \qquad u_0 \in C(\mathbb{R}^N) \cap L^\infty(\mathbb{R}^N) \text{ and } u_0 \geq 0 \text{ in } \mathbb{R}^N.$$

Equation (1) is of degenerate parabolic type: at points where $u = 0$ it loses its parabolicity. As a first consequence, Problem I does not always have classical solutions and we have to define solutions in a weaker sense.

DEFINITION 1. $u \in L^\infty(\mathbb{R}^N \times \mathbb{R}^+) \cap L^2_{\text{loc}}([0,\infty) : H^2_{\text{loc}}(\mathbb{R}^N))$ *is called a solution of* Problem 1 *if for* a.e. $t \geq 0$

$$\int_{\mathbb{R}^N} u(t)\psi(t)\,dx = \int_{\mathbb{R}^N} u_0\psi(0)\,dx + \\ + \int_0^t \int_{\mathbb{R}^N} \{u\psi_t - u\nabla u \cdot \nabla\psi - (\gamma+1)|\nabla u|^2\psi\}\,dx\,dt$$

for every test function $\psi \in C^{2,1}(\mathbb{R}^N \times [0,\infty))$ *with compact support in* $\mathbb{R}^N \times [0,\infty)$.

Below we shall discuss more dramatic consequences of the degeneracy of equation (1) in the case that $\gamma \geq 0$. To put them in the right perspective we sketch first the situation if $\gamma < 0$.

If $\gamma < 0$, equation (1) is sometimes called the *pressure equation*, related to another degenerate parabolic equation, the so-called *porous medium equation*

$$\text{(2)} \qquad v_t = \text{div}(v^{m-1}\nabla v)$$

where $m > 1$. In particular, if we substitute into (1)

$$u(x,t) = v^{m-1}(x,t) \text{ and } \gamma = -(m-1)^{-1}, \tag{3}$$

we arrive, at least formally, at equation (2).

The porous medium equation (and, more generally, a whole class of degenerate parabolic equations of similar type) is studied extensively during the last decades. The most striking consequence of the degeneracy is the fact that if the (nonnegative) initial function has compact support, the corresponding solution has compact support (in space) for all later times $t > 0$. However, apart from this "hyperbolic" phenomenon, which we could describe as finite speed of propagation, the porous medium equation has properties which remind us of parabolic equations. In particular, if we prescribe the initial function, there exists a *unique* (weak) solution. In addition, the following *local regularity* result holds:

(4) *Solutions $v_n (n = 1, 2, \ldots)$ of equation (2) which are locally uniformly bounded (i.e. in $L^\infty_{\text{loc}}(\mathbb{R}^N \times \mathbb{R}^+)$), are locally equicontinuous.*

For more general results and their proofs, we refer to Di Benedetto [**DB**].

Returning to Problem 1 we can say the following if $\gamma < 0$. For every initial function u_0, a solution $u(x,t)$ can be constructed by using the transformation (3) and solving the porous medium equation (2) with corresponding initial function. In addition, sequences of solutions which are obtained through this transformation satisfy a local regularity property similar to (4).

Observe that we did not study Problem 1 directly for $\gamma < 0$. For example, a uniqueness result is missing (see Remark 5 below). Instead we study here Problem 1 for nonnegative values of γ.

REMARK 1: Also if $\gamma \geq 0$ transformation (3) yields a relationship between equations (1) and (2). In that case, however, since $m - 1 < 0$, the transformation becomes singular, which makes it less useful for our purposes.

$\gamma \geq 0$: Existence of viscosity solutions.

Using the well-known viscosity method, we first shall construct a solution. Let $\varepsilon > 0$ and let $u_\varepsilon(x,t)$ be the unique classical solution of the uniformly parabolic problem

$$(I_\varepsilon)\begin{cases} u_t = (u+\varepsilon)\Delta u - \gamma|\nabla u|^2 & \text{in } \mathbb{R}^N \times \mathbb{R}^+ \\ u(x,0) = u_0(x) & x \in \mathbb{R}^N. \end{cases}$$

PROPOSITION 1. *For all* $(x,t) \in \mathbb{R}^N \times [0,\infty)$ *the limit*

$$u(x,t) = \lim_{\varepsilon \searrow 0} u_\varepsilon(x,t)$$

exists and the limit function u *is a solution of Problem 1. We call* u *the* viscosity solution *of Problem 1.*

REMARK 2: Our "definition" of the viscosity solution is, although quite natural, rather ad hoc. It is based on the very special property of equation (1) that it is *a priori* clear that all of the sequence $\{u_\varepsilon\}$ converges to the same limit u; this follows from the fact that for all $\varepsilon > 0$, $w_\varepsilon \equiv u_\varepsilon + \varepsilon$ is a positive and classical solution of equation (1) with initial function $u_0 + \varepsilon$, and hence, by comparison, w_ε is pointwise decreasing as $\varepsilon \searrow 0$. But we like to point out that one cannot avoid this ad hoc definition by using the definition of viscosity solutions given by P.L. Lions [**L**], which is a direct generalization of the definition given in the case of first-order Hamilton-Jacobi equations by Crandall and Lions [**C-L**]; see also [**C-E-L**]. The reason is that in general we would not have uniqueness in such a class of viscosity solutions (see Remark 6 below). In addition, we shall construct discontinuous viscosity solutions (see Theorem 3 below), while viscosity solutions in the sense of Lions are continuous by definition.

Positivity properties of viscosity solutions.

Certain positivity properties of the viscosity solutions turn out to be crucial for the study of regularity and nonuniqueness properties for equation (1).

PROPOSITION 2 ([**B-U**]). *Let* $\gamma \geq 0$ *and* u_0 *satisfy hypothesis* H_1. *Let* u *denote the viscosity solution of Problem I. Then*

(5) $$u(x_0,t_0) > 0 \Longrightarrow u(x_0,t) > 0 \text{ for all } t \geq t_0$$

where $x_0 \in \mathbb{R}^N$ *and* $t_0 \geq 0$, *and*

(6) $$\operatorname{supp} u(t) = \operatorname{supp} u_0 \text{ for all } t \geq 0.$$

This result tells us that the set where $u > 0$ is almost completely determined by the set where $u_0 > 0$. However, if for example u_0 has an isolated zero, say at $x = 0$, we cannot decide from Proposition 2 if $u(0,t) = 0$ or $u(0,t) > 0$ for $t > 0$. To describe this situation more precisely, we define

$$(7) \qquad t^* = \sup\Big\{t \geq 0 \mid \operatorname{essinf}\{u(y,t) : |y| \leq r\} = 0 \text{ for all } r > 0\Big\}.$$

Observe that if u is continuous

$$t^* = \sup\{t \geq 0 \mid u(0,t) = 0\},$$

and for this reason we call t^* the *waiting-time* of u at $x = 0$. It gives information about the positivity of u at $x = 0$: using the monotonicity of $u_\varepsilon + \varepsilon$ (see Remark 2) it can be easily shown that

$$t > t^* \Longrightarrow u(0,t) > 0,$$

and, if u is continuous, it follows from (5) that

$$0 \leq t \leq t^* \Longrightarrow u(0,t) = 0.$$

More in general, it can be shown that

$$0 \leq t < t^* \Longrightarrow \operatorname{essinf}\{u(y,t) : |y| \leq r\} = 0 \text{ for all } r > 0.$$

THEOREM 1. *Let u_0 satisfy H_1 and let $x = 0$ be an isolated zero of u_0, i.e.*

$$(\mathrm{H}_2) \qquad u_0(0) = 0 \text{ and } u_0 > 0 \text{ in } B_{2\varsigma}(0) \setminus \{0\} \text{ for some } \varsigma > 0,$$

where $B_\varsigma(y) = \{x \in \mathbb{R}^N : |x - y| < \varsigma\}$. Let t^* be defined by (7).

(i) *If $\gamma > \frac{1}{2}N$, then*

$$\int_{B_\varsigma(0)} u_0^{-\gamma}(x)dx < \infty \Longrightarrow t^* = 0$$

and

$$(8) \qquad \int_{B_\varsigma(0)} u_0^{-\gamma}(x)dx = \infty \Longrightarrow t^* = \infty$$

(ii) *If $N > 2$ and $0 < \gamma < 1$, then*

$$\limsup_{y\to 0} \int_{B_\varsigma(0)} G_\varsigma(|x-y|)u_0^{-\gamma}(x)dx = \infty \iff t^* = \infty,$$

where the Green's function G_ς is defined by

$$G_\varsigma(r) = \begin{cases} r^{-N+2} - \varsigma^{-N+2} & \text{if } N \geq 3 \\ \log(\varsigma/r) & \text{if } N = 2. \end{cases}$$

(iii) *If $N = 1$, then for all $\gamma > 0$*

$$\int_{B_\varsigma(0)} u_0^{-\gamma}(x)dx \iff t^* = \infty.$$

REMARK 3: Similar results hold for $\gamma = 0$, with $u_0^{-\gamma}$ replaced by $-\log u_0$.

The proof of Theorem 1 is based on the fact that formally u satisfies the equation

$$(u^{-\gamma})_t = -\frac{\gamma}{1-\gamma}\Delta(u^{1-\gamma}) \tag{9}$$

provided $\gamma \neq 0, 1$ (if $\gamma = 0$ or 1 similar equations are satisfied). From (9) we derive at once the formal relations

$$\int_{B_\varsigma(0)} u^{-\gamma}(x,t)dx = \int_{B_\varsigma(0)} u_0^{-\gamma}(x)dx - \frac{\gamma}{1-\gamma}\int_0^t \int_{\partial B_\varsigma(0)} \frac{\partial}{\partial \nu}(u^{1-\gamma}) \tag{10}$$

and, for $y \in B_\varsigma(0)$ and $N \geq 2$,

$$\int_0^t u^{1-\gamma}(y,\tau)d\tau = a_N\varsigma^{-N+1}\int_0^t \int_{\partial B_\varsigma(y)} u^{1-\gamma}dx\,d\tau + \tag{11}$$

$$+ b_N \frac{1-\gamma}{\gamma}\int_{B_\varsigma(y)} G_\varsigma(|x-y|)\{u^{-\gamma}(x,t) - u_0^{-\gamma}(x)\}dx,$$

for certain constants a_N and b_N. If for example $u_0^{-\gamma} \notin L^1(B_\varsigma)$, then (10) implies that $u^{-\gamma}(t) \notin L^1(B_\varsigma)$ for all $t > 0$ (here we have used that the boundary term in (10) is bounded, since $u > 0$ in a neighborhood of $\partial B_\varsigma \times [0,t]$ and hence u is smooth there). Since, by (H_2) and (5), $u > 0$ in $B_{2\varsigma} \setminus \{0\}$, this implies at once that $t^* = \infty$. Similarly, if $0 < \gamma < 1$ and $N \geq 2$, part (ii) ($\Longrightarrow$) follows from (11).

For the remaining proofs (and also a rigorous treatment of the arguments given here) we refer to [**B-U**].

Combining Theorem 1 with some straightforward arguments based on the comparison principle (from the construction of viscosity solutions it follows at once that they satisfy a comparison principle: if two viscosity solutions are initially ordered, they stay ordered for all later times $t > 0$), we can give a rather complete picture for the value of t^*. To give an idea, we give here the result in the special case that $u_0(x) = |x|^\mu$ near $x = 0$.

THEOREM 1′ ([**B-U**]). *Let $u_0(x) = |x|^\mu$ in a neighborhood of $x = 0$ for some $\mu > 0$. Then the value of t^* can be obtained from the following diagrams:*

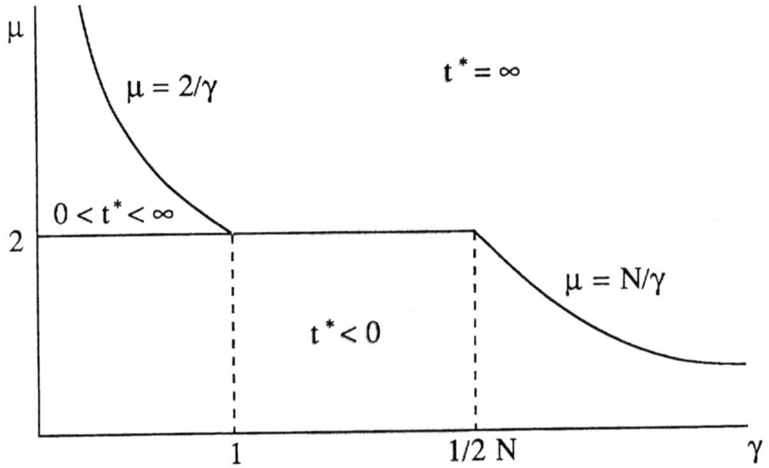

Figure 1: $N \geq 2$

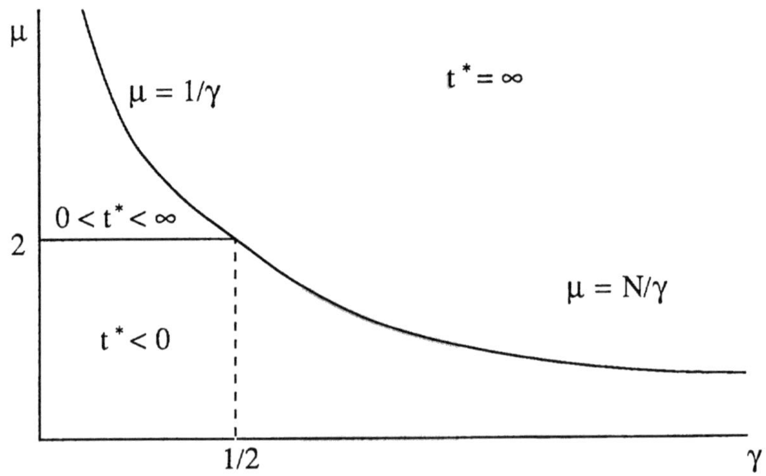

Figure 2: $N = 1$

Regularity properties of viscosity solutions.

The following theorem shows that a local regularity result similar to (4), which is typical for parabolic equations, is violated for some values of γ if $N \geq 2$.

THEOREM 2 ([**B-DP-U**]).

(i) *If* $\gamma > \frac{1}{2}N$, *then*

(12) *viscosity solutions* $\{u_n\}_{n=1,2,\ldots}$ *of equation (1) which are uniformly bounded in* $L^\infty_{\mathrm{loc}}(\mathbb{R}^N \times \mathbb{R}^+)$ *are locally equicontinuous in* $\mathbb{R}^N \times \mathbb{R}^+$.

(ii) *If* $N = 1$, *(12) is satisfied for all* $\gamma \geq 0$.

(iii) *If* $N \geq 2$ *and* $0 \leq \gamma < \frac{1}{2}N$, *there exist counterexamples to (12).*

SKETCH OF THE PROOF: Deriving equations for u_{nt} and Δu_n, it follows from the maximum principle that

$$u_{nt} \geq -\frac{1}{t}u_n \text{ in } \mathcal{D}'(\mathbb{R}^N \times \mathbb{R}^+) \tag{13}$$

and

$$\text{if } \gamma > \frac{N}{2} \text{ then } \Delta u_n \leq \frac{N}{(2\gamma - N)t} \text{ in } \mathcal{D}'(\mathbb{R}^N \times \mathbb{R}^+).$$

Combining this with equation (1) it follows that if $\gamma > \frac{1}{2}N$, then u_{nt}, $|\nabla u_n|$ and Δu_n are locally uniformly bounded in $\mathbb{R}^N \times \mathbb{R}^+$ and the first part of Theorem 2 follows.

If $N = 1$, (13) implies that u_{nxx} is locally bounded from below and hence u_n is locally uniformly Lipschitz continuous with respect to x. A result by Gilding [G] yields then local Hölder continuity with respect to t and the second part of Theorem 2 follows.

To prove part (iii) we consider the initial-boundary value problem

$$\text{(II)} \qquad \begin{cases} u_t = u\Delta u - \gamma|\nabla u|^2 & \text{in } B_1 \times \mathbb{R}^+ \\ u = 1 & \text{on } \partial B_1 \times \mathbb{R}^+ \\ u(x,0) = u_0(x) & x \in B_1 \end{cases}$$

where $B_1 \subset \mathbb{R}^N$ is the unit ball. Again Problem II possesses a viscosity solution $u(x,t)$.

For each $n = 1, 2, \ldots$ we define

$$u_n(x,t) = u(x, t+n) \quad x \in \overline{B_1}, 0 \le t \le 1.$$

To prove the result it is sufficient to choose an initial function u_0 such that the functions u_n are not equicontinuous on $B_{\frac{1}{2}} \times [\frac{1}{2}, 1]$. Therefore it is enough to choose u_0 in such a way that the viscosity solution $u(x,t)$ satisfies

$$\text{(14)} \qquad \operatorname{essinf}\{u(y,t) : |y| \le r\} = 0 \text{ for all } r > 0 \text{ and } t \ge 0$$

and, on the other hand,

$$\text{(15)} \quad u(x,t) \to 1 \text{ as } t \to \infty \text{ uniformly on compact subsets of } \overline{B_1} \setminus \{0\}.$$

It turns out that if $\gamma \neq 0$, this is the case if we choose

$$u_0(x) = |x|^\alpha, \quad x \in \overline{B_1},$$

with

$$\text{(16)} \qquad \alpha > \max\{\frac{2}{\gamma}, 2\}$$

and, if $1 < \gamma < \frac{1}{2}N$ and $N \ge 3$,

$$\text{(17)} \qquad \alpha < \frac{N-2}{\gamma - 1}.$$

Observe that, by Theorem 1′, the lowerbound (16) of α implies (14). From the upper bound (17) we derive that $\Delta(u_0^{1-\gamma}) \ge 0 (\not\equiv 0)$ in B_1 and hence $u(x,t)$ is nondecreasing with respect to t. Now (15) follows essentially from the fact that $\bar{u} \equiv 1$ is the unique steady-state solutions of Problem II which is larger than u_0. A detailed proof will be given in [**B-DP-U**].

REMARK 4: Also if $\gamma < 0$ the local regularity property (12) is satisfied. This follows from the discussion in the beginning of this paper and the fact that, for $\gamma < 0$, the solutions of Problem I obtained through the transformation (3) are precisely the viscosity solutions of Problem I. Hence, the local regularity result (12) holds both for $\gamma < 0$ and $\gamma > \frac{1}{2}N$. However, if we look at initial-*boundary* value problems we find a difference between $\gamma < 0$ and $\gamma > \frac{1}{2}N$. For example, if we consider the problem

$$\begin{cases} u_t = u\Delta u - \gamma|\nabla u|^2 & \text{in } \Omega \times \mathbb{R}^+ \\ u = 1 & \text{on } \partial\Omega \times \mathbb{R}^+ \\ u(x,0) = 0 & x \in \Omega, \end{cases}$$

where $\Omega \subset \mathbb{R}^N$ is a smooth bounded domain, then, if $\gamma < 0$, there exists a viscosity solutions which is continuous at $\{\bar{\Omega} \times [0,\infty)\} \setminus \{\partial\Omega \times \{0\}\}$. This can be proved directly, but can also be considered as a consequence of the more general fact that, if $\gamma < 0$, (12) holds up to the lateral boundary $\partial\Omega \times \mathbb{R}^+$ [**DB**]. If $\gamma \geq 0$, the situation is completely different: the viscosity method yields a limit function

$$u(x,t) = \begin{cases} 0 & \text{if } x \in \Omega, t \geq 0 \\ 1 & \text{if } x \in \partial\Omega, t \geq 0 \end{cases}$$

(this follows from the fact that property (6) continues to hold) and we cannot expect a global regularity result up to $\partial\Omega \times \mathbb{R}^+$, even if $N = 1$.

Theorem 2(i)–(ii) implies that, if $\gamma > \frac{1}{2}N$ or $N = 1$, the viscosity solution is continuous. However, if $N \geq 2$ and $0 \leq \gamma < \frac{1}{2}N$, it does not follow from Theorem 2(iii) that there actually exist discontinuous viscosity solutions of Problem 1. For $1 \leq \gamma < \frac{1}{2}N$ this is an open problem, but if $0 \leq \gamma < 1$ it turns out that discontinuous solutions do exist.

THEOREM 3. *If $N \geq 2$ and $0 \leq \gamma < 1$, there exists a class of initial function u_0 which satisfy H_{1-2} such that the corresponding viscosity solutions $u(x,t)$ of Problem I are not continuous.*

Again the proof relies basically on the positivity properties of solutions: we construct an initial function u_0 such that the viscosity solution u has a waiting-time $t^* > 0$ at $x = 0$ and such that for some $0 < t_0 < t^*$

$$u(0, t_0) > 0,$$

which implies immediately that $u(x,t_0)$ is not continuous with respect to x at $x = 0$. For the proof we refer to [**B-DP-U**]. Here we only remark that the proof, if $\gamma \neq 0$, is based on formula (11), which reduces the proof to a property of Green's functions: we construct a function u_0 such that

$$\limsup_{y \to 0} \int_{B_\varsigma(y)} G_\varsigma(|x-y|)u_0^{-\gamma}(x)dx > \int_{B_\varsigma(0)} G_\varsigma(|x|)u_0^{-\gamma}(x)dx.$$

Nonuniqueness.

Until now we have only considered the viscosity solution $u(x,t)$ of Problem I. It turns out that this is not the only solution. This nonuniqueness phenomenon was discovered independently by Dal Passo and Luckhaus [**DP-L**] and Ughi [**U**], both in the case that $\gamma = 0$.

First we discuss the result by Dal Passo and Luckhaus.

THEOREM 4. *Let $u_0 \not\equiv 0$ satisfy H_1 and let u_0 have compact support. Let $N \geq 1$ and $\gamma \geq 0$. Then there exist infinitely many solutions of Problem I, which do not satisfy the positivity properties (5) and (6).*

The proof by Dal Passo and Luckhaus in the case $\gamma = 0$ carries over immediately to the more general case $\gamma \geq 0$.

REMARK 5: Adapting the proof in [**DP-L**] a little, it is also possible to show that if $-1 < \gamma < 0$ there exists more than one solution.

The nonuniqueness phenomenon in Theorem 4 reminds us of first-order Hamilton-Jacobi equations. In particular it can be shown that the solutions constructed in Theorem 4 are not viscosity solutions in the sense of Lions [**L**], even it they are continuous.

It turns out that for certain values of γ and initial functions u_0 there also exist infinitely many solutions which do satisfy the positivity properties (5) and (6).

THEOREM 5 [**B-DP-U**]). *Let u_0 satisfy H_1 and let*

$$\gamma > \tfrac{1}{2}N \text{ if } N \geq 2$$

respectively

$$\gamma \geq 0 \text{ if } N = 1.$$

Let $u(x,t)$ be the continuous viscosity solution of Problem I

(i) *There exists a continuous solution $u^*(x,t)$ of Problem I which satisfies for all $x \in \mathbb{R}^N$ and $0 \leq t_1 < t_2$*

$$u^*(x,t_1) = 0 \iff u^*(x,t_2) = 0.$$

(ii) *If $u_0(0)$ and $t^* < \infty$, i.e. $u(0,t) > 0$ for some $t > 0$, there exists a one-parameter family (which is continuous in the topology of*

$C_{\text{loc}}(\mathbb{R}^N \times [0,\infty)))$ of continuous solutions $u^\alpha(x,t), 0 \le \alpha \le 1$, of Problem I such that

$$u^0 \equiv u^* \text{ and } u^1 \equiv u \text{ in } \mathbb{R}^N \times \mathbb{R}^+,$$

and if $0 \le \alpha < \beta \le 1$, then

$$u^\alpha \le u^\beta \text{ and } u^\alpha \not\equiv u^\beta \text{ in } \mathbb{R}^N \times \mathbb{R}^+.$$

If $\gamma > \frac{1}{2}N$, then

$$u^\alpha(0,t) > 0 \text{ for all } t > 0 \text{ and } \alpha \in [0,\infty).$$

The last property has a striking consequence. If $u_0 > 0$ in $\mathbb{R}^N \setminus \{0\}$ and $u^\alpha(0,t) > 0$ for $t > 0$, then u^α is bounded away from zero in compact subsets of $\mathbb{R}^N \times \mathbb{R}^+$. By standard parabolic theory, u^α is then a smooth solution of equation (1) and we arrive at the following result.

COROLLARY. *If $\gamma > \frac{1}{2}N$ there exists an initial function which satisfies hypotheses H_{1-2} for which Problem I has infinitely many classical solutions belonging to $C^{2,1}(\mathbb{R}^N \times \mathbb{R}^+) \cap C(\mathbb{R}^N \times [0,\infty))$.*

REMARK 6: All solutions in Theorem 5 are viscosity solutions in the sense of Lions [**L**]. For the classical solutions mentioned in the corollary this is trivial. Hence Theorem 5 implies that in general viscosity solutions in the sense of Lions are not unique.

REMARK 7: The nonuniqueness result by Ughi [**U**] was based on the construction of u^* in the case $\gamma = 0$ and $N = 1$. Her construction cannot be generalized to the higher dimensional case.

The proof of Theorem 5 is based on the local equicontinuity of uniformly bounded viscosity solutions (Theorem 2(i)–(ii)). Here we shall only sketch the construction of $u^*(x,t)$.

Let

$$\Omega = \{x \in \mathbb{R}^N : u_0(x) = 0\}$$

and define a nested sequence of open neighborhoods Ω_n of Ω:

$$\Omega_n = \{x \in \mathbb{R}^N : \text{dist}(x,\Omega) < \frac{1}{n}\}.$$

Then there exists a sequence of continuous initial functions $u_{0n}(x)$ such that

$$u_{0n} \leq u_{0n+1} \text{ in } \mathbb{R}^N,$$

$u_{0n} \to u_0$ uniformly on compact subsets of $\mathbb{R}^N$ as $n \to \infty$ and

$$u_{0n} > 0 \text{ in } \mathbb{R}^N \setminus \Omega_n; \quad u_{0n} = 0 \text{ in } \Omega_n.$$

Let $u_n(x,t)$ be the viscosity solution of Problem I with u_0 replaced by u_{0n}. Then, by (5),

$$u_n > 0 \text{ in } \{\mathbb{R}^N \setminus \Omega_n\} \times \mathbb{R}^+,$$

and, by (6) and the fact that $\Omega \subset \operatorname{supp} u_{0n}$,

$$u_n \equiv 0 \text{ in } \Omega \times \mathbb{R}^+.$$

Since $u_{n+1} \geq u_n$ in $\mathbb{R}^N \times \mathbb{R}^+$ and u_n is uniformly bounded, we may define for all $x \in \mathbb{R}^N, t \geq 0$

$$u^*(x,t) = \lim_{n \to \infty} u_n(x,t).$$

Then clearly

$$u^* > 0 \text{ in } \{\mathbb{R}^N \setminus \Omega\} \times \mathbb{R}^+$$

and

$$u^* \equiv 0 \text{ in } \Omega \times \mathbb{R}^+.$$

By Theorem 2(i)–(ii), $u^* \in C(\mathbb{R}^N \times \mathbb{R}^+)$ and it can be shown that u^* is indeed a solution of Problem I.

REMARK 8: It can be shown that in the class of continuous solutions the viscosity solution is the maximal one, and, in the class of continuous solutions which satisfy the positivity property (5) u^* is the minimal solution.

Department of Mathematics, University of Leiden
Leiden, THE NETHERLANDS

Istituto per le Applicazioni del Calcolo, CNR, Rome, ITALY

Istituto di Scienze delle Costruzioni, Trieste, ITALY

Research at MSRI supported in part by NSF Grant 812079-05.

REFERENCES

[**B-DP-U**] M. Bertsch, R. Dal Passo and M. Ughi, in preparation.

[**B-U**] M. Bertsch and M. Ughi, *Positivity properties of solutions of a degenerate parabolic equation*, to appear.

[**C-E-L**] M.G. Crandall, L.C. Evans and P.L. Lions, *Some properties of viscosity solutions of Hamilton-Jacobi equations*, Trans. Amer. Math. Soc. **282** (1984), 487–502.

[**C-L**] M.G. Crandall and P.L. Lions, *Viscosity solutions of Hamilton-Jacobi equations*, Trans. Amer. Math. Soc. **277** (1983), 1–42.

[**DP-L**] R. Dal Passo and S. Luckhaus, *On a degenerate diffusion problem not in divergence form*, to appear in J. Diff. Equat.

[**DB**] E. Di Benedetto, *Continuity of weak solutions to a general porous media equation*, Indiana Univ. Math. J. **32** (1983), 83–118.

[**G**] B.H. Gilding, *Hölder continuity of solutions of parabolic equations*, J. London Math. Soc. **13** (1976), 103–106.

[**L**] P.L. Lions, *Optimal control of diffusion processes and Hamilton-Jacobi-Bellman equations, part 2: viscosity solutions and uniqueness*, Comm. Part. Diff. Equat. **8** (1983), 1229–1276.

[**U**] M. Ughi, *A degenerate parabolic equation modelling the spread of an epidemic*, Annali Math. Pura Appl. **143** (1986), 385–400.

Existence and Meyers estimates for solutions of a nonlinear parabolic variational inequality

MARCO BIROLI

§1. Introduction and results.

Much attention has recently been given to nonlinear parabolic variational inequalities, particularly in connection with applications to stochastic control [**1**].

Three methods to prove existence theorems have been introduced: (1) the use of suitable results on the modulus of continuity of the solutions; (2) the use of a Meyers inequality; (3) the use of dual inequalities. Existence results have been proved by method (1) in [**4**] and [**14**], and can be proved by method (3) using the dual inequalities given in [**14**]. Method (2) has been used in the elliptic case in [**9**] and in the parabolic case (with linear principal part) in [**3**], [**4**], [**5**], and [**6**].

Existence theorems for the case of *equations* have been obtained in [**12**] and [**13**] by a suitable choice of test functions. Moreover, a proof of the Meyers estimate for solutions of nonlinear parabolic systems with linear principal part has been given in [**11**]; this result has been extended to *equations* by L. Boccardo and F. Murat (personal communication).

In this paper, following earlier work of the author [**5**], [**6**], we will prove a Meyers inequality for strongly nonlinear parabolic variational inequalities with a principal part having polynomial growth of order $p \geq 2$. As a consequence we are able to prove a corresponding existence theorem for these variational inequalities.

Let Ω be a bounded open set with boundary $\partial\Omega$, $Q = \Omega \times (0,T)$, $T > 0$, $\Sigma = \partial\Omega \times (0,T)$; we use the following terminology:

$$z = (x,t) \quad z_0 = (x_0,t_0) \quad x, x_0 \in \mathbb{R}^N \quad t, t_0 \in \mathbb{R}$$

$$B(R,x_0) = \{x; |x - x_0| < R\}$$

$$B(R,x_0;t) = B(R,x_0) \times \{t\}$$

$$Q(R,z_0) = B(R,x_0) \times (t_0 - R^p, t_0)$$

$$\Lambda(t_0,R) = (t_0 - R^p, t_0).$$

Let $A_i(x,t,r,\xi) : Q \times \mathbb{R} \times \mathbb{R}^N \to \mathbb{R}$ be a Carathéodory function such that

$$\sum_{i=1}^{N} A_i(x,t,r,\xi)\xi_i \geq \alpha|\xi|^p, \quad \alpha > 0 \tag{1.1}$$

for a.e. $(x,t) \in Q, \forall (r,\xi) \in \mathbb{R} \times \mathbb{R}^N$,

$$|A_i(x,t,r,\xi)| \leq \alpha^{-1}\left[|r|^{p-1} + |\xi|^{p-1}\right] \tag{1.2}$$

for a.e. $(x,t) \in Q, \forall (r,\xi) \in \mathbb{R} \times \mathbb{R}^N$, and

$$\sum_{i=1}^{N} \left[A_i(x,t,r,\xi) - A_i(x,t,r,\xi')\right]\left[\xi - \xi'\right] > 0 \tag{1.3}$$

for a.e. $(x,t) \in Q, \forall r \in \mathbb{R}, \quad \xi, \xi' \in \mathbb{R}^N, \xi \neq \xi'$.

Finally, let $H(x,t,r,\xi) : Q \times \mathbb{R} \times \mathbb{R}^N \to \mathbb{R}$ be a Carathéodory function such that

$$|H(x,t,r,\xi)| \leq b(|r|)(1 + |\xi|^p) \tag{1.4}$$

for a.e. $(x,t) \in Q, \forall (r,\xi) \in \mathbb{R} \times \mathbb{R}^N$, where $b(\cdot)$ is a continuous increasing function on $\mathbb{R}$.

We consider the following variational inequality

$$\begin{aligned}
&\int_0^t < D_t v(s), \varphi(s)\sigma'(x(s) - u(s)) >_{H^{-1,p'},H_0^{1,p}} ds + \\
&+\sum_{i=1}^{N} \int_0^t \int_\Omega A_i(x,s,u(x,s),Du(x,s)) D_i(\varphi(x,s)\sigma'(v(x,s) - u(x,s)))\, dx\, ds + \\
&\quad + \sum_{i=1}^{N} \int_0^t \int_\Omega H(x,s,u(x,s),Du(x,s))\varphi(x,s)\sigma'(v(x,s) - u(x,s)))\, dx\, ds + \\
&\qquad + \sum_{i=1}^{N} \int_0^t \int_\Omega f_i(x,s) D_i(\varphi(x,s)\sigma'(v(x,s) - u(x,s)))\, dx\, ds + \\
&\qquad\quad + \int_0^t \int_\Omega D_t\varphi(x,s)\sigma'(v(x,s) - u(x,s))\, dx\, ds + \\
&\qquad\qquad + \|\varphi\sigma(v - u_0)\|_{L^1}(0) - \|\varphi\sigma(v - u)\|_{L^1}(t) \geq 0
\end{aligned}$$

$\forall \sigma \in C^2(\mathbb{R})$, σ convex, $\sigma'(0) = 0$, $\varphi \in C^\infty(\bar{Q})$, $\varphi \geq 0$;
$\forall v \in H^{1,p'}(0,T;H^{-1,p'}(\Omega)) \cap L^p(0,T;H^{1,p}(\Omega)) \cap L^\infty(Q)$, $v \in H_0^{1,p}(\Omega)$ a.e.

in t if $\varphi(.,t) \notin C_0^\infty(\Omega)$ for some $t \in (0,T)$, with $v \leq \psi$ a.e. in $\operatorname{supp}\varphi$ and $u \in C(0,T;L^2(\Omega)) \cap L^p(0,T;H_0^{1,p}(\Omega)) \cap L^\infty(Q)$, $u \leq \psi$ a.e. in Q, where $u_0 \in L^\infty(Q)$ and $u_0 \leq \psi(.,0)$ a.e. in Ω (D = gradient in space variables, D_i = derivative with respect to the variable x_i).

In the following we suppose that $\psi \in H^{1,q''}(0,T;H^{-1,q''}(\Omega)) \cap L^{q''}(0,T;H^{1,q''}(\Omega)) \cap L^\infty(Q)$ with $q'' > p$ and that there exists $v_0 \in H^{1,q''}(0,T;H^{-1,q''}(\Omega)) \cap L^{q''}(0,T;H_0^{1,q''}(\Omega)) \cap L^\infty(Q)$ with $v_0 \leq \psi$ a.e. in Q and $v_0(.,0) \leq \psi(.,0)$ a.e. in Ω.

THEOREM 1. *Let $f_0 \in L^{q''/p}(Q)$ and $f_i \in L^{q''}(Q)$, $i = 1,2,\dots,N$, and u be a solution to (1.5); there exists $q \in (p,q'')$ such that $Du \in L^q_{\rm loc}((0,T]\times\Omega)$ and $\|Du\|_{L^q_{\rm loc}((0,T]\times\Omega)}$ depends only on α, f_0, f_i, ψ. Moreover, if we can choose v_0 with $v_0(.,0) = u_0(.)$ we have $Du \in L^q(Q)$.*

REMARK 1: A local version of Theorem 1 can also be proved for a local solution of (1.5) (i.e. $\varphi \in C_0^\infty(Q)$).

A function $\Phi \in L^\infty(0,T;H^{1,\infty}(\Omega))$ with $D_t\Phi \in L^{p'}(0,T;H^{-1,p'}(\Omega))$ is a *subsolution* of our problem if

$$D_t\Phi + \sum_{i=1}^N D_iA_i(.,.,\ ,D\Phi) + H(.,.,\ ,D\Phi) + f_0 + \sum_{i=1}^N f_i \leq 0$$

in the distribution sense, $\Phi \leq 0$ a.e. in Σ, $\Phi(.,0) \leqq u_0(.)$ and $\Phi \leq \psi$ a.e. in Q.

THEOREM 2. *Let $f_0, f_i \in L^q(Q)$, u_0 as in the second part of Theorem 1 and suppose there exists a subsolution of our problem; then there exists a solution to (1.5).*

REMARK 2: If A_i does not depend on r and

$$|H(x,t,r,\xi)| \leq K_1 + K_2(|r|)|\xi|^p, \quad f_0 = f_i = 0$$

where $K_2(.)$ is a continuous increasing function on $\mathbb{R}_+$, a subsolution of our problem can easily be exhibited.

In §2 we give a proof of Theorem 1 which is a refinement of the proof given for $p = 2$ in [**5**] and [**6**]. In §3 and §4 we give a proof of Theorem 2 by regularization on H; the method used in this proof is totally different from the one used in [**5**] for $p = 2$ or for the equations in [**12**] and [**13**].

REMARK 3: The assumptions on f_0 in Theorem 2 can be weakened.

§2. The Meyers inequality.

We denote by B a ball in $\mathbb{R}^N$ and by η a function in $C_0^\infty(B)$. Let $v \in L^\infty(B)$. We denote by c_v or simply by c the "average" of v defined by the relation

$$(2.1) \qquad \int_B (v - c_v) \exp(\mu |v - c_v|^p) \eta^2 dx = 0, \quad \mu > 0.$$

The existence and uniqueness of c_v follows easily from the zeroes theorem for continuous functions.

For the "average" c the following properties hold:

LEMMA 1.

(a) *Let* $v \geq 0$, *then* $c \geq 0$.

(b) *Let* $\eta = 1$ *in* $B' \subset\subset B$, $v \in H^{1,p}(B) \cap L^\infty(B)$; *then*

$$\int_{B'} |v - c|^p dx \leq C \int_B |Du|^p dx.$$

Moreover, if $p < N$ *we also have*

$$\left(\int_{B'} |v - c|^{p^*} dx \right)^{1/p*} \leq C \left(\int_B |v - c|^p dx \right)^{1/p}$$

where $p^* = pN(N-p)^{-1}$; *if* $p > N$ *the relation above holds for every* $p^* < +\infty$.

(c) *Let* $v \in L^\infty(B \times (0,T)) \cap C(0,T; L^2(B))$; *denote by* $c(t)$ *the average of the function* $v(.,t)$, *then* $c(t)$ *is in* $C(0,T)$.

For the proof of Lemma 1 we refer to [**5**].

We now give the proof of the Meyers inequality. In the proof, for the sake of simplicity we suppose $\psi = 0$; we can also assume

$$(2.2) \qquad |H(x,t,r,\xi)| \leq K(x,t) + K^2 |\xi|^p$$

where $K \in L^{1+\varepsilon}(Q)$, $\varepsilon > 0$.

Let $\eta \in C_0^\infty(B(2,0))$ with $\eta = 1$ in $B(1,0)$ and $|D\eta| \leq 4N^{1/2}$; we define

$$\eta_R = \eta(R^{-1}(x - x_0)).$$

Let u be a local solution of our variational inequality and assume $Q(4R, z_0) \subset\subset Q$; we denote by $c_R(t)$ the average (as defined by (2.1)) relative to $u(.,t), \mu$ and with η replaced by η_R.

LEMMA 2. *The following estimate holds:*

$$\int_{Q(R,z_0)} |Du|^p \, dx\, dt \leq C_1 R^{-p} \int_{Q(2R,z_0)} (|u - c_{2R}|^p + g) \, dx\, dt$$

where $g = C_2(K + |f_0|^p + \sum_{i=1}^N |f_i|^p)$ *is in* $L^{1+\varepsilon}(Q)$, *for a suitable* $\varepsilon > 0$ *(Caccioppoli's inequality).*

We denote by $c_{n,2R}(t)$ the function defined by the problem

$$n^{-1}(c_{n,2R}(t))' + c_{n,2R}(t) = c_{2R}(t)$$
$$c_{n,2R}(t_0 - (2R)^p) = c_{2R}(t_0 - (2R)^p).$$

We observe that from $c_{2R} \leq 0$ it follows $c_{n,2R} \leq 0$; moreover by Lemma 1 c_{2R} is in $C(t_0 - (2R)^p, t_0)$, thus

$$\lim_{n\to\infty} c_{n,2R}(t) = c_{2R}(t) \text{ in } C(t_0 - (2R)^p, t_0).$$

We choose in the variational inequality

$$\sigma(y) = \int_0^y r \exp(\mu|r|^p) \, dr$$

and $v = c_{n,2R}, \varphi = (\eta_{2R})^2(\tau_{2R})^2$, where $\tau = \tau(t)$ is such that $\tau \in C_0^\infty(\mathbb{R})$, $\tau = 1$ for $t \geq -1$, $\tau = 0$ for $t \leq -2$, $|D_t\tau| \leq 2$, and we denote $\tau_R = \tau(R^{-p}(t - t_0))$.

We obtain

(2.5)

$$\int_{Q(2R,z_0)} D_t c_{n,2R}(u - c_{n,2R}) \exp(\mu|u - c_{2R}|^p)(\eta_{2R})^2(\tau_{2R})^2 \, dx\, dt+$$
$$+\sum_{i=1}^N \int_{Q(2R,z_0)} A_i(.,.,u,Du) D_i$$
$$\{(u - c_{n,2R}) \exp(\mu|u - c_{2R}|^p)(\eta_{2R})^2(\tau_{2R})^2 \quad dx\, dt+$$
$$+\int_{Q(2R,z_0)} H(.,.,u,Du)(u - c_{n,2R}) \exp(\mu|u - c_{2R}|^p)(\eta_{2R})^2(\tau_{2R})^2 \, dx\, dt+$$
$$+\int_{Q(2R,z_0)} f_0(u - c_{n,2R}) \exp(\mu|u - c_{2R}|^p)(\eta_{2R})^2(\tau_{2R})^2 \, dx\, dt+$$
$$+\sum_{i=1}^N \int_{Q(2R,z_0)} f_i D_i\{(u - c_{n,2R}) \exp(\mu|u - c_{2R}|^p)(\eta_{2R})^2(\tau_{2R})^2 \, dx\, dt+$$
$$+\int_{B(R,x_0;t_0)} (\eta_{2R})^2 \left(\int_0^{(u-c_{n,2R})} r \exp(\mu|r|^p) \, dr \right) dx-$$
$$-\int_{Q(2R,z_0)} \exp(\mu|u - c_{2R}|^p)(\eta_{2R})^2 \tau_{2R} D_t\tau_{2R} \, dx\, dt \leq 0$$

We consider now the first term in (2.5); we obtain

(2.6)

$$\int_{Q(2R,z_0)} D_t c_{n,2R} \exp(\mu|u-c_{2R}|^p)(\eta_{2R})^2(\tau_{2R})^2\,dx\,dt \geq$$

$$\geq \int_{Q(2R,z_0)} D_t c_{n,2R}(u-c_{2R}) \exp(\mu|u-c_{2R}|^p)(\eta_{2R})^2(\tau_{2R})^2\,dx\,dt =$$

$$= \int_{t_0-(2R)^p}^{t_0} (\tau_{2R})^2 D_t c_{n,2R}\,dt \int_{B(2R;x_0)} (u-c_{2R}) \exp(\mu|u-c_{2R}|^p)(\eta_{2R})^2\,dx =$$

Taking into account (2.6) and letting $n \to \infty$ in (2.5) we obtain

(2.7)

$$\sum_{i=1}^{N} \int_{Q(2R,z_0)} A_i(.,.,u,Du)\cdot$$

$$\cdot D_i\left\{(u-c_{2R}) \exp(\mu|u-c_{2R}|^p)(\eta_{2R})^2(\tau_{2R})^2\right\}\,dx\,dt+$$

$$+\sum \int_{Q(2R,z_0)} H(.,.,u,Du)(u-c_{2R}) \exp(\mu|u-c_{2R}|^p)(\eta_{2R})^2(\tau_{2R})^2\,dx\,dt+$$

$$+\int_{Q(2R,z_0)} f_0(u-c_{2R}) \exp(\mu|u-c_{2R}|^p)(\eta_{2R})^2(\tau_{2R})^2\,dx\,dt+$$

$$+\sum_{i=1}^{N} \int_{Q(2R,z_0)} f_i D_i\{(u-c_{2R}) \exp(\mu|u-c_{2R}|^p)(\eta_{2R})^2(\tau_{2R})^2\,dx\,dt-$$

$$-\int_{Q(2R,z_0)} \left(\int_0^{(u-c_{2R})} \exp(\mu|r|^p)\,dr\right) (\eta_{2R})^2 \tau_{2R} D_t \tau_{2R}\,dx\,dt \leq 0$$

Let us consider the first term in (2.7); then

(2.8)

$$\sum_{i=1}^{N} \int_{Q(2R,z_0)} A_i(.,.,u,Du)\cdot$$

$$\cdot D_i\left\{(u-c_{2R}) \exp(\mu|u-c_{2R}|^p)(\eta_{2R})^2(\tau_{2R})^2\right\}\,dx\,dt =$$

$$= \sum_{i=1}^{N} \int_{Q(2R,z_0)} A_i(.,.,u,Du) D_i u \exp(\mu|u-c_{2R}|^p)(\eta_{2R})^2(\tau_{2R})^2\,dx\,dt+$$

$$+p\int_{Q(2R,z_0)} A_i(.,.,u,Du) D_i u |u-c_{2R}|^p \exp(\mu|u-c_{2R}|^p)(\eta_{2R})^2(\tau_{2R})^2\,dx\,dt+$$

$$+2\int_{Q(2R,z_0)} A_i(.,.,u,Du)(u-c_{2R}) \exp|u-c_{2R}|^p (\tau_{2R})^2 \eta_{2R} D_i \eta_{2R}\,dx\,dt.$$

From (2.8), taking into account that u, c_{2R} are bounded, we have

(2.9)

$$\sum_{i=1}^{N} \int_{Q(2R,z_0)} A_i(.,.,u,Du)\cdot$$
$$\cdot D_i\left\{(u-c_{2R})\exp(\mu|u-c_{2R}|^p)(\eta_{2R})^2(\tau_{2R})^2\right\}\,dx\,dt \leq$$
$$\leq C_3 \int_{Q(2R,z_0)} |Du|^p(\eta_{2R})^2(\tau_{2R})^2 \exp(\mu|u-c_{2R}|^p)\,dx\,dt+$$
$$+ C_4 \int_{Q(2R,z_0)} |Du|^p(\eta_{2R})^2(\tau_{2R})^2|u-c_{2R}|^p \exp(\mu|u-c_{2R}|^p)\,dx\,dt+$$
$$+ C_5 R^{-p} \int_{Q(2R,z_0)} |u-c_{2R}|^p\,dx\,dt.$$

For the second term in (2.7) we obtain

(2.10)

$$\left|\int_{Q(2R,z_0)} H(.,.,u,Du)\right.$$
$$\left.(u-c_{2R})\exp(\mu|u-c_{2R}|^p)(\eta_{2R})^2(\tau_{2R})^2\,dx\,dt\right| \leq$$
$$\leq C_6 \int_{Q(2R,z_0)} \left[K(.,.)+|Du|^{p-1}(u-c_{2R})\exp(\mu|u-c_{2R}|^p)\cdot\right.$$
$$\left.\cdot(\eta_{2R})^2(\tau_{2R})^2\right]\,dx\,dt \leq$$
$$\leq C_6 \int_{Q(2R,z_0)} K(.,.)\,dx\,dt+$$
$$+ C_7 \int_{Q(2R,z_0)} |Du|^p|u-c_{2R}|^{p/p-1}\exp(\mu|u-c_{2R}|^p)(\eta_{2R})^2(\tau_{2R})^2\,dx\,dt+$$
$$+ C_8 \int_{Q(2R,z_0)} |u-c_{2R}|^p\,dx\,dt \leq$$
$$\leq C_6 \int_{Q(2R,z_0)} K(.,.)\,dx\,dt+$$
$$+ 4^{-1}C_3 \int_{Q(2R,z_0)} |Du|^p\exp(\mu|u-c_{2R}|^p)(\eta_{2R})^2(\tau_{2R})^2\,dx\,dt+$$
$$+ C_9 \int_{Q(2R,z_0)} |Du|^p|u-c_{2R}|^p\exp(\mu|u-c_{2R}|^p)(\eta_{2R})^2(\tau_{2R})^2\,dx\,dt+$$
$$+ C_8 \int_{Q(2R,z_0)} |u-c_{2R}|^p\,dx\,dt.$$

We consider now the third term in (2.7), namely

$$\sum_{i=1}^{N}\int_{Q(2R,z_0)} f_i\cdot$$
$$\cdot D_i\left\{(u-c_{2R})\exp(\mu|u-c_{2R}|^p)(\eta_{2R})^2(\tau_{2R})^2\right\}\,dx\,dt =$$

$$(2.11)\quad =\sum_{i=1}^{N}\int_{Q(2R,z_0)} f_i\Big\{D_iu\exp(\mu|u-c_{2R}|^p)(\eta_{2R})^2(\tau_{2R})^2\,dx\,dt+$$
$$+D_iu|u-c_{2R}|^p\exp(\mu|u-c_{2R}|^p)(\eta_{2R})^2(\tau_{2R})^2+$$
$$+2(u-c_{2R})\exp(\mu|u-c_{2R}|^p)(\tau_{2R})^2\eta_{2R}D_i\eta_{2R}\Big\}\,dx\,dt.$$

Taking into account that u, c_{2R} are bounded, we obtain easily

$$\Bigg|\sum_{i=1}^{N}\int_{Q(2R,z_0)} f_i\cdot$$
$$\cdot D_i\left\{(u-c_{2R})\exp(\mu|u-c_{2R}|^p)(\eta_{2R})^2(\tau_{2R})^2\right\}\,dx\,dt\Bigg|\le$$

$$(2.12)\quad \le C_{10}\sum_{i=1}^{N}\int_{Q(2R,z_0)}|f_i|^{p'}\,dx\,dt+$$
$$+C_{11}R^{-p}\int_{Q(2R,z_0)}|u-c_{2R}|^p\,dx\,dt+$$
$$+4^{-1}C_3\int_{Q(2R,z_0)}|Du|^p\exp(\mu|u-c_{2R}|^p)(\eta_{2R})^2(\tau_{2R})^2\,dx\,dt.$$

Finally we have

$$(2.13)\quad \left|\int_{Q(2R,z_0)}\left(\int_0^{(u-c_{2R})} r\exp(\mu|r|^p)\,dr\right)(\eta_{2R})^2\tau_{2R}D_t\tau_{2R}\,dx\,dt\right|\le$$
$$\le C_{12}R^{-p}\int_{Q(2R,z_0)}|u-c_{2R}|^p\,dx\,dt.$$

Combining (2.5)(2.9)(2.12)(2.13) and choosing $\mu\ge 2(C_4)^{-1}C_9$ we obtain the result.

LEMMA 3. *The following inequality holds*

$$\mathrm{Sup}_{\Lambda(t_0,R)}\int_{B(R,x_0;t_0)}|u-c_{2R}|^p\,dx\,dt\le C_{13}\int_{Q(2R,z_0)}(|Du|^p+g)\,dx\,dt$$

By the same methods of Lemma 2, we obtain

$$(2.14)\quad \mathrm{Sup}_{\Lambda(t_0,R)} \int_{B(R,x_0;t_0)} |u-c_{2R}|^p\,dx \le C_{14} \int_{Q(2R,z_0)} (|Du|^p+g)\,dx\,dt.$$

Moreover we have

(2.15)

$$\int_{B(R,x_0;t_0)} |u-c_R|^p\,dx \le C_{15} \int_{B(R,x_0;t_0)} |u-c_{2R}|^p\,dx + |c_R - c_{2R}|^p R^N \le$$

$$\le C_{15} \int_{B(R,x_0;t_0)} |u-c_{2R}|^p\,dx +$$

$$+ R^N \left| \left(\int_{B(R,x_0;t_0)} u \exp(\mu|u-c_R|^p)(\eta_R)^2\,dx \right) \cdot \right.$$

$$\left. \cdot \left(\int_{B(R,x_0;t_0)} \exp(\mu|u-c_R|^p)(\eta_R)^2\,dx \right)^{-1} - c_{2R} \right|^p \le$$

$$\le C_{15} \int_{B(R,x_0;t_0)} |u-c_{2R}|^p\,dx +$$

$$+ R^N \left| \left(\int_{B(R,x_0;t_0)} (u-c_{2R}) \exp(\mu|u-c_R|^p)(\eta_R)^2\,dx \right) \cdot \right.$$

$$\left. \cdot \left(\int_{B(R,x_0;t_0)} \exp(\mu|u-c_R|^p)(\eta_R)^2\,dx \right)^{-1} - c_{2R} \right|^p \le$$

$$\le C_{16} \int_{B(R,x_0;t_0)} |u-c_{2R}|^p\,dx.$$

From (2.14)(2.15) the result follows.

LEMMA 4. *The following inequality holds*

$$\fint_{Q(2R,z_0)} |Du|^p\,dx\,dt \le \varepsilon \left(\fint_{Q(4R,z_0)} |Du|^p\,dx\,dt + \fint_{Q(4R,z_0)} g\,dx\,dt \right) +$$
$$+ C(\varepsilon) \left(\fint_{Q(4R,z_0)} |Du|^{p^{**}}\,dx\,dt \right)^{p/p^{**}} \qquad (\forall \varepsilon > 0)$$

where $p^{**} = pN(N+p)^{-1}$, *if* $p^{**} = p\theta(\theta+p)^{-1}$, $\theta > p$, *if* $p \ge N$ *and* $\fint$ *denotes the usual average.*

For the sake of simplicity we deal with the case $p < N$. We have

$$\int_{Q(2R,z_0)} |u - c_{2R}|^p \, dx\, dt \leq$$

$$\leq \left[\operatorname{Sup}_{\Lambda(t_0,R)} \left(\int_{B(2R,x_0;t_0)} |u - c_{2R}|^p \, dx\, dt \right)^{1/p'} \right] \cdot$$

$$\cdot \int_{\Lambda(t_0,R)} dt \left(\int_{B(2R,x_0;t_0)} |u - c_{2R}|^p \, dx \right)^{1/p} \leq$$

$$\leq C_{17} \left(\int_{Q(4R,z_0)} (|Du|^p + g) \, dx\, dt \right)^{1/p'} \cdot$$

$$\cdot \int_{\Lambda(t_0,R)} dt \left(\int_{B(2R,x_0;t_0)} |u - c_{2R}|^{p^{**}} \, dx \right)^{1/2p^{**}} \cdot$$

$$\cdot \left(\int_{B(2R,x_0;t_0)} |u - c_{2R}|^{p^{*}} \, dx \right)^{1/2p^{*}} \leq$$

$$\leq C_{18} R^{1/2} \left(\int_{Q(4R,z_0)} (|Du|^p + g) \, dx\, dt \right)^{1/p'} \cdot$$

$$\cdot \int_{\Lambda(t_0,R)} dt \left(\int_{B(4R,x_0;t_0)} |Du|^{p^{**}} \right) \left(\int_{B(4R,x_0;t_0)} |Du|^p \, dx \right)^{1/2p} \leq$$

$$\leq C_{19} R^{1/2} \left(\int_{Q(4R,z_0)} (|Du|^p + g) \, dx\, dt \right)^{1/p'} \cdot$$

$$\cdot R^{p(1-(1/2p^{**})-(1/2p))} \left(\int_{Q(4R,z_0)} |Du|^{p^{**}} \, dx\, dt \right)^{1/2p^{**}} \cdot$$

$$\cdot \left(\int_{Q(4R,z_0)} |Du|^p \, dx\, dt \right)^{1/2p} \leq$$

$$\leq \varepsilon R^p \left(\int_{Q(4R,z_0)} |Du|^p \, dx\, dt + \int_{Q(4R,z_0)} g \, dx\, dt \right) +$$

$$+ C(\varepsilon) R^{p(p/N)} \left(\int_{Q(4R,z_0)} |Du|^p \, dx\, dt \right)^{p/p^{**}} .$$

Dividing by $R^{-(2p+N)}$ we obtain the result.

For the case $p \geq N$ we can choose $p^{**} = p\theta(p+\theta)^{-1}$ and $p^* = p\theta(\theta - p)^{-1}$; repeating the calculation and taking care in applications of the Poincaré's inequality we obtain again the result.

We can also see that the result of Lemma 4 holds again for u extended by 0 to $\mathbb{R}^N \times (0,T)$ and $t_0 > 0$, when $Q(2R, z_0)$ in the first term is replaced by $Q(R, z_0)$ (we observe that if $Q(2R, z_0) \cap \Sigma \neq \emptyset$ and $t_0 - (4R)^p > 0$, due to the smoothness of $\partial\Omega$, we can suppose that the Newtonian capacity of $B(4R, x_0; t_0) \cap \Omega^c$ is greater than cR^{N-2}). The first part of Theorem 1 follows from Proposition 5.1 of [**10**]. The global result follows by extending u for $t < 0$ by $v^0(., -t)$.

REMARK 4: If H is bounded, $f_0 \in L^{q''}(Q)$ and $|A_i(x,t,r,\xi)| \leq C|\xi|^p$ we can repeat the above calculations by means of the usual average and obtain the result of Theorem 1 *without the assumption of boundedness* for u, using test functions with $\sigma = r^2/2$.

§3. The case H bounded.

Using the penalization method we can prove, by standard procedures,

LEMMA 5. *Let $A_i(x,t,r,\xi)$ be independent of r and $H = 0$; then there exists a unique solution u of the variational inequality (1.5) (without the condition $u, v \in L^\infty(Q)$ and with $\sigma'(r) \leq C|r|$).*

Define now

$$\hat{A}_i(x,t,r,\xi) = \begin{cases} A_i(x,t,r,\xi) & \text{if } \Phi(x,t) < r \leq \operatorname{Sup}_Q \psi \\ A_i(x,t,\Phi(x,t),\xi) & \text{if } r \leq \Phi(x,t) \\ A_i(x,t,\operatorname{Sup}_Q \psi,\xi) & \text{if } r > \operatorname{Sup}_Q \psi \end{cases}$$

and

$$\hat{H}(x,t,r,\xi) = \begin{cases} H(x,t,r,\xi) & \text{if } \Phi(x,t) < r \leq \operatorname{Sup}_Q \psi \\ H(x,t,\Phi(x,t),\xi) & \text{if } r \leq \Phi(x,t) \\ H(x,t,\operatorname{Sup}_Q \psi,\xi) & \text{if } r > \operatorname{Sup}_Q \psi. \end{cases}$$

We observe that $\hat{H}$ is not a Cathéodory function but H is everywhere defined in (r,ξ) and is a continuous operator from $L^p(0,T; H^{1,p}(\Omega))$ into $L^1(Q)$; moreover

$$(3.1) \qquad |H(x,t,v(x,t),Dv(x,t))| \leq C(1 + |Dv|^p)$$

a.e. in Q, $\forall v \in L^p(0,t;H^{1,p}(\Omega))$, [**9**] [**12**].

We denote by $(\widehat{1.5})$ the variational inequality (1.5), where A_i and H are replaced by $\hat{A}_i$, $\hat{H}$.

Denote

$$\hat{H}_n(x,t,r,\xi) = \begin{cases} \hat{H}(x,t,r,\xi) & \text{if } |H(x,t,r,\xi)| \le n \\ \hat{H}(x,t,r,\xi)|H(x,t,r,\xi)|^{-1} & \text{if } |H(x,t,r,\xi)| > n. \end{cases}$$

We observe that $\hat{H}_n$ is bounded and continuous with respect to r, (3.1) holds again and $\hat{H}_n$ defines a continuous operator from $L^p(0,T;H^{1,p}(\Omega))$ into $L^1(Q)$.

We denote by $(\widehat{1.5})_n$ the problem $(\widehat{1.5})$ with H replaced by H_n (without the assumption $u, v \in L^\infty(Q)$ and with $\sigma'(r) \le C|r|$). Taking into account the strict monotonicity of the $\hat{A}_i$ with respect to s and using the method of Lemma 10 in [**5**] we obtain:

LEMMA 6. *For $n > n_0$ the solution $\hat{u}_n$ of $(\widehat{1.5})_n$ satisfies*

$$\Phi \le \hat{u}_n \le \operatorname{Sup}_Q \psi. \tag{3.2}$$

LEMMA 7. *The problem $(\widehat{1.5})_n$ has a solution.*

The above lemma is proved by a Schauder fixed point method. We denote by $u(w,z)$ a solution of $(\widehat{1.5})_n$ with $A_i(x,t,r,\xi)$ replaced by $A_i(x,t,w(x,t),\xi)$ and $\hat{H}_n(x,t,r,\xi)$ replaced by $H_n(x,t,z(x,t),Dz(x,t))$, $w \in L^{3/2}(Q)$, $z \in L^{p''}(0,T;H^{1,p''}(\Omega))$, $p'' \in \varepsilon(p,q)$; by $u(z)$ we denote a solution of $(\widehat{1.5})_n$ with $\hat{H}_n(x,t,r,\xi)$ replaced by $\hat{H}_n(x,t,z(x,t),Dz(x,t))$.

Since $\hat{H}_n$ is bounded, we have

$$\|u(w,z)\|_{L^p(0,T;H^{1,p}(\Omega))} \le C_1 \tag{3.3}$$

where C_1 depends only on n. From (3.3) $\hat{H}_n$ being bounded and taking into account the Remark 1 in § 2 we obtain

$$\|u(w,z)\|_{L^q(0,T;H^{1,p}(\Omega))} \le C_2 \tag{3.4}$$

where C_2 depends only on n. Finally, we observe that $u(w,z) \in C(0,T;L^2(\Omega))$.

Now we prove that, for (w,z) in a bounded set of $L^{3/2}(Q) \times L^{p''}(0,T;H^{1,p''}(\Omega))$ the solution $u(w,z)$ is equicontinuous in $L^2_{\text{loc}}(\Omega)$ on (ε,T), $\forall \varepsilon > 0$.

Let $0 < \delta < \varepsilon$ and let t' be fixed in (ε, T), then there exists $t^* \in (t'-\delta, t')$ such that

$$\|u(t^*) - \psi(t^*)\|_{H^{1,q}(\Omega)} \le (C_3)^{1/q}.$$

If we choose $\sigma = r^2/2$, $\varphi \in C_0^\infty(\Omega)$, $v = u(t^*) + \psi(t) - \psi(t^*)$, by standard methods we obtain

$$\begin{aligned}
&\|(u(t) - u(t^*) - \psi(t) + \psi(t^*))\psi\|_{L^2}^2 \le \\
&\le \left| \int_{t^*}^{t} \langle D_i A_i(.,.,w,Du), \varphi(u(s) - u(t^*) - \psi(s) + \psi(t^*)) \rangle_{H^{-1,p'},H_0^{1,p}} \, ds \right| + \\
&+ \left| \int_{t^*}^{t} \left\langle f_0 + \sum_{i=1}^{N} D_i f_i \varphi(u(s) - u(t^*) - \psi(s) + \psi(t^*)) \right\rangle_{H^{1,p'},H_0^{1,p}} ds \right| + \\
&+ \left| \int_{t^*}^{t} \int_\Omega H_n(.,.,z,Dz), \varphi(.)(u(.,s) - u(.,t^*) - \psi(.,s) + \psi(.,t^*)) \, dx \, dt \right| + \\
&+ \left| \int_{t^*}^{t} \langle D_t \varphi(u(s) - u(t^*) - \psi(s) + \psi(t^*)) \rangle_{H^{1,p'},H_0^{1,p}} \, ds \right|,
\end{aligned}$$

where $t \in (t', t' + \delta)$. Thus

$$\begin{aligned}
&\|(u(t) - u(t^*))\psi\|_{L^2}^2 \le \|\psi(t) - \psi(t^*))\|_{L^2}^2 + \\
&+ C_4 \int_{t^*}^{t} \int_\Omega \Big[|Du|^p + |D\psi|^p + |Dz|^p + |f_0|^p + \\
&\qquad + \sum_{i=1}^{N} |f_i|^p + |u - \psi|^p \Big] \, dx \, ds + \\
&+ \int_{t^*}^{t} \|D_t \psi\|_{H^{-1,p}}^{p'} \, ds + \|u(t^*) - \psi(t^*)\|_{H^{1,q}}^{p} \le \\
&\le \|\psi(t) - \psi(t^*)\|_{L^2} + C_5(\delta^{p/q} + \delta^{1-(p/q)}).
\end{aligned} \tag{3.5}$$

From (3.5) the result easily follows.

Consider now the map $w \to u(w,z)$ from $L^{3/2}(Q)$ into $L^{3/2}(Q)$; when w is in a bounded set of $L^{3/2}(Q)$, $u(w,z)$ is equicontinuous in $L^2_{\text{loc}}(\Omega)$ on $(\varepsilon, T], \forall \varepsilon > 0$, and, from (3.3), $|u(w,z)|^{3/2}$ is equiintegrable in Q.

From (3.4) and the equicontinuity, $u(w,z)$ is relatively compact in $C(\varepsilon, T; L^2_{\text{loc}}(\Omega))$, then, from the equiintegrability on Q, is relatively compact in $L^{3/2}(Q)$.

Thus we can apply Schauder's fixed point theorem and prove the existence of a solution $u(z)$. We observe that (3.3) and (3.4) still hold for $u(z)$.

Let now $\{z_m\}$ be a sequence such that

$$\lim_{m\to\infty} z_m = z \text{ in } L^{p''}(0,T;H^{1,p''}(\Omega)). \tag{3.6}$$

Since (3.4) holds for $u(z_m)$ uniformly in m, then we can suppose

$$w - \lim_{m\to\infty} u(z_m) = \chi \text{ in } L^q(0,T;H^{1/q}(\Omega)). \tag{3.7}$$

By the same methods used in the first part we can prove that $u(z_m)$ are locally equicontinuous on $(0,T]$ in $L^2_{\mathrm{loc}}(\Omega)$, then from (3.7) we obtain

$$\lim_{m\to\infty} u(z_m) = \chi \tag{3.8}$$

in $L^2_{\mathrm{loc}}(\Omega)$ uniformly on $(\varepsilon,T], \forall\varepsilon > 0$, then a.e. in Q.

From (3.4) $|u(z_m)|^s$ are equiintegrable for $s < q$, then from (3.8) we obtain

$$\lim_{m\to\infty} u(z_m) = \chi \text{ in } L^s(Q). \tag{3.9}$$

The following inequality can be obtained by standard methods

$$\sum_{i=1}^{N}\int_Q \Big(\hat{A}_i(.,.,u(z_m),Du(z_m)) - \hat{A}_i(.,.,u(z_{m'}),Du(z_{m'}))\Big)\cdot$$
$$\cdot D_i((u(z_m)-u(z_{m'}))\varphi)\,dx\,ds \leq$$
$$\leq \int_Q \Big(\hat{H}_n(.,.,z_m,Dz_m) - \hat{H}_n(.,.,z_{m'},Dz_{m'})(u(z_m)-u(z_{m'}))\Big)\,dx\,ds,$$

$\forall\varphi\in C_0^\infty(Q)$; then from (3.9) we have

$$\lim_{m,m'\to\infty}\sum_{i=1}^{N}\int_Q \Big(\hat{A}_i(.,.,u(z_m),Du(z_m)) - \hat{A}_i(.,.,u(z_{m'}),Du(z_{m'}))\Big)\cdot$$
$$\cdot D_i((u(z_m)-u(z_{m'}))\varphi)\,dx\,ds = 0.$$

From (3.8)(3.10) and the strict monotonicity of the A_i in the variable ξ we obtain

$$\lim_{m,m'\to\infty} D(u(z_m)-u(z_{m'})) = 0 \text{ a.e. in } Q; \tag{3.11}$$

then

$$\lim_{m\to\infty} Du(z_m) = D\chi \quad \text{a.e. in } Q. \tag{3.12}$$

From (3.4) the sequence $\{|Du(z_m)|^{p''}\}$ is equiintegrable on Q; then

$$\lim_{m\to\infty} Du(z_m) = D\chi \text{ in } L^{p''}(Q) \tag{3.13}$$

We have proved that the map $z \to u(z)$ is compact from $L^{p''}(0,T;H^{1,p''}(\Omega))$ into itself; then from (3.4) we can apply Schauder's fixed point theorem and prove the existence of a fixed point $\hat{u}_n$, which is a solution of $(\widehat{1.5})_n$.

§4. The general result.

Using v_0 as the test function in $(1.5)_n$ and choosing $\varphi = 1$ and $\sigma(y) = \int_0^y r\exp(\mu|r|^p)\,dr$ (this is possible since $\tilde{u}_n$ is bounded for $n \geq n_0$) with $\mu > 0$ a suitable constant, obtain, for $n \geq n_0$,

$$\|\hat{u}_n\|_{L^q(0,T;H^{1,q}(\Omega))} \leq C_1. \tag{4.1}$$

Then, from the Meyers estimate (since $\{\tilde{u}_n\}$ are uniformly bounded for $n \geq n_0$), we have

$$\|\hat{u}_n\|_{L^q(0,T;H^{1,q}(\Omega))} \leq C_2. \tag{4.2}$$

From (4.2) we can suppose

$$w - \lim_{n\to\infty} \hat{u}_n = \hat{u} \text{ in } L^q(0,T;H^{1,q}(\Omega)) \tag{4.3}$$

As in §3. we obtain from (4.3) the local equicontinuity of $\{\tilde{u}_n\}$ in $L^2_{\text{loc}}(\Omega)$ on $(0,T]$; then, the sequence $\{\tilde{u}_n\}$ being uniformly bounded for $n \geq n_0$ (see lemma 6), we have

$$\lim_{n\to\infty} \hat{u}_n = \hat{u} \text{ in } L^s(Q), \quad \forall s < +\infty$$
$$\hat{u} \in C((0,T];L^2(\Omega)).$$

By (4.3)(4.4) and the same method as those applied to obtain (3.13), we prove

$$\lim_{n\to\infty} \hat{u}_n = \hat{u} \text{ in } L^{p''}(0,T,H^{1,p''}(\Omega)), \quad p'' \in (p,q) \tag{4.5}$$

From (4.4)(4.5) we easily obtain that $\hat{u}$ is a solution of $(\widehat{1.5})$ with the additional assumption $\sigma'(r) \leq C|r|$; then, $\hat{u}$ being bounded, $\hat{u}$ is a solution of (1.5) without any restrictions.

We observe that from Lemma 6 we have $\Phi \leq \hat{u} \leq \operatorname{Sup}_Q \psi$, thus $\hat{u}$ is also a solution to (1.5).

Finally, we observe that a standard regularization method gives the continuity of $\hat{u}$ at $t = 0$ in $L^2_{\text{loc}}(\Omega)$, then, due to the boundedness of $\hat{u}$, in $L^2(\Omega)$.

Department of Mathematics, Politecnico di Milano, ITALY

References

1. A. Bensoussan and J.L. Lions, "Applications des inéquations variationnelles en contrôle stochastique," Dunod, 1978.
2. M. Biroli, *Un estimation L^p du gradient de la solution d'une inéquation elliptique du 2° ordre*, C.R.A.S. Paris **A288** (1979), 453–455.
3. M. Biroli, *Existence and Meyers estimate for the solution of some nonlinear parabolic unilateral problems*, Ric. di Mat. **XXXII, No. 1** (1983), 63–73.
4. M. Biroli, *Nonlinear parabolic variational inequalities*, Comm. Math. Univ. Carolinae **26, No. 1** (1985), 23–39.
5. M. Biroli, *Existence and Meyers estimate for nonlinear parabolic variational inequalities*, Ric. di Mat., in print.
6. M. Biroli, *Existence et estimation de Meyers pour des problèmes d'obstacle paraboliques quasi-linéaires*, C.R.A.S. Paris **303, No. 12** (1986), 543–546.
7. L. Boccardo, *An L^s estimate for the gradient of solutions of some strongly nonlinear unilateral problems*, Comm. in P.D.E., in print.
8. L. Boccardo, F. Murat and J.P. Puel, *Existence results for some quasilinear parabolic equations*, preprint, Lab. An. Num. Univ. Paris **VI**.
9. L. Boccardo, F. Murat and J.P. Puel, *Résultats d'existence pour certaines problèmes elliptiques quasi-linéaires*, Ann. Sc. Norm. Sup. Pisa **2** (1984), 213–237.
10. M. Giaquinta and G. Modica, *Regularity results for some classes of higher order nonlinear elliptic systems*, J. Reine Angew. Math. **311/312** (1979), 145–169.
11. M. Giaquinta and M. Struwe, *On the partial regularity of weak solutions of parabolic systems*, Math. Zeit. **179** (1982), 437–451.
12. A. Mokrane, *Existence de solutions pour certaines problèmes quasi-linéaires elliptiques et paraboliques*, Thèse, Univ. Paris VI.
13. J.P. Puel, *A compactness theorem in quasilinear parabolic problems and applications to an existence result*, in "Nonlinear parabolic equations: qualitative properties of the solutions," Research Notes in Mathematics, Pitman, Rome, 1985.
14. M.A. Vivaldi, *Nonlinear parabolic variational inequalities: existence of weak solutions and regularity propertites*, Boll. U.M.I. **VII, I-B, 1** (1987), 259–275.

Convergence to Traveling Waves for Systems of Kolmogorov-like Parabolic Equations

MAURY BRAMSON

We are interested in systems of equations of the form

$$
\begin{aligned}
u_t &= u_{xx} + g(u,v)u \\
v_t &= v_{xx} + h(u,v)v
\end{aligned}
\tag{1}
$$

on $-\infty < x < \infty$ and $t \geq 0$. Here $g(u,v)$ and $h(u,v)$ are C^1 functions and satisfy

$$
\begin{aligned}
g(u,v) &> 0 \text{ for } 0 \leq u < 1, 0 \leq v \leq 1 \\
h(u,v) &> 0 \text{ for } 0 \leq u \leq 1, 0 \leq v < 1
\end{aligned}
\tag{2a}
$$

with

$$
\begin{aligned}
&g(1,v) = 0, h(u,1) = 0 \\
&g(u,v), h(u,v) \downarrow \text{ as } u, v \uparrow .
\end{aligned}
\tag{2b}
$$

The variables $u(x,t)$ and $v(x,t)$ can be thought of as the densities of two competing species which migrate at a fixed rate. By (2b) their growth rates go down as the density of either increases, until saturation is reached at $u = v = 1$. We assume that the initial data is measurable with

$$
0 \leq u(x,0), v(x,0) \leq 1. \tag{3}
$$

Consequently,

$$
0 \leq u(x,t), v(x,t) \leq 1
$$

for all t.

We are interested in the asymptotic behavior of (u,v) as $t \to \infty$. In particular, we ask when u and v converge to traveling waves under initial data which satisfy

$$
\begin{aligned}
u(x,0), v(x,0) &\to 1 \text{ as } x \to -\infty \\
&\to 0 \text{ as } x \to \infty
\end{aligned}
\tag{4}
$$

and appropriate additional conditions. That is, do there exist $m^1(t)$, $m^2(t)$ and $U(x), V(x)$ with

(5)
$$\begin{aligned} U(x), V(x) &\to 1 \text{ as } x \to -\infty \\ &\to 0 \text{ as } x \to \infty \end{aligned}$$

such that

(6)
$$\begin{aligned} u(x+m^1(t),t) &\to U(x) \\ v(x+m^2(t),t) &\to V(x) \end{aligned}$$

as $t \to \infty$? We will distinguish between two basic cases:

I. Where one can choose $m^1(t) = m^2(t)$ and (u,v) converges to a system of traveling waves, and
II. Where $m^1(t) - m^2(t) \to \infty$ and (u,v) decouples as $t \to \infty$ to the scalar equations

(7a)
$$u_t = u_{xx} + g(u,0)u$$

and

(7b)
$$v_t = v_{xx} + h(1,v)v.$$

To investigate this asymptotic behavior, we find it instructive to first review the behavior of the corresponding scalar equation.

Scalar Analog.

The solution $u(x,t)$ of the equation

(8)
$$u_t = u_{xx} + F(u),$$

$-\infty < x < \infty$, $t \geq 0$, with

(9a)
$$F(0) = F(1) = 0, \quad F(u) > 0 \text{ for } 0 < u < 1$$

and

(9b)
$$F'(0) = 1, \quad F'(u) \leq 1 \text{ for } 0 < u \leq 1$$

has been studied in the mathematical literature since 1937. Fisher [1] and Kolmogorov-Petrovsky-Piscounov [2] studied (8)–(9) as a model for the spread of an advantageous gene through a population. More recently (8)–(9) has been considered in the context of branching Brownian motion (McKean [3] and Skorohod [4]). In that setting $u(x,t)$ can be interpreted as measuring the spread of cosmic rays through the atmosphere. Of course, $u(x,t)$ can also be interpreted as measuring the spread of a single species throughout an environment. Typically, $0 \leq u(x,0) \leq 1$ is assumed.

In order to investigate when u converges to a traveling wave as $t \to \infty$, one needs to examine solutions of

$$U^\lambda_{xx} + \lambda U^\lambda_x + f(U^\lambda) = 0 \tag{10}$$

with $0 \leq U(x) \leq 1$. Since the initial data will be assume to satisfy

$$\begin{aligned} u(x,0) &\to 1 \text{ as } x \to -\infty \\ &\to 0 \text{ as } x \to \infty, \end{aligned} \tag{11}$$

we also require that

$$\begin{aligned} U^\lambda(x) &\to 1 \text{ as } x \to -\infty \\ &\to 0 \text{ as } x \to \infty. \end{aligned} \tag{12}$$

Using standard phase plane arguments, one can show that such solutions exist for $\lambda \geq 2$, in which case they are unique (up to translation).

Using this information and an extended maximum principle, Kolmogorov et. al. showed that under Heaviside initial data

$$\begin{aligned} u(x,0) &= 1 \text{ for } x \leq 0 \\ &= 0 \text{ for } x > 0, \end{aligned} \tag{13}$$

$m(t)$ satisfying $\dot{m}(t) \to 2$ as $t \to \infty$ can be chosen so that

$$u(x+m(t),t) \to U^2(x) \text{ as } t \to \infty. \tag{14}$$

More recently, general conditions have been given on the initial data under which

$$u(x+m(t),t) \to U^\lambda(x) \text{ as } t \to \infty \tag{15}$$

for $\lambda \geq 2$ and appropriate $m(t)$ (Bramson [5], Lau [6], Uchiyama [7]). In particular, it was shown in [5] that (14) holds if and only if

$$\overline{\lim_{y\to\infty}} \frac{1}{y} \log \left[\int_y^{y(1+h)} u(x,0)dx \right] \leq -1 \tag{16}$$

for some $h > 0$. Not surprisingly,

$$\dot{m}(t) \to 2 \text{ as } t \to \infty$$

under (16). Note that the theory for $F(u)$ chosen as in (9) is different than for $F(u)$ of Kanel type, where for some $0 < \alpha < 1$,

$$\begin{aligned} &F(u) < 0 \text{ for } 0 < u < \alpha \\ &F(u) > 0 \text{ for } \alpha < u < 1. \end{aligned} \tag{17}$$

Then, a solution to (10) and (12) exists for a unique λ_0. Convergence occurs as in (15), but for a much wider class of initial data (Kanel [8] and Fife-McLeod [9]). Also, one can set $m(t) = \lambda_0 t$, whereas in the present case this is typically not possible.

Basic techniques.

Since similar techniques will be employed for the system of equations (1)–(2), we briefly outline the techniques used to obtain (16). The methodology is largely probabilistic and uses the Feynman-Kac formula together with estimates for Brownian bridge. For general background and more detail see Bramson [5].

Let $Z_{x,y}$ denote Brownian bridge, that is Brownian motion defined on $0 \leq s \leq t$, where $Z_{x,y}(0) = y$ and $Z_{x,y}(t) = x$ are specified. Here, we scale Brownian motion so that it has variance $2s$ at time s. The Feynman-Kac formula states that for

$$u_t = u_{xx} + k(x,t)u, \tag{18}$$

then under reasonable conditions on k (for instance, k is smooth and bounded),

$$u(x,t) = \int_{-\infty}^{\infty} u(y,0)\Phi(x-y,t)E\left[\exp\left\{\int_0^t k(Z_{x,y}(s),s)ds\right\}\right] dy, \tag{19}$$

where

$$\Phi(y,t) = e^{-y^2/4t}/\sqrt{4\pi t}.$$

Here, $E[\cdot]$ denotes expectation, that is, the average over all paths $Z_{x,y}$.

In our case, we set

$$k(x,t) = F(u(x,t))/u(x,t). \tag{20}$$

Although $k(x,t)$ is not explicitly known, one can use

$$\begin{aligned} u(x,t) &\to 1 \text{ as } x \to -\infty \\ &\to 0 \text{ as } x \to \infty \end{aligned} \tag{21}$$

to conclude that

$$\begin{aligned} k(x,t) &\to 0 \text{ as } x \to -\infty \\ &\to 1 \text{ as } x \to \infty. \end{aligned} \tag{22}$$

Let $z_{x,y}(s), s \in [0,t]$, denote a continuous function with $z_{x,y}(0) = y$ and $z_{x,y}(t) = x$. (22) implies that if $z_{x,y}(s)$ is large enough for most $s \in [0,t]$, then

$$\exp\left\{\int_0^t k(z_{x,y}(s),s)ds\right\} \approx e^t, \tag{23}$$

whereas if $z_{x,y}(s)$ is small for a substantial portion of time, then

$$\exp\left\{\int_0^t k(z_{x,y}(s),s)ds\right\} \ll e^t. \tag{24}$$

One can show that Brownian bridge paths of the second type can be omitted when computing (19). Moreover, for appropriate $m(t)$ (for instance, $u(m(t),t) = \frac{1}{2}$), one can restrict consideration to paths satisfying

$$z_{x,y}(s) > m(s) \text{ for } s \in [N,t] \tag{25}$$

for $x > m(t)$. (As $z_{x,y}(s) - m(s)$ and N increase, the accuracy of the corresponding estimates improves.) For these paths, one can justify the substitution of (23) into (19). One therefore obtains

$$u(x,t) \approx e^t \int_{-\infty}^{\infty} u(y,0)\Phi(x-y,t)P\big[Z_{x,y}(s) > m(s), s \in [N,t]\big]dy. \tag{26}$$

One can show that for a large class of initial data, $m(s)$ is "almost" the straight line $\ell_t(s) = sm(t)/t$, $s \in [0,t]$. The above probability $P[\cdot]$ is explicitly computable if ℓ_t is substituted for m. (The general case requires more effort.) One can then show from (26) after some estimation that

$$u(x,t) \approx C_1(x - m(t)) \exp\{-(x - m(t))\} \tag{27}$$

for large t, where the estimate improves as $x - m(t)$ increases. The right side of (27) gives the right tail of $U^2(x - m(t))$. Application of the maximum principle enables one to show that this tail in fact determines the shape of $u(x + m(t), t)$ as $t \to \infty$, which must in face be $U^2(x)$. The limit (14) follows. One can also derive formulas for $m(t)$.

Systems of equations.

We return to the systems given in (1)–(2). As mentioned earlier, two possible types of behavior as $t \to \infty$ occur when

I. The solutions travel at the same basic rate, and

II. They travel at different rates.

To specify which occurs when, we set

$$g(0,0) = C, \quad h(0,0) = D, \quad h(1,0) = \hat{D}. \tag{28}$$

In analogy with (16), we also assume that

$$\begin{aligned} &\overline{\lim_{y\to\infty}} \frac{1}{y} \log\left[\int_y^{y(1+h)} u(x,0)\,dx\right] \le -\sqrt{C} \\ &\overline{\lim_{y\to\infty}} \frac{1}{y} \log\left[\int_y^{y(1+h)} v(x,0)\,dx\right] \le -\sqrt{D}. \end{aligned} \tag{29}$$

I. $C = D$.

The equations

$$\bar{u}_t = \bar{u}_{xx} + g(\bar{u}, 0)\bar{u} \tag{30}$$

and

$$\bar{v}_t = \bar{v}_{xx} + h(0, \bar{v})\bar{v} \tag{31}$$

can each be written in the form (8)–(9) after rescaling. Since the conditions (29) reduce to (16) in this setting, there are centering terms $\bar{m}^1(t)$ and $\bar{m}^2(t)$ corresponding to $\bar{u}$ and $\bar{v}$, respectively, under which convergence as in (15) holds. By

$$(u(x+\alpha(t),t), v(x+\beta(t),t)) \tag{32}$$

being tight, we will mean that for each $\varepsilon > 0$ there are t_0 and N so that for all $t \geq t_0$,

$$u(x+\alpha(t),t) \geq 1-\varepsilon, \quad v(x+\beta(t),t) \geq 1-\varepsilon$$

if $x \leq -N$ and

$$u(x+\alpha(t),t) \leq \varepsilon, \quad v(x+\beta(t),t) \leq \varepsilon$$

if $x \geq N$. The conditions

$$C(u_2-u_1) \geq [g(u_2,v)u_2 - g(u_1,v)u_1] - [h(u_2,v)-h(u_1,v)]\,v \tag{33}$$

and

$$D(v_2-v_1) \geq [h(u,v_2)v_2 - h(u,v_1)v_1] - [g(u,v_2)-g(u,v_1)]\,u \tag{34}$$

for $u_2 > u_1$ and $v_2 > v_1$ in effect say that each species is more sensitive to the effects of its own saturation than to those of the other species.

THEOREM 1. *Assume that (1)–(4) and (29) hold. If $C = D$ and*

$$\lim_{t\to\infty} [\bar{m}^1(t) - \bar{m}^2(t)] \tag{35}$$

exists, then (32) is tight with $\alpha(t) = \beta(t) = \bar{m}^1(t)$. If in addition (33)–(34) hold, then

$$\begin{aligned} u(x+\bar{m}^1(t),t) &\to U(x) \\ v(x+\bar{m}^1(t),t) &\to V(x) \end{aligned} \tag{36}$$

as $t \to \infty$ for appropriate $U(x)$ and $V(x)$ satisfying (5). In any case,

$$\dot{\bar{m}}^1(t) \to 2\sqrt{C} \text{ as } t \to \infty. \tag{37}$$

It is not surprising that (U, V) forms a system of traveling waves moving at rate $2\sqrt{C}$. The existence of such systems was first shown in Tang-Fife [10]. One can show that if (35) fails, then (36) cannot hold. Conditions (33) and (34) are presumably not necessary to demonstrate (36), but the author does not see how to proceed without them.

II. $C > D$.

The problem breaks down into two subcases, depending on whether

$$D \leq C + 2\hat{D} - 2\sqrt{C\hat{D}} \tag{38}$$

or

$$C + 2\hat{D} - 2\sqrt{C\hat{D}} < D < C. \tag{39}$$

Here $U(x)$ and $V(x)$ denote solutions of

$$U_{xx} + 2\sqrt{C}U_x + g(U,0)U = 0 \tag{40}$$

and

$$V_{xx} + \lambda^* V_x + h(1,V)V = 0 \tag{41}$$

with $\lambda^* \geq 2\sqrt{\hat{D}}$, and which satisfy (5).

THEOREM 2. *Assume that (1)–(4) and (29) hold. If $C > D$ and (38) is satisfied, then*

$$\begin{aligned} u(x + m^1(t), t) &\to U(x) \\ v(x + m^2(t), t) &\to V(x) \end{aligned} \tag{42}$$

as $t \to \infty$ for $U(x)$ and $V(x)$ satisfying (40)–(41) with $\lambda^* = 2\sqrt{\hat{D}}$, and appropriate $m^1(t)$ and $m^2(t)$ with

$$\begin{aligned} \dot{m}^1(t) &\to 2\sqrt{C} \\ \dot{m}^2(t) &\to 2\sqrt{\hat{D}} \end{aligned} \tag{43}$$

as $t \to \infty$.

The behavior under (39) is similar except that $\lambda^* > 2\sqrt{\hat{D}}$.

THEOREM 3. *Assume that (1)–(4) and (29) hold. If (39) is satisfied, then (42) holds as above but with*

$$2\sqrt{\hat{D}} < \lambda^* < 2\sqrt{D} \tag{44}$$

and

$$\begin{aligned} \dot{m}^1(t) &\to 2\sqrt{C} \\ \dot{m}^2(t) &\to \lambda^* \end{aligned} \tag{45}$$

as $t \to \infty$.

Basic techniques.

The basic techniques used to obtain Theorems 1–3 are extensions of the procedure outlined in (20)–(27) for the scalar equation. If we set

$$\begin{aligned} k_1(x,t) &= g(u(x,t), v(x,t)) \\ k_2(x,t) &= h(u(x,t), v(x,t)), \end{aligned} \tag{46}$$

then (1) becomes

$$\begin{aligned} u_t &= u_{xx} + k_1(t,x)u \\ v_t &= v_{xx} + k_2(t,x)v. \end{aligned} \tag{47}$$

Here, u and v satisfy the analogs of the Feynman-Kac formula (19) with k_1 and k_2 substituted in for k. In analogy with (22),

$$\begin{aligned} k_1(x,t) &\to 0 \text{ as } x \to -\infty \\ &\to C \text{ as } x \to \infty \end{aligned} \tag{48}$$

and

$$\begin{aligned} k_2(x,t) &\to 0 \text{ as } x \to -\infty \\ &\to D \text{ as } x \to \infty. \end{aligned} \tag{49}$$

As before, analysis of (19) involves deciding which paths $z_{x,y}(s)$, $0 \le s \le t$, should be kept and evaluated, and which can be thrown out. The case where $C = D$ is considerably simplified by the comparisons

$$\begin{aligned} u(x,t) &\le \bar{u}(x,t) \\ v(x,t) &\le \bar{v}(x,t). \end{aligned} \tag{50}$$

Under (35), (50) implies that

$$g(u,v), h(u,v) \approx C \tag{51}$$

for $z - \bar{m}^1(s)$ large enough. The main contribution in (19) for $\bar{u}$ and $\bar{v}$ comes from such paths if $x - \bar{m}^1(t)$ is large. Moreover,

$$g(\bar{u},\bar{v}), h(\bar{u},\bar{v}) \leq C.$$

Therefore,

$$\begin{aligned} u(x,t) &\approx \bar{u}(x,t) \\ v(x,t) &\approx \bar{v}(x,t) \end{aligned} \tag{52}$$

if $x - \bar{m}^1(t)$ is large. (52) is enough to demonstrate tightness in Theorem 1, since growth in the right tail can be estimated. If one assumes (33)–(34), then reasoning similar to that after (27) implies (36).

For both Theorems 2 and 3 one needs to go back and forth between upper and lower bounds on u and v to obtain suitable estimates. Upper bounds for u and v can be obtained as in (50). (One can also substitute C and D for g and h and solve explicitly.) Since $C > D$,

$$\bar{m}^2(s) \ll \bar{m}^1(s) \tag{53}$$

for large s. Far enough above $\bar{m}^2(s)$,

$$g(u,v) \approx g(u,0). \tag{54}$$

By making (53)–(54) more precise, one can in (19) justify throwing out those paths which violate (54). One thus obtains

$$u(x,t) \approx \bar{u}(x,t) \tag{55}$$

for $x - \bar{m}^2(t)$ large. The first half of (42) and (43) in both Theorems 2 and 3 follows from (55).

Under (38), one can show that paths $z_{x,y}(s)$ for v spending much time above $m^1(s)$ can be thrown out, as there are few such paths and D is too small to adequately compensate. The remaining paths remain in the region where $u \approx 1$, and therefore

$$h(u,v) \approx h(1,v) \tag{56}$$

for our purposes. The substitution of $h(1,v)$ for $h(u,v)$ in (1) decouples v from u. Reasoning as below (30)–(31), we can obtain the remaining half of Theorem 2.

Estimation of v under (39) requires more work. In effect, the following behavior occurs. For large t, the main contribution in (19) comes from paths which remain above $m^1(s)$ until approximately time γt, appropriate $\gamma \in (0,1)$, and then "head towards" $x \approx m^2(t)$ while quickly falling below $m^1(s)$. The correct choice of γ is that which balances the greater heating exhibited above $m^1(s)$ (D versus $\hat{D}$) against the scarcity of paths which end at x and remain high for so long. Carrying through this reasoning, one can obtain the remaining half of Theorem 3. The reader requesting more detail for the arguments behind Theorems 1–3 is referred to Bramson [**11**].

School of Mathematics, University of Minnesota, Minneapolis MN 55455

REFERENCES

1. R. A. Fisher, *The advance of advantageous genes*, Ann. of Eugenics **7** (1937), 355–369.
2. A. Kolmogorov, I. Petrovsky and N. Piscounov, *Étude de l'équation de la diffusion avec croissance de la quantité de matière et son application à un problème biologique*, Moscow University Bull. Math. **1** (1937), 1–25.
3. H. P. McKean, *Application of Brownian motion to the equation of Kolmogorov-Petrovskii-Piskunov*, Comm. Pure Appl. Math. **28** (1975), 323–331.
4. V. V. Skorohod, *Branching diffusion processes*, Theory of Probability and its Applications **9** (1964), 445–449.
5. M. Bramson, *Convergence of solutions of the Kolmogorov equation to travelling waves*, Memoirs of the AMS **285** (1983), 1–190.
6. K. S. Lau, *On the nonlinear diffusion equation of Kolmogorov, Petrovsky and Piscounov*, J. Diff. Eq. **59** (1985), 44–70.
7. K. Uchiyama, *The behavior of solutions of some nonlinear diffusion equations for large time*, J. Math. Kyoto Univ. **18** (1978), 453–508.
8. Ya. I. Kanel, *The behavior of solutions of the Cauchy problem when the time tends to infinity in the case of quasilinear equations arising in the theory of combustion*, Soviet Math. Dokl. **1** (1960), 533–536.
9. P. C. Fife and J. B. McLeod, *The approach of solutions of nonlinear diffusion equations to travelling wave front solutions*, Arch. Rat. Mech. Anal. **65** (1977), 335–362.
10. M. M. Tang and P. C. Fife, *Propagating fronts for competing species equations with diffusion*, Arch. Rat. Mech. Anal. **73** (1980), 69–77.
11. M. Bramson, *Convergence to travelling waves for certain competition models*, in preparation.

Symmetry Breaking in Semilinear Elliptic Equations with Critical Exponents

CHRIS BUDD AND JOHN NORBURY

1. Introduction.

In this paper we examine the symmetry breaking bifurcations which occur on the radially symmetric solution branches of the following semilinear elliptic equation: for Δ the usual Laplacian operator and $0 < \lambda \in \mathbb{R}$

$$(1.1) \qquad \begin{cases} \Delta v + \lambda(v + v^5) = 0, & \text{for } \underset{\sim}{r} \in B, \\ v = 0, & \text{for } \underset{\sim}{r} \in \partial B. \end{cases}$$

Here B is the unit ball (with boundary ∂B) in the space $\mathbb{R}^3$ and we do not require that any radially symmetric function, $v(\underset{\sim}{r}) \equiv v(r)$ for $r = |\underset{\sim}{r}|$, is positive throughout B. Indeed it is a necessary requirement for the existence of a symmetry breaking bifurcation that $v(r)$ has at least one zero for $r \in (0,1)$.

The set of radially symmetric solutions $v(r)$ of problem (1.1) has recently been the subject of considerable research (see this volume). Rather less is known, however, about solutions which do not have all the symmetries of the ball, and in this paper we shall discuss those solutions of problem (1.1) which bifurcate from branches of radially symmetric solutions.

DEFINITION 1.1. *A radially symmetric solution branch for problem (1.1) (henceforth denoted by Γ) is a connected set of solutions $(\lambda, v) \in \mathbb{R} \times C^2(B)$ such that the function $v(\underset{\sim}{r})$ is invariant under the action of the group O_3 (and thus $v(\underset{\sim}{r}) \equiv v(r)$). A nonsymmetric solution branch for problem (1.1) is any connected set of solutions (λ, v) not possessing this invariance property.*

DEFINITION 1.2. *A symmetry breaking bifurcation point (an SBB) for Γ is any point $((\lambda, v)) \in \Gamma$ which is a limit point of a nonsymmetric solution branch of problem (1.1).*

The symmetry breaking bifurcation points (henceforth SBB's) play a significant role in the study of the stability of the solutions of problem (1.1).

A particular instance of this is the change in solutions if the domain B is perturbed. In general the function $v(\underset{\sim}{r})$ will perturb smoothly, but in a neighborhood of an SBB more complex behavior is possible. Some of this is described in Budd [4] and in Chillingworth [9]. We are further motivated to study the location of SBB's of problem (1.1) if we wish to calculate numerically the solution branches. An effective numerical method for locating the nonsymmetric solution branches is to first compute Γ and then to locate the SBB's. Both of these calculations involve fairly simple systems of ordinary differential equations in r, and details are given in Budd [6]. We may then search for nonsymmetric solutions in a neighborhood of the SBB with further local information being given by a Liapounov-Schmidt reduction of the system.

A necessary condition for the existence of an SBB is that for $(\lambda, v) \in \Gamma$ there is a solution $\bar{\psi}(\underset{\sim}{r})$ of the following linear elliptic boundary value problem:

$$(1.2) \qquad \begin{cases} \Delta\bar{\psi} + \lambda(1+5v^4)\bar{\psi} = 0, & \text{for } \underset{\sim}{r} \in B, \\ \bar{\psi} = 0, & \text{for } \underset{\sim}{r} \in \partial B. \end{cases}$$

DEFINITION 1.3 (SMOLLER AND WASSERMAN [12]). *If problem (1.2) has a solution $\bar{\psi}(\underset{\sim}{r})$ at some point $(\lambda, v) \in \Gamma$ then there is an infinitesimal symmetry breaking bifurcation (henceforth denoted by ISB) at this point.*

The existence of an ISB at $(\lambda, v) \in \Gamma$ is certainly not sufficient for the existence of an SBB at (λ, v) and we shall turn to this in section 4. We now state the main result of this paper.

THEOREM 1.1.

(i) *If for $(\lambda, v) \in \Gamma$ the function $v(r)$ has precisely one zero for $r \in (0,1)$ then there is at least one point $(\lambda, v) \in \Gamma$ at which problem (1.2) has a solution $\bar{\psi}(\underset{\sim}{r})$. Further,*

$$\bar{\psi}(\underset{\sim}{r}) = R(r)Y_{2,m}(\theta, \phi)$$

where $\underset{\sim}{r} = (r, \theta, \phi)$ are the usual spherical coordinates, $Y_{2,m}(\theta, \phi)$ is a second order spherical harmonic and $R(r)$ is positive in $(0,1)$.

(ii) *There is a branch of axisymmetric solutions of problem (1.1) which bifurcates from point $(\lambda, v) \in \Gamma$.*

Our method of proof for this theorem is to use a priori bounds on the function $v(r)$ to establish (i), and then to use an argument based upon topological degree to establish (ii).

A study of the symmetry breaking bifurcations from positive radially symmetric solution branches of problems which include (1.1) has been made by Smoller and Wasserman [**12**]. They show, in particular, that there can be no SBB's from any branch of positive solutions of problem (1.1). We thus restrict our discussion to the study of the non-positive solution branches Γ. To simplify the calculations we employ a rescaling described in Budd and Norbury [**7**] and in Smoller and Wasserman [**12**]; namely we set

$$\underset{\sim}{s} = \lambda^{\frac{1}{2}} \underset{\sim}{r} \text{ and } u(\underset{\sim}{s}) = v(\lambda^{-\frac{1}{2}} \underset{\sim}{s}).$$

The function $u(\underset{\sim}{s})$ then satisfies the following elliptic differential equation problem:

$$(1.3) \qquad \begin{cases} \Delta u + u + u^5 = 0, & \text{for } \underset{\sim}{s} \in B\mu, \\ u = 0, & \text{for } \underset{\sim}{s} \in \partial B\mu; \end{cases}$$

where $B\mu = \{\underset{\sim}{s} \in R^3 : s \equiv |\underset{\sim}{s}| < \mu \equiv \lambda^{\frac{1}{2}}\}$.

The radially symmetric solutions $u(s) \equiv u(\underset{\sim}{s})$ of (1.3) then satisfy the following ordinary differential equation problem:

$$(1.4) \qquad \begin{cases} u_{ss} + \frac{2}{s} u_s + u + u^5 = 0, \text{for } s \in (0, \mu) \\ u_s(0) = 0, \text{ and } u(\mu) = 0, \end{cases}$$

where $s \equiv |\underset{\sim}{s}|$ and $\mu \equiv \lambda^{\frac{1}{2}}$. It proves analytically and numerically convenient to pose (1.4) as an initial value problem. Thus we set $u(0) = N$ and shoot from the origin to find the values of μ at which $u(s)$ vanishes (as a function of N). If μ_j is the j-th zero of $u(s)$ then we say that the corresponding function $v(r)$ lies on the j-th solution branch of problem (1.1).

Similarly we set $\psi(\underset{\sim}{s}) = \psi(\lambda^{-\frac{1}{2}} \underset{\sim}{s})$ so that problem (1.2) transforms to

$$(1.5) \qquad \begin{cases} \Delta \psi + (1 + 5u^4)\psi = 0, & \text{for } \underset{\sim}{s} \in B\mu, \\ \psi = 0, & \text{for } \underset{\sim}{s} \in \partial B\mu. \end{cases}$$

In section 2 we seek solutions $\psi(s)$ of problem (1.5) in the form

$$\psi(\underset{\sim}{s}) = S_\ell(s) Y_{\ell m}(\theta, \phi), \tag{1.6}$$

where $\underset{\sim}{s} = (s, \theta, \phi)$ and $Y_{\ell m}(\theta, \phi)$ is an ℓ-th order spherical harmonic. From the completeness of the spherical harmonics and the smoothness of solutions of (1.5) we note that all such solutions must comprise a linear combination of functions of the form (1.6) for integer values of the parameter ℓ. The forms that this particular solution can take are limited by the following lemma which we shall prove in section 4.

LEMMA 1.2. *Let $(\lambda, v) \in \Gamma$ be such that $\lambda^{\frac{1}{2}} = \mu_j$ is the j-th zero of the solution $u(s)$ of (1.3). Then there is a value $L(j) < \infty$ such that, if $\ell > L(j)$, problem (1.5) has no solution $\phi(\underset{\sim}{s})$ of the form (1.6).*

Smoller and Wasserman [**12**] show that there is no solution of problem (1.5) of the form (1.6) with $\ell = 1$. The particular case of $\ell = 0$ corresponds to the case of bifurcations from radially symmetric solutions of problem (1.1) which do not cause a change in the symmetry of the solution. It is shown in Budd [**5**] that these points must be fold bifurcations of the form described by Jepson and Spence [**11**]. Numerical and asymptotic evidence strongly implies that no such bifurcations occur for solutions of problem (1.1). If an SBB occurs, however, then there must exist a branch of axisymmetric solutions and this is what we will prove exists in section 4. In order to do this, we need to establish inequalities on the zeros of eigenvalues of the solutions of the linearized problem at the SBB. We develop in section 2 some comparison lemmas which we then apply to our problem, using a key a priori bound on $u(s)$ developed in Budd [**3**].

We conclude this introduction by stating some numerical and asymptotic calculations which have been made for problems (1.1) and (1.5) for differing values of ℓ (for details see Budd [**6**]). By solving the initial value problem (1.3) for varying values of $u(0) \equiv N$ we numerically determine $\mu_j(N)$ and a graph of this for $j = 1, 2$ is given in Figure 1. We have indicated the location of the SBB whose existence is proved in Theorem 1.1. (The asymptotic form of this graph is discussed in Budd [**5**] and in McLeod [**13**].)

Figure 1 also shows what happens when we replace (1.1) and (1.2) by the following problem:

$$\begin{cases} \Delta v + \lambda(v + v|v|^{p-1}) = 0 & \text{for } \underset{\sim}{r} \in B \\ v = 0 & \text{for } \underset{\sim}{r} \in \partial B, \end{cases} \tag{1.7}$$

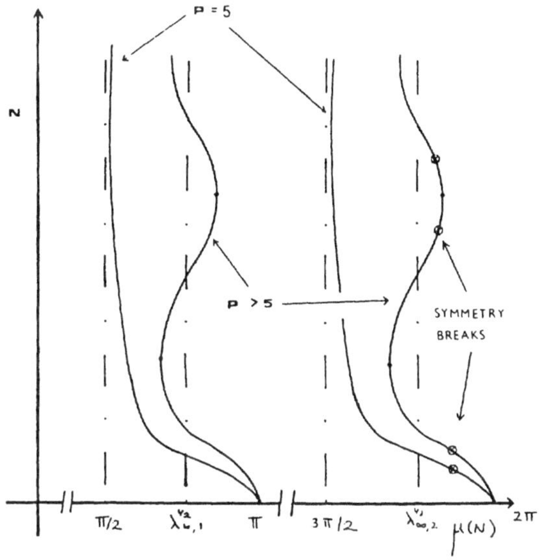

Figure 1

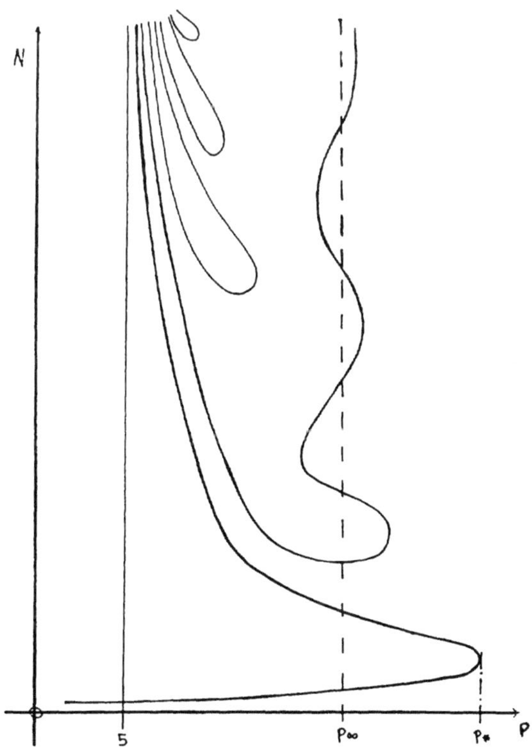

Figure 2

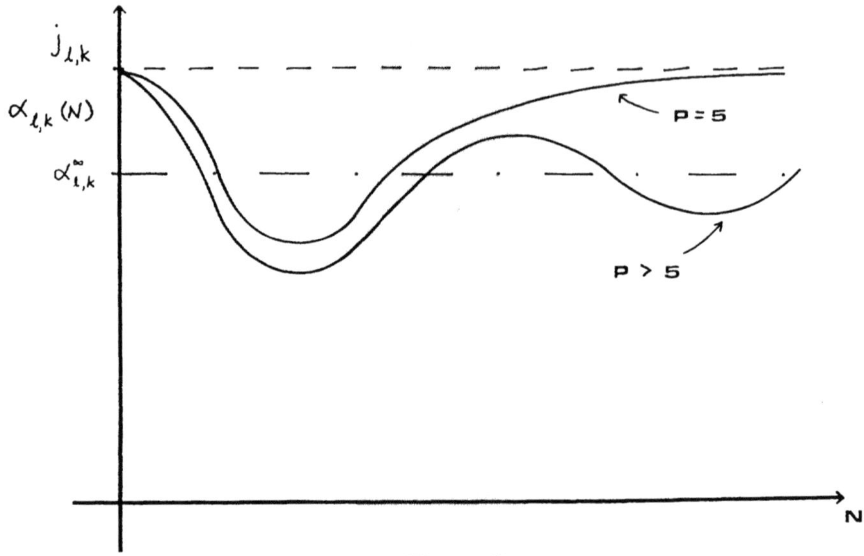

Figure 3

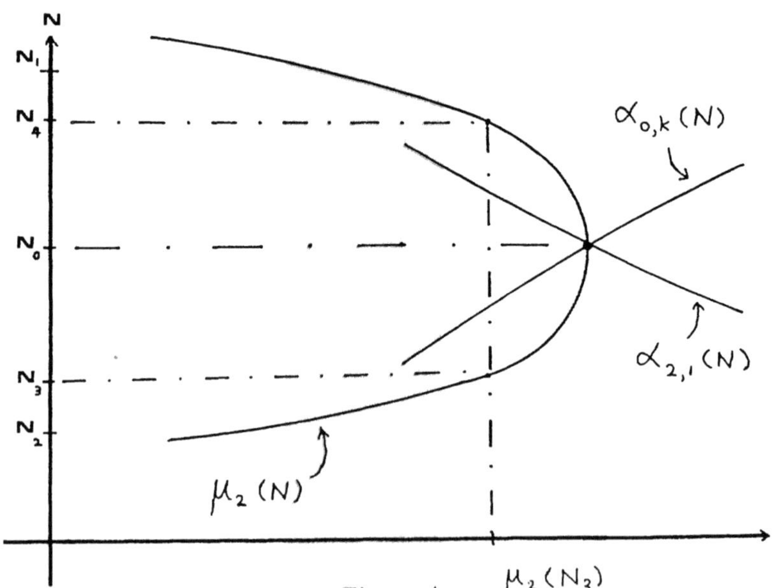

Figure 4

together with

$$(1.8)\qquad \begin{cases} \Delta\bar{\psi} + \lambda(1 + p|v|^{p-1})\bar{\psi} = 0 & \text{for } \underset{\sim}{r} \in B, \\ \bar{\psi} = 0 & \text{for } \underset{\sim}{r} \in \partial B. \end{cases}$$

It is shown in Budd and Norbury [**7**] that if $p > 5$ there are values of $\lambda \equiv \lambda_{\infty,j}$ such that problem (1.7) has an infinite number of solutions and an infinite number of fold bifurcations with a limit point at $\lambda_{\infty,j}$. Numerical calculations for the case $p = 7$ and the location of the SBB for $\ell = 2$ are shown in Figure 1. We note that only the SBB point with the smallest value of N may be obtained from a continuous perturbation in p of the point shown to exist in Theorem 1.1. We may in fact follow the locations of the SBB's as we change the value of p in problem (1.7). Of particular interest is the location of these points as p passes through the value 5 and a graph showing this is given in Figure 2 where we have plotted the value of N at which a bifurcation occurs against p for the particular case of $\ell = 2$. The qualitative form of this graph is consistent with formal asymptotic calculations made in Budd [**4**]. We note that there are a large number of SBB's as p tends to 5 from above, the existence of only one of which has been proved in Theorem 1.1.

2. The existence of infinitesimal symmetry breaking bifurcations.

We establish in this section the existence of a point $(\lambda, u) \in \mathbb{R} \times C^2(B)$ at which there is an ISB on a radially symmetric solution branch Γ of problem (1.1). The full result is stated in Theorem 1.1 (i). First we state a priori bounds on $u(s)$, henceforth defined by (1.3) and $u(0) = N$.

LEMMA 2.1 (BUDD [**3**]).

(i) *For* $0 < s < \mu_1(N)$

$$(2.1)\qquad 0 < u(s) < N(1 + N^4 s^2/3)^{-\frac{1}{2}}.$$

(ii) *If* $s > \mu_1(N)$ *there is a constant* A *independent of* N *such that*

$$(2.2)\qquad |u(s)| < AN^{-1}.$$

These key a priori bounds underline our results in that they enable us to constrast the behavior of zeros as $N \to \infty$ with the familiar behavior as $N \to 0$.

LEMMA 2.2. *Let $R_\ell(s)$ be the solution for $s > 0$ of the differential equation*

$$(R_\ell)_{ss} + 2(R_\ell)_s/s + (1 + 5u^4 - \ell(\ell+1)/s^2)R_\ell = 0 \tag{2.3}$$

which satisfies

$$R_\ell \sim s^\ell \text{ as } s \to 0.$$

Further, let $\alpha_{\ell,k}(N)$ be the k-th zero of $R_\ell(s)$ where $\mu_j(N)$ denotes the j-th zero of $u(s)$.

(i) *$\alpha_{\ell,k}(N)$ and $\mu_j(N)$ are analytic functions of N.*
(ii) *If for some value of the triple (j, k, ℓ) there is a value N such that*

$$\alpha_{\ell,k}(N) = \mu_j(N) \tag{2.4}$$

then there is an ISB for problem (1.1) with $\lambda = \mu_j(N)^2$.

PROOF: The result (i) has been established by Smoller [**14**]. To prove (ii) we examine solutions of (1.5) and substitute

$$\psi(s,\theta,\phi) = S_\ell(s)Y_{\ell m}(\theta,\phi), \tag{2.5}$$

where $Y_{\ell m}(\theta,\phi)$ is a spherical harmonic of degree ℓ. The function $S_\ell(s)$ then satisfies the ordinary differential equation problem (2.3) together with the boundary condition

$$S_\ell(\mu_j(N)) = 0.$$

Thus $R_\ell(s)$ equals $S_\ell(s)$ if the condition (2.4) holds.

The condition (2.4) proves to be a convenient formula for the numerical calculation of an ISB for problem (1.1) and details of this calculation are given in Budd [**6**].

We now determine properties of the curves $\alpha_{\ell,k}(N)$ and $\mu_j(N)$ which allow us to establish whether (2.4) occurs.

LEMMA 2.3. *As $N \to 0$*

$$\alpha_{\ell,k}(N) \to j_{\ell,k} \text{ and } \mu_j(N) \to j\pi \tag{2.6}$$

where $j_{\ell,k}$ is the k-th zero of the Bessel function $J_{\ell+\frac{1}{2}}(s)$.

PROOF: Let $u(s)$ be the solution of (1.4). If we consider the function

$$H(s) = u_s^2 + u^2 + u^6/3$$

then a simple calculation shows that $H(s)$ is monotone decreasing for $s > 0$. Thus $u^2 + u^6/3$ is bounded above by $N^2 + N^6/3$ for all $s > 0$. Each of (2.6) then follows from a simple application of the Sturmian Comparison Theorem to problems (2.3) and (1.5) respectively.

LEMMA 2.4. *As $N \to \infty$*

$$(2.7) \qquad \begin{cases} (i)\ \mu_j(N) \to (j - \frac{1}{2})\pi \text{ and} \\ (ii)\ \text{if } \ell \geq 2 \text{ then } \alpha_{\ell,k}(N) \to j_{\ell,k}. \end{cases}$$

PROOF: The result (i) is proved in McLeod [**13**] and in Budd [**5**]. We return to the proof of (ii) after the following, which establishes Theorem 1.1 (i).

LEMMA 2.5. *Let $j = 2$, $k = 1$, and $\ell = 2$. Then there exists a finite value $N_* > 0$ such that*

$$(2.8) \qquad \alpha_{2,1}(N_*) = \mu_2(N_*)$$

and thus there is an ISB at this point.

PROOF: From Lemmas 2.3 and 2.4 we have

$$\mu_2(N) \to 2\pi \text{ as } N \to 0, \quad \mu_2(N) \to 3\pi/2 \text{ as } N \to \infty,$$

and

$$\alpha_{2,1} \to j_{2,1} = 5.763459\ldots \text{ as } N \to 0 \text{ and } N \to \infty.$$

Since $3\pi/2 < j_{2,1} < 2\pi$ the continuity of $\alpha_{\ell,k}(N)$ and $\mu_k(N)$ as functions of N shows that there is a value N_* such that (2.8) holds.

PROOF OF LEMMA 2.4 (ii): We first prove two comparison Lemmas.

LEMMA 2.6. *Let the function $F(s)$ be continuous for $s \geq 0$ with $F(0) < 0$. Further, let $F(s)$ satisfy the inequality*

$$(2.9) \qquad F(s) > s^2[1 + 5u^4] - \ell(\ell+1).$$

Let $\psi(s)$ be a solution which is regular at the origin of

$$(s^2\psi_s)_s + F(s)\psi = 0. \tag{2.10}$$

Let β_k be the k-th positive zero of $\psi(s)$. Then

$$\beta_k < \alpha_{\ell,k} < j_{\ell,k}. \tag{2.11}$$

PROOF: The left hand inequality in (2.11) follows from an application of the Sturmian Comparison Theorem to the functions $\psi(s)$ and $R_\ell(s)$. (The existence of a suitable function $\psi(s)$ is guaranteed by the behavior of $F(s)$ at the origin.) The right-hand inequality follows from comparing $\psi(s)$ with the function $s^{-\frac{1}{2}}J_{\ell+\frac{1}{2}}(s)$ which satisfies (2.3) with u set to zero.

We now construct a suitable function $F(s)$ for which β_k is forced to $j_{\ell,k}$ as we increase N, thus making $\alpha_{\ell,k}$ do the same.

LEMMA 2.7. *Define the function $F(s) \equiv F(\varepsilon, N; s)$ as follows:*

(i) *If $s \le N^{-\frac{1}{2}}$ then*

$$F(\varepsilon, N; s) = s^2 5N^4(1 + \tfrac{N^4 s^2}{3})^{-2} + \varepsilon - \ell(\ell+1) \tag{2.12}$$

(where we insist that $\varepsilon < \ell(\ell+1)$).

(ii) *If $s > N^{-\frac{1}{2}}$ then*

$$F(\varepsilon, N; s) = s^2[1 + 5N^4(1 + N^3/3)^{-2}] + \varepsilon - N^{-1} - \ell(\ell+1). \tag{2.13}$$

Then $F(\varepsilon, N; s)$ is a continuous function for $s > 0$; further there is a constant B independent of N and ε such that if $N > \varepsilon^{-1} > B$ then the inequality (2.9) is satisfied.

PROOF: We shall first consider the range $s < N^{-\frac{1}{2}}$. If $N^{-\frac{1}{2}} < \pi/2$ then a result of Brezis and Nirenberg [2] shows that the function $u(s)$ is positive in $s \in (0, \pi/2)$. For s in this range we have, from Lemma 2.1, that

$$u^4 < N^4(1 + s^2N^4/3)^{-2}.$$

Further, for $N > \varepsilon^{-1} > 1$

$$1 - \ell(\ell+1)s^{-2} > \varepsilon s^{-2} - \ell(\ell+1)s^{-2},$$

which establishes the inequality (2.9) if $s < N^{-\frac{1}{2}}$. To prove the inequality (2.9) for $N^{-\frac{1}{2}} < s < \mu_1(N)$ we note that $u(s)$ is monotone decreasing in this interval, and is consequently bounded above by $N(1+N^3/3)^{-\frac{1}{2}}$. For $s > \mu_1(N)$ we use Lemma 2.1 (ii) to obtain

$$u^4 < A^4 N^{-4} < N^4(1+N^3/3)^{-2},$$

provided that

$$N^8 > A^4(1+N^3/3)^2$$

which is certainly true if $N > \varepsilon^{-1}$ and ε is taken smaller than $\varepsilon(A)$ where $\varepsilon(A)$ depends only upon the value A above. The continuity of $F(s)$ and $F(0) < 0$ are evident from the definition.

We shall now state Lemma 3.7, which locate the zeros of the function $\psi(s)$ defined by (2.10) where the function $F(\varepsilon, N; s)$ is defined by Lemma 2.7.

LEMMA 3.7. *For $\ell \geq 2$, let $\beta_k(\varepsilon, N)$ be the k-th positive zero of $\psi(s)$. Then there is a function $\delta(\varepsilon) \to 0$ as $\varepsilon \to 0$ and a value $N_0(\varepsilon)$ such that if $N > N_0(\varepsilon)$ then*

$$j_{\ell,k} - \delta(\varepsilon) < \beta_k(\varepsilon, N) < j_{\ell,k}. \tag{2.14}$$

The proof of this lemma is given in section 3.

Hence Lemmas 2.6, 2.7 and the above allow us to deduce Lemma 2.4 (ii) and thus we have proved Lemma 2.5.

The estimate for $\alpha_{\ell,k}(N)$ given by (2.11) and (2.14) appear pessimistic. A sharper bound is suggested by using formal asymptotic techniques (Budd [4]) and is consistent with numerical calculations.

PROPOSITION 2.8. *Let $\alpha_{\ell,k}(N)$ be defined as in Lemma 2.2 and let $\ell \geq 2$. Then there is a constant $A_{\ell,k} > 0$ such that as $N \to \infty$*

$$\alpha_{\ell,k}(N) \sim j_{\ell,k} - A_{\ell,k} N^{-4}. \tag{2.15}$$

The behavior of the function $\alpha_{\ell,k}(N)$ is thus of the form indicated by Figure 3.

It is of interest to contrast this behavior with that of the similarly defined function $\alpha_{\ell,k}(N)$ where we now consider solutions $u(s)$ of the supercritical problem defined in (1.7) where $p > 5$.

PROPOSITION 2.9. *Let $\alpha_{\ell,k}(N)$ be the k-th zero of the function $R_\ell(s)$ where*

$$(1.5)\qquad \begin{aligned}&(R_\ell)_{ss} + 2(R_\ell)_s/s + (1 + p|u|^{p-1} - \ell(\ell+1)s^{-2})R_\ell = 0\\ &R_\ell \sim s^\ell \text{ as } s \to 0 \text{ and } p > 5.\end{aligned}$$

Here $u(s)$ is the function $u(s) \equiv v(\lambda^{-\frac{1}{2}}s)$ and $v(s)$ is defined in (1.7). Then, if $\ell \geq 2$, there is a point $\alpha^\infty_{\ell,k}$ such that

$$(2.17)\qquad \begin{aligned}&(i)\ \left|j_{\ell,k} - \alpha^\infty_{\ell,k}\right| \ll 1 \text{ and}\\ &(ii)\ \alpha_{\ell,k} \sim \alpha^\infty_{\ell,k} + \varepsilon A \sin[\omega \ln N + B] \text{ as } N \to \infty,\end{aligned}$$

where $\varepsilon = N^{-(p-5)/4}$ and A, B, ω are suitable constants.

The asymptotic formula (2.17) is formally established in Budd [**4**] and is consistent with some numerical experiments, in which A is found to be small. We indicate the behavior of $\alpha_{\ell,k}(N)$ in this case in Figure 3.

3. The proof of Lemma 3.7.

This section proves Lemma 3.7. We construct solutions of the differential equation (2.10) which are valid in the regions $s < N^{-\frac{1}{2}}$ and $s > N^{-\frac{1}{2}}$ respectively and then match these solutions at the point $N^{-\frac{1}{2}}$. First we examine the "inner" region $s < N^{-\frac{1}{2}}$.

LEMMA 3.1. *Let the function $\psi(s)$ be the regular solution of (2.10) with the function $F(s) \equiv F(\varepsilon, N; s)$ defined by Lemma 2.7 and (2.12).*

(i) *If $\ell = 1$ and $\varepsilon = 0$ then, up to a multiplicative constant*

$$(3.1)\qquad \psi(s) = -N^5 s / \left[3(1 + N^4 s^2/3)^{\frac{3}{2}}\right].$$

(ii) *If $\ell \geq 2$ then there are constants $\gamma, B > 0$ defined by (3.8) and (3.19) respectively such that if $t \equiv N^2 s \gg 1$ there are function $f(s)$, $g(s)$ bounded independently of s and N with*

$$(3.2)\qquad \begin{cases} (2\gamma + 1)\psi(s) = s^\gamma[B + f(s)N^{-4}s^{-2}],\\ (2\gamma + 1)\psi_s(s) = s^{\gamma-1}[\gamma B + g(s)N^{-4}s^{-2}].\end{cases}$$

PROOF:

(i) The problem

$$w_{ss} + \tfrac{2}{s} w_s + w^5 = 0, \qquad w(0) = N, \quad w_s(0) = 0, \tag{3.3}$$

has the exact solution

$$w(s) = N(1 + N^4 s^2/3)^{-\frac{1}{2}}. \tag{3.4}$$

It follows from the definition of $\psi(s)$ and $F(\varepsilon, N; s)$ that and exact formula for $\psi(s)$ in the case $\varepsilon = 0$ is given by

$$\psi(s) \equiv w_s(s).$$

(ii) To prove this result we first make the substitutions

$$t = N^2 s \text{ and } \tilde{\psi}(t) = \psi(N^{-2}t).$$

The system (2.10), (2.12) is now transformed, for $0 < t < N^{\frac{3}{2}}$, into

$$\tilde{\psi}_{tt} + \tfrac{2}{t}\tilde{\psi}_t + \left(5(1 + t^2/3)^{-2} - (\ell(\ell+1) - \varepsilon)t^{-2}\right)\tilde{\psi} = 0, \tag{3.5}$$

with $\tilde{\psi}(t) \to 0$ as $t \to 0$.

We observe that the equation (3.5) is independent of N, a result crucial to the further development of the theory, since we may take $t \to \infty$ independently of N. It t is either very small or very large then the problem (3.5) is approximated by the equation

$$\hat{\psi}_{tt} + \tfrac{2}{t}\hat{\psi}_t - \tilde{\psi}(\ell(\ell+1) - \varepsilon)t^{-2} = 0. \tag{3.6}$$

This problem has the two independent solutions

$$\hat{\psi}(t) = t^{\gamma} \text{ and } \hat{\psi}(t) = t^{-1-\gamma}, \tag{3.7}$$

where

$$\gamma = -\tfrac{1}{2} + ((\ell + \tfrac{1}{2})^2 - \varepsilon)^{\frac{1}{2}}. \tag{3.8}$$

For t small we shall thus take the regular solution

$$\hat{\psi}(t) \sim t^{\gamma} \text{ as } t \to 0. \tag{3.9}$$

We shall further, in (3.20), construct solutions of problem (3.6) for large t as perturbations of the functions defined in (3.7).

We first prove the Lemmas 3.2 and 3.4, which describe this qualitative behavior of $\tilde{\psi}(t)$ as $t \to \infty$.

LEMMA 3.2. *Let $\tilde{\psi}(t)$ be the solution of problem (3.5) with initial behavior (3.9), and let $\ell \geq 2$. Then*

$$\text{(3.10)} \qquad \begin{aligned} &(i)\ \tilde{\psi}(t) > 0 \text{ for all } t > 0 \text{ and} \\ &(ii)\ \tilde{\psi}(t) \to \infty \text{ as } t \to \infty. \end{aligned}$$

PROOF: If we set $\theta(t) = t\tilde{\psi}(t)$, then $\theta(t)$ satisfies

$$\text{(3.11)} \qquad \theta_{tt} = G(t)\theta \equiv \left[(\ell(\ell+1) - \varepsilon)t^{-2} - 5(1 + t^2/3)^{-2}\right]\theta,$$

and

$$\text{(3.11)} \qquad \theta(t) \sim t^{\gamma+1} \text{ and } t \to 0.$$

A simple estimate shows that

$$\text{(3.12)} \qquad G(t) > [\ell(\ell+1) - \varepsilon - 15/4]t^{-2};$$

that is, $t^2 G(t) > K > 0$ if $\ell \geq 2$ and $\varepsilon < 2$. (This estimate is no longer true if $\ell = 1$.) Thus the function $\theta(t)$ is convex and positive for all $t > 0$. It follows from the Supporting Hyperplane Theorem that there are constants $a > 0$ and b such that

$$\text{(3.13)} \qquad \theta(t) > at + b \text{ for } t > 0.$$

Thus

$$\theta_{tt} > Kat^{-1} + bt^{-2},$$

and for sufficiently large t there is a value $c > 0$ such that

$$\theta(t) > ct \log t.$$

Hence $\tilde{\psi}(t) > c \log t$ and (3.10) follows.

The next lemma now gives an upper bound for $\tilde{\psi}(t)$ valid for all $t > 0$.

LEMMA 3.3. *Let $\varsigma(t) = t^{\gamma}\tilde{\psi}(t)$. Then*

$$\text{(3.14)} \qquad 0 < \varsigma(t) < (2\gamma + 1) \text{ for all } t > 0.$$

PROOF: A simple calculation shows that $\varsigma(t)$ satisfies

$$\text{(3.15)} \qquad \begin{cases} \varsigma_{tt} + 2(\gamma+1)\varsigma_t/t + 5\varsigma(1 + t^2/3)^{-2} = 0, \\ \varsigma(0) = 1, \varsigma_t(0) = 0. \end{cases}$$

Multiplying (3.15) by t and integrating, we see that

$$t\varsigma_t + (2\gamma+1)\varsigma = (2\gamma+1) - 5\int_0^t t\varsigma(1+t^2/3)^{-2}\,dt.$$

Now, the previous lemma implies that $\varsigma(t) > 0$ and thus

$$t\varsigma_t + \varsigma < (2\gamma+1). \tag{3.16}$$

Hence, on integrating (3.16) we see that $t\varsigma < (2\gamma+1)t$ and (3.14) follows.

To proceed further we recast the differential equation (3.15) in an integral form using the Method of Variation of Constants. We see that $\varsigma(t)$ satisfies the integral identity

$$\begin{aligned}(2\gamma+1)\varsigma(t) &= B + Ct^{-1-2\gamma} + \\ &+ t^{-1-2\gamma}\int_0^t 5\varsigma(x)x^{2+2\gamma}(1+x^2/3)^{-2}\,dx + \\ &+ \int_t^\infty 5\varsigma(x)x(1+x^2/3)^{-2}\,dx,\end{aligned} \tag{3.17}$$

where the existence of each integral is guaranteed by the bound (3.14). The constants B and C remain to be determined. It is immediate, from the behavior of $\varsigma(t)$ close to $t=0$, that $C=0$. We now claim that $B \neq 0$. If $B=0$, then substituting the bound (3.14) into (3.17) shows that as $t \to \infty$

$$0 < \varsigma(t) < Kt^{-2},$$

where $K > 0$ is a suitable constant. By repeated substitution of the bounds into the formula (3.17) we show that

$$0 < \varsigma(t) < Lt^{-1-2\gamma} \text{ as } t \to \infty, \tag{3.18}$$

where L is another constant. This bound contradicts (3.10) and hence $B \neq 0$. Thus

$$\lim_{t\to\infty}(2\gamma+1)\varsigma(t) = B = (2\gamma+1) - \int_0^\infty 5\varsigma(x)x(1+x^2/3)^{-2}\,dx. \tag{3.19}$$

LEMMA 3.4. *There are function $f(t)$,$g(t)$ bounded independently of t such that, as $t \to \infty$*

$$\left\{ \begin{array}{ll} (2\gamma+1)\varsigma(t) = B + f(t)t^{-2} \text{ and} \\ (2\gamma+1)\varsigma_t(t) = \qquad g(t)t^{-2}. \end{array} \right. \tag{3.20}$$

PROOF: These estimates follow immediately on substituting (3.14) into (3.17) and into the differential of (3.17) with respect to t.

To complete the proof of Lemma 3.1 we now substitute $\psi(s) = s^\gamma \varsigma(N^2 s)$ into the identity (3.20).

We shall now construct an outer solution of the differential equation (2.10) with $F(\varepsilon, N; s)$ defined by (2.13) and which is valid for $s > N^{-\frac{1}{2}}$. Since equation (2.10) then amounts to a form of the Bessel equation, we can solve it explicitly to obtain the following result.

LEMMA 3.5.

(i) *If $\psi(s)$ is a solution of problem (2.10,13) for $s > N^{-\frac{1}{2}}$ then there are constant P and Q such that*

$$\psi(s) = Ps^{-\frac{1}{2}} J_\nu(\Lambda s) + Qs^{-\frac{1}{2}} J_{-\nu}(\Lambda s), \tag{3.21}$$

where

$$\begin{array}{l} \nu^2 = (\ell + \frac{1}{2})^2 + N^{-1} - \varepsilon \text{ and} \\ \Lambda^2 = 1 + 5N^4(1 + N^3/3)^{-2}. \end{array} \tag{3.22}$$

Here $J_\gamma(s)$ is the ν-th Bessel function normalized such that

$$J_\nu(s) \sim s^\nu \text{ as } s \to 0.$$

(ii) *As $s \to 0$ the function $\psi(s)$ has the form*

$$\psi(s) = \left[P\Lambda^\nu s^{\nu-\frac{1}{2}} + Q\Lambda^{-\nu} s^{-\nu-\frac{1}{2}}\right](1 + 0(s^2)) \tag{3.23}$$

with a similar expression for $\psi_s(s)$.

LEMMA 3.6. *For $N \gg 1$, values of P, Q of the form*

$$\left\{ \begin{array}{l} P = (2\gamma+1)^{-1}[1 + 0(N^{-1})] \text{ and} \\ Q = 0(N^{-1-\gamma}) \end{array} \right. \tag{3.24}$$

ensure that $\tilde{\psi}(s)$ *and* $\tilde{\psi}_s(s)$ *are both continuous at* $s = N^{-\frac{1}{2}}$.

PROOF: Since the function $\tilde{\psi}(s)$ and $s\tilde{\psi}_s(s)$ must be continuous at the point $s = N^{-\frac{1}{2}}$, we evaluate the expressions (3.1,23) at this point. This leads to the following equation for P and Q.

$$\begin{bmatrix} \Lambda^{\nu} N^{-\kappa/2}, & \Lambda^{-\nu} N^{\kappa/2} \\ \Lambda^{\nu}\kappa N^{-\kappa/2}, & \kappa\Lambda^{-\nu} N^{\kappa/2} \end{bmatrix} (1 + 0(N^{-1})) \begin{bmatrix} P \\ Q \end{bmatrix} = \frac{A}{(2\gamma+1)} N^{\gamma/2} \begin{bmatrix} 1 \\ \gamma \end{bmatrix} (1 + 0(N^{-3})). \tag{3.25}$$

Here κ takes the value $\nu - \frac{1}{2}$ and we observe that

$$\Lambda = 1 + 0(N^{-2}) \text{ and } \gamma = \kappa + 0(N^{-1}). \tag{3.26}$$

Thus we solve (3.25,6) to obtain (3.24).

LEMMA 3.7. *Let* $\psi(s)$ *be defined by (3.23) and let* β_k *be the k-th zero of* $\psi(s)$. *Then, there is a* $\delta(\varepsilon) \to 0$ *as* $\varepsilon \to 0$ *such that for* $N > \varepsilon^{-1}$

$$|\beta_k j_{\ell,k}| < \delta(\varepsilon)$$

where $j_{\ell,k}$ *is the k-th positive zero of* $J_{\ell+\frac{1}{2}}(s)$.

PROOF: Standard results from the theory of Bessel function (Abramowitz and Stegun [**1**]) ensure that if N is taken sufficiently large, for example $N > \varepsilon^{-1}$, then there is a value $\delta(\varepsilon) \to 0$ as $\varepsilon \to 0$ such that if $\hat{\beta}_k$ is the k-th positive zero of $J_\gamma(\Lambda s)$ then

$$\left|\hat{\beta}_k - j_{\ell,k}\right| < \delta(\varepsilon)/2.$$

We now observe from (3.24) that $Q/P = 0(N^{-1-\gamma})$ as $N \to \infty$; thus it follows immediately from (3.19) that by taking N sufficiently large

$$\left|\beta_k - \hat{\beta}_k\right| < \delta(\varepsilon)/2,$$

and the result follows.

4. Symmetry breaking.

We finally show that there is a branch of axisymmetric solutions which bifurcates from the ISB point which we found in section 2. Our main tool will be the use of topological degree methods based upon topological features of the curve $\alpha_{\ell,k}(N)$ described in section 2.

In general, the link between actual symmetry breaking bifurcations (SBB's) and ISB's is somewhat complex. It is not clear when symmetry breaking does occur at an ISB and, if so, what symmetries the bifurcating branch or branches possess. These results usually depend upon higher order terms in the local Taylor expansion of the solution of problem (1.1) which are difficult to calculate analytically. For a detailed discussion of this difficulty see Vanderbauwehde [**15**] and Golubitsky and Schaeffer [**10**]. An abstract result is:

THEOREM 4.1 (VANDERBAUWEHDE [**15**]).

(i) *Let the nonlinear operator problem*

$$M(v;\lambda)=0, \tag{4.1}$$

be equivariant under a representation G of the group 0_3. Let (λ_0, v_0) be an ISB such that the kernel K of the linear operator $M_v(v_0,\lambda_0)$ is not empty. Then, an SBB occurs if the following two conditions are satisfied:

$$\begin{cases} \text{(i) the restriction of } G \text{ to } K \text{ is irreducible and} \\ \text{(ii) } M_{\lambda v}(v_0,\lambda_0)\xi \notin \operatorname{Range} M_v(v_0,\lambda_0) \quad \forall \xi\in K. \end{cases} \tag{4.2}$$

(iii) *If a bifurcation does occur then there is a branch of axisymmetric solutions bifurcating from (λ_0, v_0).*

We shall reformulate the abstract conditions (4.2) in terms of the behavior of the zeros that are discussed in section 2. The key condition (4.5) of Theorem 4.1 amounts to saying that a sufficient condition for SBB is that a unique curve of zeros of the eigenfunctions should cross transversally.

In view of the result (iii) above we shall consider, for the remainder of this section, only the solutions of problem (4.1) which are axisymmetric about some fixed axis $\underset{\sim}{L}$. This restriction considerably reduces the complexity of the space K defined above, and typically K is of dimension one. We consider problem (1.1) in the form (4.1) as follows:

$$\begin{cases} M(\lambda, v) \equiv v + \lambda\Delta^{-1}(v + v^5) = 0, \\ v \in \beta \equiv \overset{o}{C}(B) \cap \{\text{Axisymmetric functions about } \underset{\sim}{L}\} \end{cases}$$

there Δ^{-1} is the Greens function for the Laplacian in B. A simple calculation shows that, in the Banach space β

$$M_v\psi \equiv \left[I + \lambda\Delta^{-1}(1 + 5v^4)\right]\psi, \tag{4.4}$$

so that the operator M_v is a compact perturbation of the identity. The structure of M_v is closely related to the properties of the curves $\alpha_{\ell,k}(N)$ described in section 2. Let $v(r)$ be a radially symmetric solution of problem (1.1) and thus of (4.3), and consider the eigenvalue problem

$$M_v\psi = \gamma\psi. \tag{4.5}$$

Courant and Hilbert [8] show that for some $\ell, \psi(r, \theta, \phi)$ has the form

$$\psi(r, \theta) = T(r)P_\ell(\cos\theta) \tag{4.6}$$

where $P_\ell(\cos\theta)$ is the ℓ-th Legendre Polynomial, the appropriate spherical harmonic for the axisymmetric functions. We label γ by $\gamma_{\ell,k}(N)$ where $v(0) = N$ and $T(r)$ has k zeros in the interval $(0, 1]$. The following result interprets Vanderbauwehde's Threorem explicitly for our problem, and is proved in Budd [6].

THEOREM 4.2.

(i) *A necessary condition for an ISB is that* $\gamma_{\ell,k}(N) = 0$ *for some* ℓ *and* k.

(ii) *Let* $v(r)$ *have* j *zeros in the interval* $[0, 1]$. *Let* $\alpha_{\ell,k}(N)$ *and* $\mu_j(N)$ *be defined as in Lemma 2.2. Then*

$$\begin{cases} \gamma_{\ell,k}(N) < 0 \text{ iff } \alpha_{\ell,k}(N) < \mu_j(N), \\ \gamma_{\ell,k}(N) = 0 \text{ iff } \alpha_{\ell,k}(N) = \mu_j(N), \\ \gamma_{\ell,k}(N) > 0 \text{ iff } \alpha_{\ell,k}(N) > \mu_j(N). \end{cases} \tag{4.7}$$

(iii) *Let* N *and* j *be such that* $\gamma_{\ell,k}(N) = 0$ *for some value of the pair* (ℓ, k). *Then the condition (4.2) is satisfied if the pair* (ℓ, k) *is unique and if*

$$d\alpha_{\ell,k}(N)/dN \neq d\mu_j(N)/dN; \tag{4.8}$$

that is, the ISB is an SBB.

In section 2 we established that $\alpha_{2,1}(N) = \mu_2(N)$ for some value of N. To apply Vanderbauwehde's theorem we require the condition (4.8). Because this proves difficult we will instead use ideas from the theory of topological degree. First we prove some results concerning the topology of the curves $\alpha_{\ell,k}(N)$.

LEMMA 4.3.

(i) *If $\ell \leq \ell'$ and $k \leq k'$ then $\alpha_{\ell,k}(N) \leq \alpha_{\ell' k'}(N)$ with equality only if $\ell = \ell'$ and $k = k'$.*

(ii) *$\alpha_{1,k}(N) \neq \mu_j(N)$ for all values of N, j and k.*

(iii) *There exists, for each $j > 0$, a value $L < \infty$ such that if $\ell > L$ or if $k > j$ then*

$$\alpha_{\ell,k}(N) > \mu_j(N)$$

for all $N > 0$.

PROOF: The result (i) is an immediate consequence of the Sturmian Comparison Theorem, and (ii) follows from the observation that $\alpha_{1,k}(N)$ is the k-th positive zero of du/ds. It remains, therefore, to prove (iii). We immedately observe from the Sturmian Comparison Theorem that $\alpha_{\ell,j}(N) > \mu_j(N)$ for all $\ell \geq 0$ and all $N > 0$. We shall now establish the result for L by using the upper bound for $u(s)$ given in Lemma 2.1. We have shown that an upper bound for $u(s)$ is given by the function $H(N,s)$ where

$$H(N,s) = N(1 + N^4 s^2/3)^{-\frac{1}{2}} \text{ for } s < \mu_1(N)$$

and

$$H(N,s) = \min(AN^{-1}, N) \text{ for } s > \mu_1(N).$$

If further follows, after some manipulation, that if ℓ is sufficiently large then

$$1 + 5H^4(N,s) < \ell(\ell+1)s^{-2}$$

for all $N \geq 0$ and for all $s \leq j\pi$. Thus we may deduce that the function $\ell(\ell+1)s^{-2} - (1 + 5u^4(s))$ is positive on the interval $(0, j\pi]$ and thus the function $R_\ell(s)$ defined in (2.3) is also positive in this interval. Hence

$$\mu_j(N) \leq j\pi < \alpha_{\ell,1}(N)$$

and (iii) follows.

COROLLARY 4.4. *Let N_* be such that $\alpha_{2,1}(N_*) = \mu_2(N_*)$. If $\alpha_{\ell,k}(N_*) = \mu_2(N_*)$ then $(\ell,k) = (2,1)$ or $\ell = 0$.*

PROOF: This follows immediately from Lemma 4.3 (i), (ii).

LEMMA 4.5. *There exists at least one point N_0 such that $\alpha_{2,1}(N_0) = \mu_2(N_0)$, and there exists $N_1 > N_0$ such that $N \in (N_0, N_1)$ implies $\mu_2(N) < \alpha_{2,1}(N)$ and $N_2 < N_0$ such that $N \in (N_2, N_0)$ implies that $\alpha_{2,1}(N) < \mu_2(N)$.*

PROOF: We have shown that as $N \to 0$ then $\alpha_{2,1}(N) < \mu_2(N)$ and as $N \to \infty$ then $\mu_2(N) < \alpha_{2,1}(N)$. Let N_{MAX} be defined as

$$N_{\mathrm{MAX}} = \sup\{N : \alpha_{2,1}(N) < \mu_2(N)\} \text{ and}$$
$$N_{\mathrm{MIN}} = \inf\{N : \mu_2(N) < \alpha_{2,1}(N)\}.$$

We claim that either of N_{MAX} and N_{MIN} is a candidate for N_0. To prove this we note that the function

$$e(N) \equiv \mu_2(N) - \alpha_{2,1}(N) \tag{4.9}$$

is an analytic function of N, and consequently there is an integer r such that

$$e^{(r)}(N_{\mathrm{MAX}}) \neq 0 \text{ and } e^{(p)}(N_{\mathrm{MAX}}) = 0 \text{ for } p < r.$$

If r is even it follows that $e(N)$ is one-signed in a neighborhood of N_{MAX}, which is a contradiction. Hence r is odd and the result follows. A similar argument holds for N_{MIN}.

LEMMA 4.6. *Let N_0 be the point constructed in Lemma 4.5 and suppose that $\alpha_{0,k}(N_0) \neq \mu_2(N)$ for any $k > 0$. Then there is an SBB at this point.*

PROOF: It follows from Lemmas 4.3, 4.5 and Corollary 4.4 that there is a neighborhood of N_0 defined by points (N_2, N_1) such that the contradictions of Lemma 4.5 are satisfied and in which there is no value of N such that $\alpha_{\ell,k}(N) = \mu_2(N)$ for any pair (ℓ, k) other than $(2,1)$. Let E be the set of negative eigenvalues of the operator $M_v(\lambda, N)$ defined in (4.4). It follows from (4.7) that the eigenvalue $\gamma_{2,1}(N)$ belongs to E if $N \in (N_0, N_1)$ and not if $N \in (N_2, N_0)$. Further, it follows form the above that if $(\ell,k) \neq (2,1)$ then the eigenvalue $\gamma_{\ell,k}(N)$ is either always positive or always negative in the interval $N \in (N_2, N_1)$. Let $\cap_E$ be the total number of elements of E (this is finite as M_v is a compact perturbation of the identity) where we count the eigenvalues according to their multiplicity. It follows from the above and from the simplicity of the eigenfunction $\psi(\underset{\sim}{r})$ (which we may deduce from Sturm-Liouville theory) that

$$\cap_E(N_2) = \cap_E(N_1) - 1. \tag{4.10}$$

We may now compute the topological index of the radial solution (μ_2^2, v) of the problem (4.1) at the two points N_2 and N_1. The index of this solution is defined in Krasnoselsky [**16**] and we shall denote it by $i(v)$. The next theorem allows us to calculate the value of $i(v)$.

THEOREM 4.7 (KRASNOSELSKY [**16**]). *Let (λ, v) be a solution of problem (4.3) and let $\cap_E$ and E be defined as above. Further, suppose that 0 is not an eigenvalue of the operator $M_v(\lambda, v)$. Then*

$$i(v) = (-1)^{\cap_E}. \tag{4.11}$$

PROOF: See Krasnoselsky [**16**, Chapter 2].

We may immediately deduce from (4.10,11) that the value of $i(v)$ at N_2 differs from its value at N_1. Since the operator $M(\lambda, v)$ changes continuously from N_2 to N_1 it follows from Krasnoselsky [**16**] that there is a branch of (axisymmetric) solutions of problem (4.3) bifurcating from the symmetric solution branch at $v(0) = N_0$.

We have included in the proof of Lemma 4.6 the condition $\alpha_{0,k}(N_0) \neq \mu_2(N_0)$ which amounts to prohibiting fold bifurcations on the solution branch of (4.3). There is strong numerical and asymptotic evidence that there are no fold bifurcations for this problem. However, such bifurcation points do occur for equations very similar to problem (4.3) (such as when $p > 5$), so we examine the possibility that $\alpha_{0,k}(N_0) = \mu_2(N_0)$. First we consider the local behavior of the curve $\alpha_{0,k}(N)$ noting the analyticity of α, μ in N.

LEMMA 4.8. *Let k and j be such that at N_0, $\alpha_{0,k}(N_0) = \mu_j(N_0)$. If*

$$\mu_j(N) = \mu_j(N_0) + A(N - N_0)^r + 0((N - N_0)^{r+1}) \tag{4.12}$$

for some integer $r > 1 (A \neq 0)$, then there is a constant $B \neq 0$ such that

$$\alpha_{0,k}(N_0) = \mu_j(N_0) + B(N - N_0)^{r-1} + 0((N - N_0)^r). \tag{4.13}$$

PROOF: Since $\alpha_{0,k}(N)$ is the k-th positive zero of the function $\psi(s) = \partial u(s)/\partial N$, a necessary condition for bifurcation at N_0 within the spherically symmetric functions is that

$$\partial\mu/\partial N = 0.$$

It follows from differentiating with respect to N the identities $u(\mu(N)) = 0$ and $\psi(\alpha_{0,k}(N)) = 0$ that

$$\partial\mu_j/\partial N u_s(\mu_j(N)) + \psi(\mu_j(N)) = 0 \tag{4.14}$$

and

$$\partial\alpha_{0,k}/\partial N \psi_s(\alpha_{0,k}(N)) + \partial\psi/\partial N(\alpha_{0,k}(N)) = 0. \tag{4.15}$$

If we repeatedly differentiate (4.14) and substitute the result that $\partial^p\mu_j/\partial N^p = 0$ for $p < r$ it follows that

$$\partial^q\psi/\partial N^q(\mu_j(N)) = 0 \text{ for } 0 \leq q < r-1 \tag{4.16}$$

and

$$-\partial^r\mu_j/\partial N^r u_s(\mu_j(N)) = \partial^{r-1}\psi/\partial N^{r-1}. \tag{4.17}$$

Similarly, by repeatedly differentiating (4.15) and substituting (4.16) and (4.17) we may deduce that

$$\partial^q\alpha_{0,k}/\partial N^q = 0 \text{ for } 1 \leq q < r-1$$

and

$$\partial^{r-1}\alpha_{0,k}/\partial N^{r-1} = \partial^r\mu_j/\partial N^r u_s(\mu_j)[\psi_s(\mu_j)]^{-1}.$$

Since neither $u_s(\mu_j)$ nor $\psi_s(\mu_j)$ vanish, (4.13) follows.

We may now complete the proof of Theorem 1.1 (ii).

LEMMA 4.9. *Let N_0 be the point constructed in Lemma 4.5. Then if $\alpha_{0,k}(N_0) = \mu_2(N_0)$ for some k there is an SBB at N_0.*

PROOF: First we assume that the value of r constructed in (4.12) is odd. Then, provided N_1 and N_2 are sufficiently close to N_0, the curve $\alpha_{0,k}(N_0)$ lies wholly to one side of the curve $\mu_e(N)$. Since $\alpha_{2,1}(N)$ does in fact cross $\mu_2(N)$ at N_0 the value of $\cap_E$ changes as we pass from N_2 to N_1 and the proof of Lemma 4.7 applies as before. We now consider the case of even r. Let N_1 and N_2 be constructed as before and taken sufficiently close to N_0 so that (4.12,13) apply in the interval (N_2, N_1). Let N_3 and N_4 be any two values such that $N_2 < N_3 < N_0 < N_4 < N_1$ and that $\mu_2(N_3) = \mu_2(N_4)$. It is evident from Lemma 4.7 and from the estimates (4.12,13) that both of the eigenvalues $\gamma_{0,k}(N)$ and $\gamma_{2,1}(N)$ change sign as N varies from N_3 to N_4 and that no other eigenvalue does so. Hence we may deduce from (4.11) that the value of $i(v)$ at N_3 equals that at N_4 and is nonzero.

We now let

$$U = \{v(\underset{\sim}{r}) \in \overset{o}{C}(B) : N_2 < \sup |v(\underset{\sim}{r})| < N_1\}$$

and calculate the topological degree of the operator $M(\lambda, v)$ defined by (4.3) at the point $\lambda \equiv \mu_2(N_3)^2$. We suppose that the only solutions of problem (4.1) in U are radially symmetric. If so, then the degree of v may be calculated as the sum of the values of $i(v)$ at N_3 and N_4 which, by the above argument is nonzero. We now continuously deform $M(\lambda, v)$ by varying λ in a neighborhood of $\lambda_0 \equiv \mu_2^2(N_0)$ so that the degree remains nonzero, in particular at a value λ_* where

$$(\lambda_0 - \mu_2^2(N_3))(\lambda_* - \lambda_0) > 0.$$

This calculation is illustrated in Figure 4. From the local behavior of $\mu_2(N)$ we deduce that the radially symmetric solution branch remains in U throughout this deformation and ceases to exist as λ passes through $\mu_2^2(N_0)$ and thus the degree of $M(\lambda, v)$ at λ_* is zero. This is a contradiction. Hence, there must be a branch of axisymmetric solutions which bifurcates from the point N_0, and the proof is complete.

□

Oxford University Computing Laboratory, Numerical Analysis Group
8-11 Keble Road, Oxford OX1 3QD, GREAT BRITAIN

Mathematical Institute, 24–29 St. Giles, Oxford OX1 3LB, GREAT BRITAIN

Prof. Budd's research supported by the Central Electricity Generating Board

Research at MSRI supported in part by NSF Grant DMS 812079-05.

REFERENCES

1. M. Abramowitz and I. Stegun, "Handbook of Mathematical Functions," Dover, 1964.
2. H. Brezis and L. Nirenberg, *Positive solutions of nonlinear elliptic equations involving critical Sobolev exponents*, Comm. Pure Appl. Math. **36** (1983), 437–478.
3. C. Budd, *Comparison theorems for radial solutions of semilinear elliptic equations*, to appear in J. Diff. Eqns. (1985).
4. C. Budd, *A semilinear elliptic problem*, Ph.D. thesis (1986).
5. C. Budd, *Semilinear elliptic equations with near critical growth rates*, to appear in Proc. Roy. Soc. Edinburgh (1986).
6. C. Budd, *Symmetry breaking in elliptic systems*, submitted to SIAM Appl. (1986).
7. C. Budd and J. Norbury, *Semilinear elliptic equations and and supercritical growth*, J. Diff. Eqns. **68** (1987), 169–197.
8. R. Courant and D. Hilbert, "Methods of mathematical physics," Wiley-Interscience, 1961.
9. D. Chillingworth, *Notes on some recent methods in bifurcation theory*, Banach Centre Publications **15** (1985), 161–174.
10. M. Golubitsky and D. Schaeffer, "Singularities and groups in bifurcation theory," Vol. 1, Springer-Verlag, 1985.
11. A. Jepson and A. Spence, *Folds in solutions of two parameter systems and their calculation*, SIAM J. Num. Anal. (1984).
12. J. Smoller and A. Wasserman, *Symmetry breaking from positive solutions of semilinear elliptic equations*, in "Proceedings of the 1984 Dundee Conference on Differential Equations," Springer Verlag, 1984.
13. J. B. McLeod, personal communication, 1987.
14. J. Smoller, personal communication, 1986.
15. A. Vanderbauwhede, "Local Bifurcation and Symmetry," Pitman, 1982.
16. M. Krasnoselsky, "Topological Methods in the Theory of Nonlinear Integral Equations," Pergamon, 1964.

Remarks on Saddle Points in the Calculus of Variations

KUNG-CHING CHANG

Existence and regularity of the minima of a functional have been studied extensively in the calculus of variations. It is our purpose here to present an analogy for saddle points. We shall prove two existence theorems for saddle points for certain functionals. The first is based on the Von Neumann-Sion-Ky Fan theorem [**1**], and the second is an application of a local minimax theorem due to S.Z. Shi and the author [**5**].

We study saddle points of the integral

$$J(u,v) = \int_\Omega F(x,u(x),v(x),\nabla u(x),\nabla v(x))dx,$$

where F is a Carathéodory function and u, v are vector-valued functions defined on a domain $\Omega \subset \mathbb{R}^n$.

The structure conditions for the function F are made to assure the existence and regularity of critical points.

The existence and $H^{1,p}_{\text{loc}} \times H^{1,q}_{\text{loc}}$ regularity of a saddle point for the functional J, where F is convex-concave in (u,v), are obtained in §1. Partial regularity of saddle points is studied in §2. The methods used to obtain regularity and partial regularity are based on Giaquinta and Giusti [**3**]. Avoiding the convexity of u, §3 contains an existence theorem based on a generalized Palais-Smale condition.

Nash point equilibria, which are more general than saddle points, have been studied by A. Bensoussan and J. Frehse [**2**]. But both the conditions and methods used here are different from theirs.

§1. Existence and regularity for saddle points.

The first existence theorem is a consequence of a variant of the well-known minimax theorem via weak topology [**5**], namely

THEOREM A. *Suppose that E and F are two closed convex sets of reflexive Banach spaces. We assume that*

(1) $\forall y \in F, x \to f(x,y)$ *is lsc and quasi-convex,*
(2) $\exists y_0 \in F$ *such that* $x \to f(x,y_0)$ *is bounded below and coercive,*
(3) $\forall x \in E, y \to f(x,y)$ *is usc and quasi-concave,*
(4) $\exists x_0 \in E$ *such that* $y \to -f(x_0,y)$ *is bounded below and coercive.*

Then there exists a saddle point $(\bar{x},\bar{y}) \in E \times F$ *such that*

$$f(\bar{x},y) \leqq f(\bar{x},\bar{y}) \leqq f(x,\bar{y}), \qquad \forall x \in E, \quad \forall y \in F.$$

We apply this theorem to variational integrals. Assume that $\Omega \subset \mathbb{R}^n$ is a bounded domain, and that

(I) F is a Carathéodory function defined on $\Omega \times \mathbb{R}^N \times \mathbb{R}^M \times \mathbb{R}^{nN} \times \mathbb{R}^{nM}$.

(II) $\forall (x_0, q_0, Q_0) \in \Omega \times \mathbb{R}^M \times R^{nM}$, the function $(p,P) \to F(x_0,p,q_0,P,Q_0)$, and $\forall (x_0,p_0,P_0) \in \Omega \times \mathbb{R}^N \times \mathbb{R}^{nN}$, the function $(q,Q) \to -F(x_0,p_0,q,P_0,Q)$ are convex.

(III) $\exists m, r > 1$, such that

$$|F(x,p,q,P,Q)| \leqq C\left(1 + |p|^{\bar{m}} + |q|^{\bar{r}} + |P|^m + |Q|^r\right)$$

where C is a constant, and

$$\bar{m} = \tfrac{mn}{n-m} \text{ if } m < n \quad \text{(otherwise no restriction)}$$
$$\bar{r} = \tfrac{rn}{n-r} \text{ if } r < n \quad \text{(otherwise no restriction).}$$

(IV) There exist constants $C, C_1 > 0$, and a function $\chi \in L^1(\Omega, \mathbb{R}^1)$ such that

$$F(x,p,o,P,0) \geqq C\,|P|^m - C_1\,|p|^{\hat{m}} - \chi, \quad \text{with } 1 \leqq \hat{m} < m,$$
$$-F(x,o,q,0,Q) \geqq C\,|Q|^r - C_1\,|q|^{\hat{r}} - \chi, \quad \text{with } 1 \leqq \hat{r} < r.$$

The following theorem on the existence of saddle points now follows directly from Theorem A.

THEOREM 1. *Under the assumptions* I–IV, *the functional*

$$J(u,v) = \int_\Omega F(x,u(x),v(x),\nabla u(x),\nabla v(x))dx$$

defined on $H_0^{1,m}(\Omega,\mathbb{R}^N)\times H_0^{1,r}(\Omega,\mathbb{R}^M)$ *possesses a saddle point* (u_0,v_0).

We now turn to the regularity of the saddle point (u_0,v_0). For $C^{0,\alpha}$-regularity, we may assume without loss of generality that $m,r<n$.

We next replace IV with a stronger assumption:

(IV′) There exist constants $C,C_1,C_2>0$ such that

$$F(x,p,q,P,Q)\geqq C\,|P|^m - C_1\,|Q|^r - C_2(|p|^{\hat m}+|q|^{\bar r}+1),\quad \text{with } 1\leqq \hat m<m,$$
$$-F(x,p,q,P,Q)\geqq C\,|Q|^r - C_1\,|P|^m - C_2(|p|^{\bar m}+|q|^{\hat r}+1),\quad \text{with } 1\leqq \hat r<r.$$

THEOREM 2. *Under the assumptions* I, II, III, *and* IV′, *the saddle point* (u,v) *of* J *belongs to* $H^{1,\ell}_{\mathrm{loc}}(\Omega,\mathbb{R}^N)\times H^{1,s}_{\mathrm{loc}}(\Omega,\mathbb{R}^M)$ *for some* $\ell>m$, $s>r$, *and we have*

$$\fint_{B_R(x_0)}\left(1+|\nabla u|^\ell+|\nabla v|^s+|u|^{\frac{\ell\bar m}{m}}+|v|^{\frac{s\bar r}{r}}\right)dx$$
$$\leqq C\left[\left\{\fint_{B_{2R}(x_0)}(1+|\nabla u|^m+|u|^{\bar m})dx\right\}^{\frac{\ell}{m}}+\left\{\fint_{B_{2R}(x_0)}(1+|\nabla v|^r+|v|^{\bar r})dx\right\}^{\frac{s}{r}}\right]$$

for all $x_0\in\Omega$, *and* $R<\frac12\operatorname{dist}(x_0,\partial\Omega)$.

The proof is quite similar to the correspondent proof for minima (Giaquinta-Giusti [**3**]). We need an extension of the reverse Hölder inequality. Put

$$Q_R(x_0)=\{x=(x_1,\dots,x_n)\in\mathbb{R}^n \mid |x_i-x_{0i}|\leqq R, i=1,\dots,n\},$$

where $x_0=(x_{01},\dots,x_{0n})$, $R>0$.

LEMMA 1. *Let* Q *be an* n*-cube. Suppose that*

$$\fint_{Q_R(x_0)}(g^p+h^q)dx\leqq C\left[(\fint_{Q_{2R}(x_0)}g)^p+(\fint_{Q_{2R}(x_0)}h)^q\right]$$
$$+\fint_{Q_{2R}(x_0)}(f^p+e^q)+\theta\fint_{Q_{2R}(x_0)}(g^p+h^q).$$

holds for each $x_0\in Q$ *and each* $R<\frac12\operatorname{dist}(x_0,\partial\Omega)\wedge R_0$, *where* R_0,C,θ *are constants with* $C>1$, $R_0>0$, $0\leqq\theta<2^{-(p-1)}$. *Then* $g\in L^{\tilde m}_{\mathrm{loc}}(Q)$ *and* $h\in L^{\tilde r}_{\mathrm{loc}}(Q)$ *for some* $\tilde m\in[p,p+\varepsilon)$, $\tilde r\in[q,q+\varepsilon)$, *and*

$$\text{(1.1)}\quad \fint_{Q_R}(g^{\tilde m}+h^{\tilde r})\leqq C_1\left[(\fint_{Q_{2R}}g^p)^{\frac{\tilde m}{p}}+(\fint_{Q_{2R}}h^q)^{\frac{\tilde r}{q}}+\fint_{Q_{2R}}(f^{\tilde m}+e^{\tilde r})\right]$$

for $Q_{2R} \subset Q$ and $R < R_0$, where C_1 and ε are positive constants.

PROOF: One may assume $q > p$. Let $k = h^{q/p}$ and $\ell = e^{q/p}$. Our assumption gives

$$\fint_{Q_R(x_0)} (g^p + k^p) \leqq C\left[(\fint_{Q_{2R}(x_0)} g)^p + (\fint_{Q_{2R}(x_0)} k^{\frac{p}{q}})^q\right] + \fint_{Q_{2R}(x_0)} (f^p + \ell^p) + \theta \fint_{Q_{2R}(x_0)} (g^p + k^p).$$

It then follows that

$$\fint_{Q_R(x_0)} (g+k)^p \leqq C_2\left[\fint_{Q_{2R}(x_0)} (g+k)^p + \fint_{Q_{2R}(x_0)} (f+\ell)^p\right] + 2^{p-1}\theta \fint_{Q_{2R}(x_0)} (g+k)^p.$$

Now we apply the reverse Hölder inequality (c.f. [**3**, Prop. 1.1, Ch. V]), and (1.1) is proved with $\tilde{r} = \tilde{m} \cdot q/p$.

REMARK: Carefully following the proof of Prop. 1.1, Ch. V in [**3**], we may assume $0 \leqq \theta < 1$.

PROOF OF THEOREM 2: For all $x_0 \in \Omega$, with $R < \frac{1}{2}\operatorname{dist}(x_0, \partial\Omega)$, we set

$$u_{x0,R} = \fint_{Q_{2R}(x_0)} u, \quad \text{and} \quad v_{x0,R} = \fint_{Q_{2R}(x_0)} v.$$

Let η be a C_0^∞ function on $Q_{2R}(x_0)$ satisfying

$$0 \leqq \eta \leqq 1, \qquad |\nabla \eta| \leqq \frac{2}{s-t},$$

$\eta = 1$ on $Q_t(x_0)$, $\eta = 0$ outside $Q_s(x_0)$, for $0 < t < s < 2R$. Since (u,v) is a saddle point of J, we have

$$\int_{Q_s} [F(x, \tilde{u}, v, \nabla\tilde{u}, \nabla v) - F(x, u, \tilde{v}, \nabla u, \nabla\tilde{v})]\, dx \geqq 0$$

for $\tilde{u} = (1-\eta)u + \eta u_{x0,R}$ and $\tilde{v} = (1-\eta)v + \eta v_{x0,R}$. According to the assumptions III and IV′, we obtain

$$C\int_{Q_s} (|\nabla u|^m + |\nabla v|^r)\, dx \leqq C_1 \int_{Q_s} (|\nabla\tilde{u}|^m + |\nabla\tilde{v}|^r)\, dx + \\ + C_2 \int_{Q_s} (|\tilde{u}|^{\tilde{m}} + |\tilde{v}|^{\tilde{r}})\, dx + C_2 \int_{Q_s} (|u|^{\hat{m}} + |v|^{\hat{r}})\, dx + 2C_2|Q_s|.$$

By the interpolation inequality, we have

$$\int_{Q_s} \left(|\nabla u|^m + |\nabla v|^r\right) dx \leqq$$
$$\leqq C_3 \left[\int_{Q_s} \left(|\nabla \tilde{u}|^m + |\nabla \tilde{v}|^r + |\tilde{u}|^{\bar{m}} + |\tilde{v}|^{\bar{r}}\right) dx + |Q_s|\right].$$

Therefore

$$\int_{Q_t} \left(|\nabla u|^m + |\nabla v|^r\right) dx \leqq C_3 \Bigg[\int_{Q_s \setminus Q_t} \left(|\nabla u|^m + |\nabla v|^r + |u|^{\bar{m}} + |v|^{\bar{r}}\right) dx +$$
$$\frac{1}{(s-t)^m} \int_{Q_{2R}} |u - u_{x0,R}|^m \, dx + \frac{1}{(s-t)^r} \int_{Q_{2R}} |v - v_{x0,R}|^r \, dx$$
$$+ |Q_r| \left(|u_{x0,R}|^m + |v_{x0,R}|^r + 1\right)\Bigg]$$

and

$$\int_{Q_t} \left(|\nabla u|^m + |\nabla v|^r + |u|^{\bar{m}} + |v|^{\bar{r}}\right) dx$$
$$\leqq \theta \int_{Q_s} \left(|\nabla u|^m + |\nabla v|^r + |u|^{\bar{m}} + |v|^{\bar{r}}\right) dx$$
$$+ \int_{Q_{2R}} \left(|u|^{\bar{m}} + |v|^{\bar{r}}\right) dx$$
$$+ C_4 \Bigg[\frac{1}{(s-t)^m} \int_{Q_{2R}} |u - u_{x0,R}|^m \, dx$$
$$+ \frac{1}{(s-t)^r} \int_{Q_{2R}} |v - v_{x0,R}|^r \, dx + \left(|u_{x0,R}|^{\bar{m}} + |v_{x0,R}|^{\bar{r}} + 1\right) |Q_R|\Bigg].$$

Applying Lemma 3.2 in [**2**], we obtain

$$\int_{Q_R} \left(|\nabla u|^m + |\nabla v|^r + |u|^{\bar{m}} + |v|^{\bar{r}}\right) dx$$
$$\leqq C_5 \Bigg[\frac{1}{R^m} \int_{Q_{2R}} |u - u_{x0,R}|^m \, dx + \frac{1}{R^r} \int_{Q_{2R}} |v - v_{x0,R}|^r \, dx$$
$$+ \int_{Q_{2R}} \left(|u|^{\bar{m}} + |v|^{\bar{r}}\right) dx + |Q_R|\Bigg].$$

The Sobolev-Poincaré inequality implies

$$R^{-m} \fint_{Q_{2R}} |u - u_{x0,R}|^m \, dx \leqq C_6 \left(\fint_{Q_{2R}} |\nabla u|^{\frac{mn}{m+n}} \, dx\right)^{\frac{m+n}{n}}$$

and

$$R^{-r}\fint_{Q_{2R}} |v - v_{x0,R}|^r\,dx \leqq C_6\left(\fint_{Q_{2R}} |\nabla v|^{\frac{rn}{r+n}}\,dx\right)^{\frac{r+n}{n}}.$$

On the other hand,

$$\int_{Q_{2R}} |u|^{\bar m}\,dx \leqq C_7\int_{Q_{2R}} |u - u_{x0,R}|^{\bar m}\,dx + C_7|Q_{2R}|\left|\fint_{Q_{2R}} u\,dx\right|^{\bar m}$$
$$\leqq C_8\left[\left(\int_{Q_{2R}} |\nabla u|^m\,dx\right)^{\frac{\bar m}{m}} + |Q_{2R}|\left(\fint_{Q_{2R}} |u|^{\frac{mn}{m+n}}\,dx\right)^{\frac{m+n}{n}}\right].$$

Similarly,

$$\int_{Q_{2R}} |v|^{\bar r}\,dx \leqq C_8\left[\left(\int_{Q_{2R}} |\nabla v|^r\,dx\right)^{\frac{\bar r}{r}} + |Q_{2R}|\left(\fint_{Q_{2R}} |v|^{\frac{\bar r n}{r+n}}\,dx\right)^{\frac{r+n}{n}}\right].$$

Because of the absolute continuity theorem of Lebesgue, we may choose $0 < \theta < 1$ sufficiently small (and $0 < R < R_0$ small enough) such that

$$\fint_{Q_R} \left(|\nabla u|^m + |\nabla v|^r + |u|^{\bar m} + |v|^{\bar r}\right)\,dx$$
$$\leqq C_9\left\{\left[\fint_{Q_{2R}} \left(|\nabla u|^m + |u|^{\bar m}\right)^{\frac{n}{n+m}}\,dx\right]^{\frac{n+m}{n}}\right.$$
$$\left. + \left[\fint_{Q_{2R}} \left(|\nabla v|^r + |v|^{\bar r}\right)^{\frac{n}{n+r}}\,dx\right]^{\frac{n+r}{n}} + \theta\fint_{Q_{2R}} \left(|\nabla u|^m + |\nabla v|^r\right) + 1\right\}.$$

The proof is completed by an application of Lemma 1 with

$$g = \left(|\nabla u|^m + |u|^{\bar m}\right)^{\frac{n}{n+m}}, \quad h = \left(|\nabla v|^r + |v|^{\bar r}\right)^{\frac{n}{n+r}}, \quad f = e = 0.$$

□

2. Partial Regularity.

In this section, we restrict ourselves to the following special form of F:

$$F(x,p,q,P,Q) = A_{ij}^{\alpha\beta}(x,p,q)P_\alpha^i P_\beta^j - B_{kl}^{\alpha\beta}(x,p,q)Q_\alpha^k Q_\beta^l + g(x,p,q,P,Q),$$

where we have used the convention summation with $\alpha, \beta = 1,\ldots,n$; $i,j = 1,\ldots,N$; $k,l = 1,\ldots,M$, and $n > 2$. We assume that

$$A_{ij}^{\alpha\beta}(x,p,q)\xi_\alpha\xi_\beta\eta^i\eta^j \geqq \lambda|\xi|^2|\eta|^2,$$

and

$$B_{kl}^{\alpha\beta}(x,p,q)\xi_\alpha\xi_\beta\varsigma^k\varsigma^l \geqq \lambda|\xi|^2|\varsigma|^2,$$

for $\lambda > 0$, $\forall\xi \in \mathbb{R}^n$, $\forall\eta \in \mathbb{R}^N$, $\forall\varsigma \in \mathbb{R}^M$, where $A_{ij}^{\alpha\beta}$ and $B_{kl}^{\alpha\beta}$ are uniformly continuous and bounded. We also suppose

$$|g(x,p,q,P,Q)| \leqq C\left[|p|^{\frac{2n}{n-2}} + |q|^{\frac{2n}{n-2}} + (|p|+|q|)^s(|P|+|Q|)^r\right]$$

with $r < 2$, $2 < r+s < 2+\frac{2s}{n}$.

THEOREM 3. *Let $(u,v) \in H_0^{1,2}(\Omega,\mathbb{R}^N) \times H_0^{1,2}(\Omega,\mathbb{R}^M)$ be a saddle point for the functional J. Then there exists an open set $\Omega_0 \subset \Omega$ such that $(u,v) \in C^{0,\alpha}(\Omega_0,\mathbb{R}^N) \times C^{0,\alpha}(\Omega_0,\mathbb{R}^M)$ for every $0 < \alpha < 1$. Moreover we have*

$$\Omega \setminus \Omega_0 = \left\{x_0 \in \Omega \mid \underline{\lim}_{R\to+0} R^{2-n}\int_{B_R(x_0)} \left(|\nabla u|^2 + |\nabla v|^2\right) > \varepsilon_0\right\}$$

where $\varepsilon_0 > 0$ is a constant independent of u and v, and

$$\mathcal{H}^{n-q}(\Omega\setminus\Omega_0) = 0$$

for some $q > 2$.

PROOF: The proof is based on the following estimate, which we shall prove in three steps,

$$\int_{B_\rho(x_0)} (1+|\nabla u|^2+|\nabla v|^2)$$
$$\leqq C\left[(\tfrac{\rho}{R})^n + \omega\left(R^2 + C_1R^{2-n}\int_{B_R(x_0)}(|\nabla u|^2+|\nabla v|^2)\right)^{1-\frac{2}{q}}\right]\cdot$$
$$\cdot\int_{B_{2R}(x_0)}(1+|\nabla u|^2+|\nabla v|^2) + C_2R^n\left(|u_{x0,R}|+|v_{x0,R}|\right)^{\frac{2n}{n-2}} + C_3R^n$$

for each $x_0 \in \Omega$, and for every $\rho < R < \frac{1}{2}\,\mathrm{dist}(x_0,\partial\Omega)$, where $C, C_1, C_2, C_3 > 0$ are constants and where $\omega : \mathbb{R}^+ \to \mathbb{R}^+$ is the continuity modulus of $A_{ij}^{\alpha\beta}$ and $B_{kl}^{\alpha\beta}$. Here ω is increasing, concave, continuous, and satisfies $\omega(0) = 0$, $\omega(t) \leqq M$ and

$$\left|A_{ij}^{\alpha\beta}(x,p,q) - A_{ij}^{\alpha\beta}(x',p',q')\right| \leqq \omega(|x-x'|^2+|p-p'|^2+|q-q'|^2),$$

and

$$\left|B_{kl}^{\alpha\beta}(x,p,q) - B_{kl}^{\alpha\beta}(x',p',q')\right| \leqq \omega(|x-x'|^2+|p-p'|^2+|q-q'|^2).$$

Step 1. Put

$$A^{\alpha\beta}_{ij,0} = A^{\alpha\beta}_{ij}(x_0, u_{x0,R}, v_{x0,R}),$$

and

$$B^{\alpha\beta}_{kl,0} = B^{\alpha\beta}_{kl}(x_0, u_{x0,R}, v_{x0,R}).$$

Let $(\bar{u}, \bar{v})$ be a solution of the following variational problems,

$$\int_{B_R(x_0)} A^{\alpha\beta}_{ij,0} D_\alpha \bar{u}^i D_\beta \bar{u}^j = \min \int_{B_R(x_0)} A^{\alpha\beta}_{ij,0} D_\alpha w^i D_\beta w^j$$
$$w|_{\partial B_R(x_0)} = \bar{u}|_{\partial B_R(x_0)} = u|_{\partial B_R(x_0)}$$

and

$$\int_{B_R(x_0)} B^{\alpha\beta}_{kl,0} D_\alpha \bar{v}^k D_\beta \bar{v}^l = \min \int_{B_R(x_0)} B^{\alpha\beta}_{kl,0} D_\alpha Z^k D_\beta Z^l$$
$$Z|_{\partial B_R(x_0)} = \bar{v}|_{\partial B_R(x_0)} = v|_{\partial B_R(x_0)}.$$

The Legendre-Hadamard condition and Theorem 2.1 of [**3**, Chapter III] imply that there are constants C_4 and $C(p) > 0$ such that

$$\int_{B_\rho(x_0)} (|\nabla \bar{u}|^2 + |\nabla \bar{v}|^2) \leqq C_4 (\tfrac{\rho}{R})^n \int_{B_R(x_0)} (|\nabla \bar{u}|^2 + |\nabla \bar{v}|^2) \tag{2.1}$$

and

$$\int_{B_R(x_0)} (|\nabla \bar{u}|^p + |\nabla \bar{v}|^p) \leqq C_{(p)} \int_{B_R(x_0)} (|\nabla u|^p + |\nabla v|^p) \tag{2.2}$$

for $1 < p < \infty$.

Let $w = u - \bar{u}$, and $Z = v - \bar{v}$. By the Garding inequality and the Legendre-Hadamard condition, we obtain

$$\lambda \int_{B_R(x_0)} (|\nabla w|^2 + |\nabla Z|^2) \leqq \int_{B_R(x_0)} \left(A^{\alpha\beta}_{ij,0} D_\alpha w^i D_\beta w^j + B^{\alpha\beta}_{kl,0} D_\alpha Z^k D_\beta Z^l \right).$$

Since

$$\int_{B_R(x_0)} A^{\alpha\beta}_{ij,0} D_\alpha \bar{u}^i D_\beta w^j = \int_{B_R(x_0)} B^{\alpha\beta}_{kl,0} D_\alpha \bar{v}^k D_\beta Z^l = 0,$$

it follows that

$$\int_{B_R} A_{ij,0}^{\alpha\beta} D_\alpha w^i D_\beta w^j + B_{kl,0}^{\alpha\beta} D_\alpha Z^k D_\beta Z^l$$
$$= \int A_{ij,0}^{\alpha\beta} D_\alpha u^i D_\beta w^j + B_{kl,0}^{\alpha\beta} D_\alpha v^k D_\beta Z^l$$
$$= \int_{B_R} \left[A_{ij,0}^{\alpha\beta} - A_{ij}^{\alpha\beta}(x,u,\bar v)\right] D_\alpha(u^i + \bar u^i) D_\beta w^j$$
$$+ \left[B_{kl,0}^{\alpha\beta} - B_{kl}^{\alpha\beta}(x,\bar u,v)\right] D_\alpha(v^k + \bar v^k) D_\beta Z^l$$
$$+ \int_{B_R} \left[A_{ij}^{\alpha\beta}(x,\bar u,v) - A_{ij}^{\alpha\beta}(x,u,\bar v)\right] D_\alpha \bar u^i D_\beta \bar u^j$$
$$+ \left[B_{kl}^{\alpha\beta}(x,u,\bar v) - B_{kl}^{\alpha\beta}(x,\bar u,v)\right] D_\alpha \bar v^k D_\beta \bar v^l$$
$$+ \int_{B_R} A_{ij}^{\alpha\beta}(x,u,\bar v) D_\alpha u^i D_\beta u^j + B_{kl}^{\alpha\beta}(x,\bar u,v) D_\alpha v^k D_\beta v^l$$
$$- A_{ij}^{\alpha\beta}(x,\bar u,v) D_\alpha \bar u^i D_\beta \bar u^j + B_{kl}^{\alpha\beta}(x,u,\bar v) D_\alpha \bar v^k D_\beta \bar v^l$$
$$= I_1 + I_2 + I_3 + I_4.$$

Now if we extend $(\bar u, \bar v) = (u, v)$ outside the ball $B_R(x_0)$, then

$$I_3 + I_4 = -J(\bar u, v) + J(u, \bar v) - \int_{B_R} [g(x,u,\bar v,\nabla u,\nabla \bar v) - g(x,\bar u,v,\nabla \bar u,\nabla v)]$$
$$\leqq \int_{B_R} [g(x,\bar u,v,\nabla \bar u,\nabla v) - g(x,u,\bar v,\nabla u,\nabla \bar v)] = II,$$

by the saddle point property. Thus

(2.3)
$$\int_{B_R} (|\nabla w|^2 + |\nabla z|^2)$$
$$\leqq C_5 \int_{B_R} \Big[\omega^2\left(R^2 + |u - u_{x0,R}|^2 + |\bar v - v_{x0,r}|^2\right)\left(|\nabla u|^2 + |\nabla \bar u|^2\right)$$
$$+ \omega^2\left(R^2 + |\bar u - u_{x0,R}|^2 + |v - v_{x0,R}|^2\right)\left(|\nabla v|^2 + |\nabla \bar v|^2\right)$$
$$+ \omega(|w|^2 + |z|^2)(|\nabla \bar u|^2 + |\nabla \bar v|^2)\Big] + II$$
$$\leqq C_5 \int_{B_R} \Big[\omega^2\left(R^2 + 2|u - u_{x0,R}|^2 + 2|v - v_{x0,R}|^2 + 2|w|^2 + 2|Z|^2\right)$$
$$\cdot\left(|\nabla u|^2 + |\nabla v|^2 + |\nabla \bar u|^2 + |\nabla \bar v|^2\right)$$
$$+ \omega(|w|^2 + |Z|^2)(|\nabla \bar u|^2 + |\nabla \bar v|^2)\Big] + II$$

Step 2. We next estimate the various terms on the RHS of (2.3). Noticing that ω is bounded, we have for some $q > 2$

$$\begin{aligned}
&\int_{B_R} \omega^2(|\nabla u|^2 + |\nabla v|^2) \\
&\qquad \leqq C_6 \int_{B_R} (|\nabla u|^q + |\nabla v|^q)^{\frac{2}{q}} (\int_{B_R} \omega)^{1-\frac{2}{q}} \\
&\qquad \leqq C_6 \int_{B_{2R}} (1 + |\nabla u|^2 + |\nabla v|^2) \cdot (\fint_{B_R} \omega)^{1-\frac{2}{q}},
\end{aligned}$$

by Theorem 2. By the concavity of ω we have

$$\begin{aligned}
&\fint_{B_R} \omega \left(R^2 + 2|u - u_{x0,R}|^2 + 2|v - v_{x0,R}|^2 + 2|w|^2 + 2|Z|^2\right) \\
&\quad \leqq \omega \left(2 \fint_{B_R} (|u - u_{x0,R}|^2 + |v - v_{x0,R}|^2) + 2 \fint_{B_R} (|w|^2 + |Z|^2) + R^2\right) \\
&\quad \leqq \omega \left(R^2 + C_7 R^{2-n} \int_{B_R} |\nabla u|^2 + |\nabla v|^2\right).
\end{aligned}$$

The last inequality follows from the Poincaré inequality and the L^2 estimate (2.2).

For the same reason, for $q > 2$ we have

$$\begin{aligned}
&\int_{B_R} \omega^2(|\nabla \bar{u}|^2 + |\nabla \bar{v}|^2) \\
&\qquad \leqq C_7 \left(\int_{B_R} (|\nabla \bar{u}|^q + |\nabla \bar{v}|^q)^{\frac{2}{q}}\right)^{\frac{2}{q}} \left(\int_{B_R} \omega\right)^{1-\frac{2}{q}} \\
&\qquad \leqq C_8 \left[\int_{B_R} (|\nabla u|^q + |\nabla v|^q)\right]^{\frac{2}{q}} (\int_{B_R} \omega)^{1-\frac{2}{q}} \\
&\qquad \leqq C_9 \int_{B_{2R}} (1 + |\nabla u|^2 + |\nabla v|^2) (\fint_{B_R} \omega)^{1-\frac{2}{q}}.
\end{aligned}$$

We next estimate

$$\begin{aligned}
II &= \left|\int_{B_R} [g(x, \bar{u}, v, \nabla \bar{u}, \nabla v) - g(x, u, \bar{v}, \nabla u, \nabla \bar{v})]\right| \\
&\leqq C_{10} \int_{B_R} \left(|u|^{\bar{2}} + |v|^{\bar{2}}\right) + \left(|\bar{u}|^{\bar{2}} + |\bar{v}|^{\bar{2}}\right) \\
&+ C_{10} \int_{B_R} (|u| + |v| + |\bar{u}| + |\bar{v}|)^s (|\nabla u| + |\nabla \bar{u}| + |\nabla v| + |\nabla \bar{v}|)^r = II_1 + II_2.
\end{aligned}$$

From the Poincaré inequality it follows that

$$
\begin{aligned}
II_1 &\leqq C_{11}\left[\int_{B_R}\left(|u|^{\bar 2}+|v|^{\bar 2}\right)+\int_{B_R}\left(|w|^{\bar 2}+|Z|^{\bar 2}\right)\right]\\
&\leqq C_{12}\left[\int_{B_R}\left(|u-u_{x0,R}|^{\bar 2}+|v-v_{x0,R}|^{\bar 2}+|w|^{\bar 2}+|Z|^{\bar 2}\right)\right.\\
&\qquad\left.+|B_R|\left(|u_{x0,R}|^{\bar 2}+|v_{x0,R}|^{\bar 2}\right)\right]\\
&\leqq C_{13}\left[\left(\int_{B_R}|\nabla u|^2+|\nabla v|^2+|\nabla w|^2+|\nabla Z|^2\right)^{\frac{\bar 2}{2}}\right.\\
&\qquad\left.+|B_R|\left(|u_{x0,R}|^2+|v_{x0,R}|^2\right)\right]\\
&\leqq C_{14}\left[\left(\int_{B_R}|\nabla u|^2+|\nabla v|^2)^{\frac{\bar 2}{2}}\right)+|B_R|\left(|u_{x0,R}|+|v_{x0,R}|\right)^{\bar 2}\right],
\end{aligned}
$$

and

$$
\begin{aligned}
II_2 &\leqq C_{15}\left(\int_{B_R}(|u|+|v|+|\bar u|+|\bar v|)^{\frac{2s}{2-r}}\right)^{1-\frac{r}{2}}\\
&\quad\cdot\left(\int_{B_R}(|\nabla u|^2+|\nabla v|^2+|\nabla\bar u|^2+|\nabla\bar v|^2)\right)^{\frac{r}{2}}\\
&\leqq C_{16}\left[\int_{B_R}(|u-u_{x0,R}|+|v-v_{x0,R}|+|w|+|Z|)^{\frac{2s}{2-r}}\right.\\
&\quad\left.+|B_R|\left(|u_{x0,R}|+|v_{x0,R}|\right)^{\frac{2s}{2-r}}\right]^{1-\frac{r}{2}}\cdot\left(\int_{B_R}|\nabla u|^2+|\nabla v|^2\right)^{\frac{r}{2}}\\
&\leqq C_{17}\left[\left(\int_{B_R}|\nabla u|^2+|\nabla v|^2)^{\frac{r+s}{2}}\right)+|B_R|\left(1+|u_{x0,R}|+|v_{x0,R}|\right)^{s+r}\right].
\end{aligned}
$$

Finally, we obtain

(2.4)

$$
\begin{aligned}
&\int_{B_R}(|\nabla w|^2+|\nabla z|^2)\leqq C_{18}\left(\int_{B_{2R}}(1+|\nabla u|^2+|\nabla v|^2)\cdot\right.\\
&\quad\left.\left[\omega^{1-\frac{2}{q}}\left(R^2+C_7R^{2-n}\int_{B_R}(|\nabla u|^2+|\nabla v|^2)\right)+\left(\int_{B_R}|\nabla u|^2+|\nabla v|^2\right)^{\frac{2}{n-2}}\right]\right.\\
&\quad+C_{18}R^n\left(|u_{x0,R}|+|v_{x0,R}|\right)^{\frac{2n}{n-2}}+C_{18}R^n.
\end{aligned}
$$

Step 3. Using the relation

$$
(2.5)\quad \int_{B_\rho}(|\nabla u|^2+|\nabla v|^2)\leqq 2\int_{B_\rho}(|\nabla\bar u|^2+|\nabla\bar v|^2)+2\int_{B_\rho}(|\nabla w|^2+|\nabla Z|^2),
$$

together with (2.1) and (2.4), we obtain

$$\int_{B_\rho} (|\nabla u|^2 + |\nabla v|^2) \leqq$$

$$C_{19}\Big[(\tfrac{\rho}{R})^n + \omega^{1-\frac{2}{q}} \Big(R^2 + C_7 R^{2-n} \textstyle\int_{B_R}(|\nabla u|^2 + |\nabla v|^2)\Big)$$

$$+ \int_{B_R} (|\nabla u|^2 + |\nabla v|^2)^{\frac{2}{n-2}}\Big].$$

$$\int_{B_{2R}} (1 + |\nabla u|^2 + |\nabla v|^2) + C_{19}R^n \left(|u_{x0,R}| + |v_{x0,R}|\right)^{\frac{2n}{n-2}} + C_{19}R^n.$$

That is the estimate stated at the beginning. The rest of the proof is essentially the same as Theorem 1.1 in [**3**, Chapter IV] and Theorem 1.1 in [**3**, Chapter VI].

3. A critical point theorem.

The above existence theorem depends heavily on the given convexity-concavity property of the function F. The following theorem in [**5**] avoids the convexity assumption.

THEOREM B. *Let E, F be two closed convex sets of Banach spaces, and let $f \in C^1(E \times F, \mathbb{R}^1)$ be a function satisfying*

(1) *For every $x \in E$, the function $y \to -f(x,y)$ is bounded below and quasi-convex, and satisfies (GPS),*
(2) *There exists a $y_0 \in F$ such that $x \to f(x, y_0)$ is bounded below,*
(3) *(TPS) holds for f.*

Then there exists an element $(\bar{x}, \bar{y}) \in E \times F$ such that

(1) $f(\bar{x},\bar{y}) = \max_{y\in F} f(\bar{x}, y) = \min_{x\in E} \max_{y\in F} f(x,y)$,
(2) $\langle f'_x(\bar{x},\bar{y}), h\rangle \geqq 0$ *for* $h \in E - \bar{x}$.

Moreover, if $x \to f(x,\bar{y})$ is locally pseudo-convex at x, then $(\bar{x},\bar{y})$ is a saddle point of f.

For the conditions (GPS) and (TPS), we refer the reader to [**5**]. Since we are only interested in the case where E and F are Banach spaces, (GPS) is just the Palais-Smale condition, and (TPS) reads as follows:

Any sequence $\{(x_n, y_n)\} \subset E \times F$, along which $f(x_n, y_n) = \mathrm{Sup}_{y\in F} f(x_n, y)$ is bounded and $\|f'_x(x_n, y_n)\|_{E^*} \to 0$, possesses a convergent subsequence.

We note that in case E and F are Banach spaces, even though $(\bar{x}, \bar{y})$ is not a saddle point, it is a critial point of f.

We shall make the following assumptions:

I. The function $F \in C^1(\bar{\Omega} \times \mathsf{R}^n \times \mathsf{R}^M \times \mathsf{R}^{nN} \times \mathsf{R}^{nM})$ satisfies the conditions

$$
\begin{aligned}
|F_P| &\leqq C_1 \left(1 + |P|^{m-1} + |Q|^{\frac{r}{\bar{m}'}}\right) + C_1 \left(|p|^{\frac{\bar{m}}{m'}} + |q|^{\frac{r}{m'}}\right)^s \\
|F_Q| &\leqq C_1 \left(1 + |P|^{\frac{m}{r'}} + |Q|^{r-1}\right) + C_1 \left(|p|^{\frac{\bar{m}}{r'}} + |q|^{\frac{r}{r'}}\right)^s \\
|F_p| &\leqq C_1 \left(1 + |P|^{\frac{m}{\bar{m}'}} + |p|^{\frac{\bar{m}}{\bar{m}'}} + |Q|^{\frac{r}{\bar{m}'}} + |q|^{\frac{r}{\bar{m}'}}\right)^s \\
|F_q| &\leqq C_1 \left(1 + |P|^{\frac{m}{\bar{r}'}} + |p|^{\frac{\bar{m}}{\bar{r}'}}\right)^s + C_1(|Q| + |q|)^\tau,
\end{aligned}
$$

where $0 < s < 1$, $0 < \tau < r - 1$, $m > 1$, $\frac{1}{m} + \frac{1}{m'} = 1$, and $\bar{m}'$, r' and $\bar{r}'$ are defined similarly. And we assume

$$
|F_p(x, p, o, P, O) \leqq C_1 \left(1 + |P|^{\frac{m\hat{s}}{\bar{m}'}} + |p|^{\hat{m}-1}\right),
$$

with $0 < \hat{m} < m$, and $0 < \hat{s} < \frac{\bar{m}'}{m'}$.

II.

$$
\begin{aligned}
&\left(F_P(x, p, q, P, Q) - F_P(x, p, q, \bar{P}, \bar{Q})\right)(P - \bar{P}) \\
-&\left(F_Q(x, p, q, P, Q) - F_Q(x, p, q, \bar{P}, \bar{Q})\right)(Q - \bar{Q}) > 0 \text{ if } (P, Q) \neq (\bar{P}, \bar{Q}).
\end{aligned}
$$

III.

$$
\begin{aligned}
|P|^m &\leqq F_P \cdot P + C_2 \left(|P|^{\hat{m}} + |p|^{\hat{m}} + 1\right) \text{ with } 0 < \hat{m} < m, \\
|Q|^r &\leqq -F_Q \cdot Q + C_2 \left(|P|^{\hat{m}} + |p|^\sigma + |Q|^{\hat{r}} + |q|^{\hat{r}} + 1\right)
\end{aligned}
$$

with $\hat{r} = \tau + 1$ and $\sigma < \frac{\bar{m}}{\bar{m}'}$.

IV. There exists $\theta \in (0, \frac{1}{m})$ such that

$$
\tfrac{1}{r} F_q \cdot q + \theta F_p \cdot p + \tfrac{1}{m} F_P \cdot P + \tfrac{1}{r} F_Q \cdot Q \leqq F + C_3(1 + |p|^{\hat{m}} + |P|^{\hat{m}}),
$$

for large $|p| + |P|$.

V.

$$
(q, Q) \to F(x_0, p_0, q, P_0, Q) \text{ is concave for each } (x_0, p_0, P_0).
$$

LEMMA 2. *Under the assumptions I and III, we have*

$$
\begin{aligned}
&- F(x,p,q,P,Q) \geqq \\
&\tfrac{1}{r}|Q|^r - C_4\left[\left(|P|^{\frac{m}{r'}} + |p|^{\frac{\bar m}{r'}}\right)|q| + |Q|^{\tau+1} + |q|^{\tau+1} + 1\right] - F(x,p,o,P,O).
\end{aligned}
$$

PROOF: Since

$$-F(x,p,q,P,Q) = -\int_0^1 \frac{d}{dt}F(x,p,tq,P,tQ)\,dt - F(x,p,o,P,O)$$

and

$$
\begin{aligned}
&-\int_0^1 \frac{d}{dt}F(x,p,tq,P,tQ)\,dt \geqq \tfrac{1}{r}|Q|^r - C_2\left(|P|^{\hat m} + |p|^{\bar m} + \tfrac{1}{\hat r}|Q|^{\hat r} + \tfrac{1}{\hat r}|q|^{\hat r} + 1\right) \\
&- C_1\left(|P|^{\frac{m}{r'}}|q| + |p|^{\frac{\bar m}{r'}}|q| + \tfrac{1}{\tau}|Q|^{\tau}|q| + \tfrac{1}{\tau}|q|^{\tau+1} + 1\right),
\end{aligned}
$$

the conclusion follows from an interpolation inequality. As an example, we turn to study of the functional

$$J(u,v) = \int_\Omega F(x,u(x),v(x),\nabla u(x),\nabla v(x))\,dx$$

on the space $H_0^{1,m}(\Omega,\mathbb{R}^N) \times H_0^{1,r}(\Omega,\mathbb{R}^M)$.

Verification of (GPS) for $v \to J(u_0,v)$. Let $u_0 \in H_0^{1,m}(\Omega,\mathbb{R}^N)$ be fixed. Assume that $J(u_0,v_k) \to C$ for $\{v_k\} \subset H_0^{1,r}(\Omega,\mathbb{R}^M)$. We shall first prove the $\{v_k\}$ is bounded in $H_0^{1,r}(\Omega,\mathbb{R}^M)$.

In fact, by Lemma 2 we obtain

$$
\begin{aligned}
&\tfrac{1}{r}\int_\Omega |\nabla v_k|^r\,dx \leqq \\
&-J(u_0,v_k) + C_4\int_\Omega \left(|\nabla u_0|^m + |u_0|^{\bar m} + |\nabla v_k|^{\hat r} + |v_k|^{\hat r} + 1\right)dx + J(u_0,0).
\end{aligned}
$$

This implies that $\{v_k\}$ is bounded in $H_0^{1,r}(\Omega,\mathbb{R}^M)$, which in turn implies that

$$
\begin{aligned}
A_Q^k(x) &= F_Q(x,u_0(x),v_k(x),\nabla u_0(x),\nabla v_k(x)) &&\rightharpoonup A_Q \text{ in } L^{r'}(\Omega,\mathbb{R}^{nM}), \\
A_q^k(x) &= F_q(x,u_0(x),v_k(x),\nabla u_0(x),\nabla v_k(x)) &&\rightharpoonup A_q \text{ in } L^{\hat r'}(\Omega,\mathbb{R}^{M}).
\end{aligned}
$$

For the strong convergence of a subsequence of $\{v_k\}$, we refer the reader to [**4**]. Actually, the proof is quite similar to, and considerably simpler than, steps 2 and 3 in the verification of (TPS) below.

Verification of (TPS). Suppose that $\{u_k, v_k\} \subset H_0^{1,m} \times H_0^{1,r}$ is a sequence along which

(1) $J(u_k, v_k)$ is bounded,

(2)

$$\langle J_v'(u_k, v_k), \psi\rangle = \int_\Omega F_{qk}\psi + F_{Qk}\nabla\psi = 0, \quad \forall \psi \in H_0^{1,r}(\Omega, \mathbb{R}^M),$$

and

(3)

$$\langle J_u'(u_k, v_k), \phi\rangle = \int_\Omega F_{pk}\phi + F_{Pk}\nabla\phi = 0(\|\phi\|), \quad \forall \phi \in H_0^{1,m}(\Omega, \mathbb{R}^N),$$

where $F_{q,k} = F_q(x, u_k, v_k, \nabla u_k, \nabla v_k)$. We define $F_{Q,k}$, $F_{p,k}$, and $F_{P,k}$ similarly.

Step 1. $\{u_k, v_k\}$ is bounded. According to assumption III,

$$(\tfrac{1}{m} - \theta)|\nabla u_k|^m \leqq (\tfrac{1}{m} - \theta)F_{Pk}\cdot\nabla u_k + C_2\left(|\nabla u_k|^{\hat{m}} + |u_k|^{\hat{m}} + 1\right).$$

By IV, (2) and (3), we obtain

$$\begin{aligned}
(\tfrac{1}{m} - \theta)\,\|u_k\|_{H_0^{1,m}}^m &\leqq \tfrac{1}{m}\int_\Omega F_{Pk}\cdot\nabla u_k - \theta\int_\Omega F_{Pk}\cdot\nabla u_k \\
&\quad + C_2\int_\Omega \left(|\nabla u_k|^{\hat{m}} + |u_k|^{\hat{m}} + 1\right) \\
&= \tfrac{1}{m}\int_\Omega F_{Pk}\cdot\nabla u_k + \tfrac{1}{r}\int_\Omega (F_{Qk}\cdot\nabla v_k + F_{qk}\cdot v_k) \\
&\quad + \theta\int_\Omega F_{pk}\cdot u_k + 0(\|u_k\|) \\
&\quad + C_2\int_\Omega \left(|\nabla u_k|^{\hat{m}} + |u_k|^{\hat{m}} + 1\right) \\
&\leqq J(u_k, v_k) + C_2\,\|u_k\|_{H_0^{1,m}}^{\hat{m}} + 0(\|u_k\|_{H_0^{1,m}}).
\end{aligned}$$

Thus $\{u_k\}$ is bounded. Again, by assumptions III and I we have

$$\begin{aligned}
\|v_k\|_{H_0^{1,r}}^r &\leqq -\int_\Omega F_{Qk}\cdot\nabla v_k + C_2\int_\Omega \left(|\nabla u_k|^{\hat{m}} + |u_k|^{\sigma} + |\nabla v_k|^{\hat{r}} + |v_k|^{\hat{r}} + 1\right) \\
&= -\int_\Omega F_{qk}\cdot v_k + C_5\,\|v_k\|_{H_0^{1,r}}^{\hat{r}} + C_6 \\
&\leqq c\int_\Omega \left(|\nabla u_k| + |u_k| + |\nabla v_k|^{\tau} + |v_k|^{\tau}\right)|v_k| + C_5\,\|v_k\|^{\hat{r}} + C_6.
\end{aligned}$$

Thus $\{v_k\}$ is also bounded.

Step 2. Almost everywhere convergence of the sequences $\nabla u_k(x)$ and $\nabla v_k(x)$. According to assumption I and the boundedness of the sequences u_k, v_k we see that

$$(u_k, v_k, F_{pk}, F_{qk}, F_{Pk}, F_{Qk}) \rightharpoonup (u^*, v^*, A_p, A_q, A_P, A_Q)$$

in the space $H_0^{1,m} \times H_0^{1,r} \times L^{\bar{m}'} \times L^{\bar{r}'} \times L^{m'} \times L^{r'}$, and that the following two equalities hold,

$$\int_\Omega A_q\psi + A_Q\nabla\psi = 0 \qquad \forall\psi \in H_0^{1,r} \tag{3.1}$$

$$\int_\Omega A_p\phi + A_p\nabla\phi = 0 \qquad \forall\phi \in H_0^{1,m} \tag{3.2}$$

Define

$$P_k(x) = (F_{Pk} - F_{P^\bullet})(\nabla u_k - \nabla u^*) - (F_{Qk} - F_{Q^\bullet})(\nabla v_k - \nabla v^*)$$

where $F_{P^\bullet} = F_P(x, u_k, v_k, \nabla u^*, \nabla v^*)$ and $F_{Q^\bullet} = F_Q(\cdot)$.

We claim that $P_k(x) \to 0$ a.e. Let

$$I_k = \int_\Omega P_k(x)\,dx = I_k^1 + I_k^2 + I_{k'}^3 \tag{3.3}$$

where

$$\begin{aligned} I_k^3 &= \int_\Omega [(F_{Qk} - F_{Q^\bullet})\nabla v^* - (F_{Pk} - F_{P^\bullet})\nabla u^*]\,dx \\ &\to \int_\Omega [(A_Q - F_Q^*)\nabla v^* - (A_P - F_P^*)\nabla u^*]\,dx; \end{aligned} \tag{3.4}$$

here we have put $F_P^* = F_P(x, u^*, v^*, \nabla u^*, \nabla v^*)$ and similarly for F_Q^*. Now (3.5)

$$\begin{aligned} I_k^1 &= \int_\Omega (F_{Pk}\nabla u_k - F_{Qk}\nabla v_k)\,dx \\ &= \int_\Omega (F_{Pk}\nabla u_k + F_{pk}u_k - F_{Qk}\nabla v_k - F_{qk}v_k - F_{pk}u_k + F_{qk}v_k)\,dx \\ &= -\int_\Omega [(F_{pk} - A_p)u^* - (F_{qk} - A_q)v^*]\,dx - \int_\Omega (A_p u^* - A_q v^*)\,dx \\ &\quad - \int_\Omega [F_{pk}(u - u^*) - F_{qk}(v_k - v^*)] + 0(\|u_k\|) \to \int_\Omega (A_q v^* - A_p u^*)\,dx \end{aligned}$$

by the assumption I. Similarly, we have

$$(3.6)\qquad I_k^2 = \int_{\Omega} (F_{Q^*} \nabla v_k - F_{P^*} \nabla u_k)\, dx \to \int_{\Omega} (F_Q^* \nabla v^* - F_P^* \nabla u^*)\, dx.$$

Combining (3.1), (3.2), (3.3), (3,4), (3,5) and (3.6), we obtain $I_k \to 0$. Therefore $P_k(x) \to 0$ a.e. by assumption II. Since $|\nabla u_k|^m$ and $|\nabla v_k|^r$ are summable, it follows that

$$\nabla u_k(x) \to \nabla u^*(x) \text{ and } \nabla v_k(x) \to \nabla v^*(x) \text{ a.e.}$$

Step 3. Equiabsolute continuity of the integrals of $|\nabla u_k|^m$ and $|\nabla v_k|^r$. According to assumption III, we have for each measurable set $\Omega_0 \subset \Omega$

$$\begin{aligned}
&\int_{\Omega_0} (|\nabla u_k|^m + |\nabla v_k|^r)\, dx \\
&\quad \leqq \int_{\Omega_0} \big[(F_{Pk} \nabla u_k - F_{Qk} \nabla v_k) \\
&\qquad + C_2 \big(|\nabla u_k|^{\hat{m}} + |u_k|^{\sigma} + |\nabla v_k|^{\hat{r}} + |v_k|^{\hat{r}} + 1 \big) \big]\, dx \\
&\quad = \int_{\Omega_0} [(F_{Pk} - F_{P^*})(\nabla u_k - \nabla u^*) - (F_{Qk} - F_{Q^*})(\nabla v_k - \nabla v^*)]\, dx \\
&\quad + \int_{\Omega_0} [F_{P^*}(\nabla u_k - \nabla u^*) - F_{Q^*}(\nabla v_k - \nabla v^*)]\, dx \\
&\quad + \int_{\Omega_0} (F_{Pk} \nabla u^* - F_{Qk} \nabla v^*)\, dx + \tfrac{1}{2} \int_{\Omega_0} (|\nabla u_k|^m + |\nabla v_k|^r)\, dx \\
&\qquad + C_6 |\Omega_0| + C_2 \int_{\Omega_0} |u_k|^{\sigma}\, dx.
\end{aligned}$$

Thus

$$\begin{aligned}
\int_{\Omega_0} (|\nabla u_k|^m + |\nabla v_k|^r)\, dx \leqq I_k + \left(\int_{\Omega_0} |\nabla u^*|^m\, dx \right)^{\frac{m-1}{m}} \left(\int_{\Omega_0} |\nabla u_k|^m \right)^{\frac{1}{m}} + \\
+ \left(\int_{\Omega_0} |\nabla v^*|^r \right)^{\frac{1}{m'}} \left(\int_{\Omega_0} |\nabla u_k|^m \right)^{\frac{1}{m}} + \cdots + \int_{\Omega_0} |u_k|^{\sigma}\, dx.
\end{aligned}$$

Since $u_k \to u^* \quad (L^{\sigma}(\Omega, \mathbb{R}^N))$, this proves that for all $\varepsilon > 0$ there exists a $\delta > 0$ such that, for $|\Omega_0| < \delta$,

$$\int_{\Omega_0} (|\nabla u_k|^m + |\nabla v_k|^r)\, dx < \varepsilon.$$

Finally, combining steps 1, 2, and 3, we get by Egorov's theorem

$$\int_\Omega \left(|\nabla u_k - \nabla u^*|^m + |\nabla v_k - \nabla v^*|^r\right) dx \to 0 \qquad \text{as } k \to \infty.$$

It follows directly from Lemma 2 that the functional $v \to J(u_0, v)$ is bounded from above, where $u_0 \in H_0^{1,m}(\Omega, \mathbb{R}^N)$ is arbitrary but fixed. Thus, we arrive at

THEOREM 4. *Under the assumptions I through V in this section, the functional J has a critical point (u_0, v_0). Moreover, if $u \to J(u, v_0)$ is locally pseudo-convex at u_0 then (u_0, v_0) is a saddle point of J.*

PROOF: We only need verify that the functional $J(u, 0)$ is bounded from below. This may be proved as follows. Indeed

$$\begin{aligned} &F(x,p,o,P,O) = \\ &\quad F(x,o,o,O,O) + \int_0^1 \left[F_p(x,tp,o,tP,o)p + F_P(x,tp,o,tP,o)P\right] dt \\ &\qquad \geqq F_0(x) + \tfrac{1}{m}|P|^m - C_{20}\left(\tfrac{1}{\hat m}|P|^{\hat m} + \tfrac{1}{\hat m}|p|^{\hat m} + 1\right) - C_{21}|P|^{\frac{\hat s m}{m'}}|p| \end{aligned}$$

By assumptions I and III, where $F_0(x) = F(x,o,o,o,o)$. Hence

$$\begin{aligned} &\int_\Omega F(x,u,o,\nabla u,o)\,dx \geqq \\ &C_{21} + \tfrac{1}{m}\int_\Omega |\nabla u|^m\,dx - C_{22}\left[\left(\int_\Omega |\nabla u|^m\,dx\right)^{\frac{\hat s}{m'} + \frac{1}{m}} + \int_\Omega |\nabla u|^{\hat m}\,dx\right]. \end{aligned}$$

Department of Mathematics, Peking University
& Nankai Institute of Mathematics

Research at MSRI supported in part by NSF Grant DMS 812079-05.

REFERENCES

1. J.P. Aubin and I. Ekeland, "Applied Nonlinear Analysis," Wiley-Interscience, New York, 1984.
2. A. Bensoussan and J. Frehse, *Nash point equilibria for variational integrals*, in "Nonlinear Analysis and Optimization," Springer Lecture Notes in Math, vol. 1107, C. Vinti, ed., 1984.
3. M. Giaquinta and E. Giusti, *On the regularity of minima of variational integrals*, Acta Math. **148** (1982), 31–46.
4. Y.T. Sheng and Y.H. Deng, *Nontrivial solutions for high order quasilinear elliptic equations*, J. Sys. Sci. & Math. Scis. **5** (1985), 303–312.
5. S.Z. Shi and K.C. Chang, *A local minimax theorem without compactness*, in "Nonlinear and Convex Analysis: Proceedings in Honor of Ky Fan," Marcel Dekker, B. L. Lin, and S. Simons, eds., 1987, pp. 211–234.

On the Elliptic Problem
$\Delta u - |\nabla u|^q + \lambda u^p = 0$

M. CHIPOT AND F.B. WEISSLER

We consider regular solutions of the following elliptic problem,

$$
(1) \qquad \begin{cases} \Delta u - |\nabla u|^q + \lambda u^p = 0, & x \in \Omega \\ u > 0, & x \in \Omega \\ u = 0, & x \in \partial\Omega \end{cases}
$$

where $\Omega \subset \mathbb{R}^n$ is a smooth, bounded domain, $u : \bar{\Omega} \to \mathbb{R}$ is C^2, and q, p, λ are parameters satisfying $q, p > 1$ and $\lambda > 0$.

The motiviation for studying problem (1) comes from a related parabolic problem,

$$
(2) \qquad \begin{cases} v_t = \Delta v - |\nabla v|^q + v^p, & t > 0, \quad x \in \Omega \\ v(0,x) = \varphi(x) \geq 0, & x \in \Omega \\ v(t,x) = 0, & t > 0, \quad x \in \partial\Omega \end{cases}
$$

where now $v = v(t,x)$. In [**1**] we investigate whether there exist initial values φ for which the resulting solution of (2) blows up in finite time. It turns out that a natural candidate for such a φ is a solution of (1). More precisely, assume either

$$
(3) \qquad \begin{cases} 1 < q < 2p/(p+1) \\ n/2 < (p+1)/(p-1) \\ \lambda > 0 \text{ is sufficiently small,} \end{cases}
$$

or

$$
(4) \qquad \begin{cases} q = 2p/(p+1) \\ p \text{ is sufficiently large} \\ 0 < \lambda \leq 2/(p+1). \end{cases}
$$

It follows that if φ is a solution of (1), then the resulting solution of (2) blows up in finite time. (See Theorem 1.2 and Corollary 4.3 in [**1**].)

Thus we are led to investigate under what conditions there exist solutions of (1) and if these solutions are unique. So far we have considered only the ball in $\mathbb{R}^n$:

$$\Omega = B_R = \{x \in \mathbb{R}^n : |x| < R\}.$$

In this case, the techniques of Gidas, Ni, and Nirenberg [**2**] show that every solution of (1) is radially symmetric. The following theorems summarize what we know at this time.

THEOREM 1. *Suppose $1 < q < 2p/(p+1)$ and $n/2 < (p+1)/(p-1)$. Then for every $\lambda > 0$ and every $R > 0$, there is a solution of (1) on $\Omega = B_R$. If in addition $n = 1$, this solution is unique.*

THEOREM 2. *Suppose $q = 2p/(p+1)$. Let*

$$\lambda_p = \frac{(2p)^p}{(p+1)^{2p+1}}.$$

If $\lambda \leq \lambda_p$, there is no solution of (1) on $\Omega = B_R$. If $n = 1$ and $\lambda > \lambda_p$, then for each $R > 0$ there is a unique solution of (1) on $\Omega = B_R$. In the remaining cases, i.e. $n \geq 2$ and $\lambda > \lambda_p$, for a fixed λ a solution of (1) on B_R exists either for all $R > 0$ or for no $R > 0$. Furthermore, any such solution is necessarily unique.

THEOREM 3. *Suppose $q > 2p/(p+1)$ and $n/2 < (p+1)/(p-1)$. Then for all sufficiently large R, perhaps depending on λ, there is a solution of (1) on $\Omega = B_R$. If in addition, $q > p$, then for all sufficiently large R, perhaps depending on λ, there are at least two distinct solutions of (1) on $\Omega = B_R$.*

THEOREM 4. *Suppose $q \geq p$. Then for all sufficiently small R, perhaps depending on λ, there is no solution of (1) on $\Omega = B_R$.*

The proofs of these theorems are contained in Sections 4 and 5 of [**1**]. Consequently, here we will just sketch some of the main ingredients.

By abuse of notation, we write $u(r)$ for $u(x)$ in the radially symmetric case, where $r = |x|$. Thus, we are led to consider the initial value problem,

$$(5) \qquad \begin{cases} u'' + (\frac{n-1}{r})u' - |u'|^q + \lambda|u|^p = 0, \quad r > 0 \\ u(0) = a > 0 \\ u'(0) = 0. \end{cases}$$

For solutions of (5), one can check readily that $u'(r) < 0$ for $r > 0$; and so $u(r)$ has at most one zero. For each $a > 0$, we let $z(a)$ denote this zero of the corresponding solution of (5) if it exists, and $z(a) = \infty$ if $u(r)$ has no zero. Consequently, we have a solution of (1) on $\Omega = B_R$ precisely when there exists $a > 0$ with $z(a) = R$.

The first main technique is a scaling argument, used earlier in [**3**]. We start with a fixed $\lambda > 0$. Let $u(r;a)$ denote the solution of (5) with initial value a, and set

$$v_a(r) = a^{-1}u\big(ra^{-(p-1)/2};a\big).$$

Then v_a satisfies

$$\text{(6)} \qquad \begin{cases} v_a'' + (\frac{n-1}{r})v_a' - a^{q(p+1)/2-p}|v_a'|^q + \lambda|v_a|^p = 0 \\ v_a(0) = 1 \\ v_a'(0) = 0. \end{cases}$$

Moreover, one can show that $v_a(r)$ and $v_a'(r)$ are uniformly bounded on $\{r : v_a(r) \geq 0\}$, independent of $a > 0$.

If now $q < 2p/(p+1)$, we let $a \to \infty$; and if $q > 2p/(p+1)$, we let $a \to 0$. Suppose as a approaches the limit, that $v_a(r) > 0$ for all $r > 0$. Then $v_a(r) \to v(r)$ uniformly on compact sets, where $v : [0,\infty) \to \mathbb{R}$ is a positive solution of

$$\text{(7)} \qquad \begin{cases} v'' + (\frac{n-1}{r})v' + \lambda v^p = 0 \\ v(0) = 1 \\ v'(0) = 0. \end{cases}$$

If $n/2 < (p+1)/(p-1)$, we get a contradiction since no such solution of (7) exists. (See pp. 293–294 of [**4**].) In other words, if $q < 2p/(p+1)$ with $n/2 < (p+1)/(p-1)$, then $z(a) < \infty$ for sufficiently large a; and if $q > 2p/(p+1)$ with $n/2 < (p+1)/(p-1)$, then $z(a) < \infty$ for sufficiently small a. This shows the existence of solutions to (1) on $\Omega = B_R$ at least for some values of $R > 0$.

There are two estimates for $z(a)$ which are very helpful:

$$\text{(8)} \qquad z(a) \geq Ca^{-(p-1)/2},$$

$$\text{(9)} \qquad z(a) \geq Ca^{1-(p/q)}.$$

To prove (8), we observe first that

$$\frac{d}{dr}\left[\frac{u'(r)^2}{2}+\frac{\lambda}{p+1}|u(r)|^p u(r)\right]=u'(r)\left[-(\frac{n-1}{r})u'(r)+|u'(r)|^q\right]<0.$$

Thus, as long as $u(r)\geq 0$,

$$\frac{1}{2}u'(r)^2\leq\frac{\lambda}{p+1}a^{p+1}.$$

Consequently, if $z(a)<\infty$,

$$\begin{aligned}a&=u(0)-u(z(a))\\&=-\int_0^{z(a)}u'(r)dr\\&\leq z(a)a^{(p+1)/2}\sqrt{2\lambda/(p+1)},\end{aligned}$$

thereby proving (8).

To prove (9), we observe first that (if $z(a)<\infty$) the maximum value of $-u'(r)$ on $[0,z(a)]$ occurs at $r=r_0$ in the interior. This follows from calculating $u''(0)$ and $u''(z(a))$ using (5). Thus $u''(r_0)=0$, and so for all $r\in[0,z(a)]$

$$\begin{aligned}(-u'(r))^q&\leq(-u'(r_0))^q\\&=\lambda u(r_0)^p+(\frac{n-1}{r_0})u'(r_0)\\&\leq\lambda a^p.\end{aligned}$$

So again we have

$$\begin{aligned}a&=u(0)-u(z(a))\\&=-\int_0^{z(a)}u'(r)dr\\&\leq z(a)\lambda^{1/q}a^{p/q},\end{aligned}$$

thereby proving (9).

In case $q<2p/(p+1)$ and $n/2<(p+1)/(p-1)$ we know from the scaling argument that $z(a)<\infty$ for a large. Indeed, a more careful analysis of that argument shows $z(a)\to 0$ as $a\to\infty$. On the other hand, (8) implies $z(a)\to\infty$ as $a\to 0$. By continuity of $z(a)$, it follows that for all $R>0$, there is an $a > 0$ with $z(a) = R$, i.e. a solution of (1) on $\Omega = B_R$. This is most of Theorem 1.

In case $q > 2p/(p+1)$ and $n/2 < (p+1)/(p-1)$ we saw above that $z(a) < \infty$ for some $a > 0$. Also, we still have by (8) that $z(a) \to \infty$ as $a \to 0$. Thus, (1) has a solution on $\Omega = B_R$ for all sufficiently large R. Furthermore, if in addition $q > p$, then (9) implies $z(a) \to \infty$ as $a \to \infty$. Hence, for R sufficiently large, there are at least two values of a with $z(a) = R$. This proves Theorem 3.

If $q \geq p$, then (8) and (9) together imply that $z(a)$ is bounded below. This proves Theorem 4.

The borderline case $q = 2p/(p+1)$ is more delicate. In this case (6) is independent of $a > 0$, and so $v_a(r) = u(r;1)$ for all $a > 0$. It follows that $z(a) = a^{-(p-1)/2} z(1)$. This proves uniqueness of solutions on $\Omega = B_R$, as well as the fact that for fixed $\lambda > 0$, a solution to (1) exists either on all B_R or on no B_R.

If $n = 1$ and $q = 2p/(p+1)$, there is a singular solution

$$U(r) = kr^{-2/(p-1)}$$

of the first line in (5) precisely when $\lambda < \lambda_p$. A translation argument then shows that $z(a)$ can never be finite if $\lambda \leq \lambda_p$.

To show existence of a (regular) solution to (1) in case $q = 2p/(p+1), n = 1$, and $\lambda > \lambda_p$, we argue by contradiction. Suppose $z(a) = \infty$. Then one can easily see that $u(r), u'(r)$, and $u''(r)$ all tend to zero as $r \to \infty$, with $u(r) > 0$ and $u'(r) < 0$ for all $r > 0$. Thus, by (5) with $n = 1$,

$$u'' = |u'|^q - \lambda u^p \leq (-u')^q, \tag{10}$$

or

$$u''(-u')^{1-q} \leq -u'.$$

Integrating from r to ∞ yields

$$(-u'(r))^{2-q} \leq (2-q)u(r).$$

Putting this into (5), with $q = 2p/(p+1)$ and $n = 1$, we see that

$$u'' = (-u')^q - \lambda u^p \leq \left[1 - \lambda(\frac{p+1}{2})^p\right](-u')^q. \tag{11}$$

In particular, if $\lambda \geq [2/(p+1)]^p$, then $u''(r) \leq 0$ for all $r > 0$. This is clearly impossible since $u(r) > 0$ and $u'(r) < 0$ for all $r > 0$.

Therefore, if $\lambda \geq [2/(p+1)]^p$, then $z(a) < \infty$ for all $a > 0$, i.e. (1) has a solution on every ball $\Omega = B_R$. To obtain this result for all $\lambda > \lambda_p$, we observe first that (11) is an improvement over (10). This allows the previous argument to be iterated indefinitely. We refer the reader to [**1**], Lemma 4.8 through Proposition 4.11, for the details.

Département de Mathématique et d'Informatique
Université de Metz, 57045 Metz Cedex, FRANCE

Department of Mathematics, Texas A&M University, College Station, TX 77843, USA

The second author is partially supported by NSF Grant DMS 8201639.

Research at MSRI supported in part by NSF Grant DMS 812079-05.

REFERENCES

1. M. Chipot and F.B. Weissler, *Some blow-up results for a nonlinear parabolic equation with a gradient term*, preprint.
2. B. Gidas, W.-M. Ni, and L. Nirenberg, *Symmetry and related properties via the maximum principle*, Commun. Math. Phys. **68** (1969), 209–243.
3. A. Haraux and F.B. Weissler, *Nonuniqueness for a semilinear initial value problem*, Ind. Univ. Math. J. **31** (1982), 167–189.
4. F.B. Weissler, *An L^∞ blow-up estimate for a nonlinear heat equation*, Commun. Pure and Appl. Math. **38** (1985), 291–295.

Nonlinear elliptic boundary value problems: Lyusternik-Schnirelman theory, nodal properties and Morse index

CHARLES V. COFFMAN

1. Introduction.

The Lyusternik-Schnirelman theory in the calculus of variations treats isoperimetric problems for even functionals on a Banach space X. A principal application is to nonlinear elliptic eigenvalue problems of the form

$$(1.1) \qquad Ly = \lambda f(x,y) \text{ in } \Omega, \quad y = 0 \text{ on } \partial\Omega,$$

where L is a second order symmetric uniformly elliptic operator (we restrict the discussion to the second order case for simplicity) and

$$(1.2) \qquad f(x,-y) = -f(x,y) \quad (x,y) \in \Omega \times \mathsf{R}.$$

The analysis proceeds by finding stationary values of the functional

$$(1.3) \qquad a(y) = \int_\Omega \int_0^{y(x)} f(x,t)\, dt\, dx,$$

subject to

$$(1.4) \qquad b(y) = \int_\Omega y\, Ly\, dx = \text{const.} > 0$$

(alternatively one can find stationary values of b subject to $a =$ const. but below we shall confine our attention to the first-mentioned problem).

In the linear case of (1.1), where $f(x,y) = p(x)y$, there are two well-known variational principles for determining the eigenvalues $0 < \lambda_1 \leq \lambda_2 \leq \lambda_3 \leq \ldots$ of the problem, the *Poincaré principle*

$$(1.5) \qquad \lambda_n^{-1} = \max_{\dim M \geq n} \min\{a(y) : y \in M, \quad b(y) = 1\}$$

and the *Courant-Weyl principle*

$$(1.6) \qquad \lambda_n^{-1} = \min_{\text{co-dim}\, M \leq n-1} \max\{a(y) : y \in M, \quad b(y) = 1\};$$

in (1.5) and (1.6) M denotes a closed subspace of the underlying Banach space X and co-dim M is the co-dimension relative to X. For the validity of these principles it is necessary that the problem be positive definite. The Lyusternik-Schnirelman theory is based on a generalization of the Poincaré principle. The right side of (1.5) can be written as

$$\max_{K \in \mathcal{M}_n} \min\{a(y) : y \in K\},$$

where $\mathcal{M}_n$ denotes the class of sets of the form

$$\{y \in M : b(y) = 1\}, \quad \dim M \geq n.$$

In the generalization of (1.5) the class $\mathcal{M}_n$ is replaced by a larger class $\mathcal{M}'_n$ and the resulting principle gives the critical values of $a(y)$ subject to $b(y) = \text{const.}$; in the general nonlinear case these critical values bear no simple relation to the corresponding eigenvalues.

This generalization to nonlinear eigenvalue problems of the Poincaré principle gives rise to a monotone sequence of critical values of the corresponding isoperimetric problem. In the specific case of the generalization of (1.5) to treat the nonlinear case of (1.1) this sequence is non-increasing. The values of $\{c_n\}$, $(c_1 \geq c_2 \geq c_3 \geq \ldots)$ given by this principle will be referred to as the *Lyusternik-Schnirelman critical values* and the n-th term c_n will be referred to as the n-th *Lyusternik-Schnirelman critical value;* the set

$$\{y : a(y) = c_n, \quad b(y) = 1\}$$

will be referred to as the n-th critical level and the index n will be referred to as the *order* of the critical value c_n or of the corresponding critical level. (It should be pointed out that the Lyusternik-Schnirelman critical values need not include all of the critical values of the problem and that the choice referred to above of the classes of sets $\mathcal{M}'_n$ is nonunique and the resulting values can be dependent upon this choice.)

The main problem with which we are concerned here is that of establishing some correlation between properties of eigenfunctions of (1.1) that are obtained from the Lyusternik-Schnirelman principle and the order of the critical values to which they correspond. A result of this type in the linear theory is the nodal theorem for the 2nd order Sturm-Liouville problem with Dirichlet boundary conditions. A quite complete generalization of this to the corresponding nonlinear problem can be obtained for certain special types of nonlinearities and these results will be described below. More generally, for second order problems in higher dimensions, positivity of the first eigenfunction remains valid in the nonlinear case and likewise the Courant nodal line theorem, see e.g. [**9**], generalizes, at least in the case of superlinear problems. We also discuss the correlation of the (generalized) Morse indices of critical points with the order of the critical level on which they lie; the specialization of such a result to the linear case of course is trivial. In connection with the investigation of the Morse index we were led also to consider the formulation of a principle complementary to the Lyusternik-Schnirelman principle and which could be regarded as a generalization of the principle (1.6). In this we have followed certain ideas of Heinz [**10**]. Also these considerations suggested that in the formulation of the Lyusternik-Schnirelman principle the "genus", a topological invariant introduced by Krasnosel'skii [**14**], be replaced by a similar invariant (due to C.-T. Yang [**22**]) which possessed certain desirable properties which the genus did not. The same replacement was made earlier by Faddell and Rabinowitz [**8**], to take advantage of another property of this invariant that is not shared by the genus.

2. The Variational Principle.

The formulation of the variational principle that generalizes (1.5) employs an integer-valued topological invariant; we shall refer to any member of the class of invariants having the requisite properties as a "genus". Let X be a Banach space, a subset K of X is said to be *balanced* if $x \in X$ implies $-x \in X$, $B(X)$ will denote the class of closed balanced subsets of $X \setminus \{0\}$, and for $K_1, K_2 \in B(X)$, $S(K_1, K_2)$ will denote the class of odd continuous mappings $f : K_1 \to K_2$. A *genus* is a function $\tau : B(X) \to \{0, 1, 2, \dots\} \cup \{\infty\}$ with the following properties; here K, K_1, K_2 denote sets in $B(X)$.

(1) If $S(K_1, K_2) \neq \phi$, in particular if $K_1 \subseteq K_2$, then $\tau(K_1) \leq \tau(K_2)$.
(2) $\tau(K_1 \cup K_2) \leq \tau(K_1) + \tau(K_2)$.

(3) If K is compact then $\tau(K) < \infty$ and there is a neighborhood U of K with $\bar{U} \in B(X)$ and

$$\tau(\bar{U}) = \tau(K).$$

(4) $\tau(\phi) = 0, \quad \tau(S^n) = n+1, \quad n = 0,1,2,\ldots.$

One example of a genus is that introduced by Krasnosel'skii [**14,15**], and which can be defined as follows:

$$\tau(\phi) = 0, \quad \tau(K) = \inf\{n : S(K, S^{n-1}) \neq \phi\}, \quad K \neq \phi, \tag{2.1}$$

for $K \in B(X)$; see [**2**].

REMARK: If the class of all mappings $\tau : B(X) \to \{0,1,2,\ldots\} \cup \{\infty\}$ having properties 1 and 4 is given the natural ordering, then the Krasnosel'skii genus is the greatest element in that class.

We now state the main result of the Lyusternik-Schnirelman theory as it applies to the problem (1.1). We make no attempt here to achieve maximum generality. We shall assume that Ω is a bounded region in R^N. We assume also that L is as above and that there is a constant $k > 0$ such that

$$k^{-2} \|y\|^2_{W_0^{1,2}(\Omega)} \leq \int_\Omega y\, Ly\, dx \leq k^2 \|y\|^2_{W_0^{1,2}(\Omega)},$$

for all $y \in W_0^{1,2}(\Omega)$. We assume that $f(x,y)$ is continuous on $\Omega \times \mathsf{R}$ and in addition to (1.2) satisfies

$$f(x,y_1) > f(x,y_2) > 0 \text{ for } y_1 > y_2 > 0, \tag{2.2}$$

and if $N \geq 2$

$$|f(x,y)| \leq C_1|y|^\alpha + C_2, \text{ for } (x,y) \in \Omega \times \mathsf{R}, \tag{2.3}$$

where

$$0 < \alpha < (N+2)/(N-2). \tag{2.4}$$

We now formulate the variational principle. Without loss of generality we can take the side condition to be $b(y) = 1$; we put

$$S = \{y \in W_0^{1,2}(\Omega) : b(y) = 1\},$$

and

$$\Sigma_n = \{K \in B(W_0^{1,2}(\Omega)) : K \subseteq S, \tau(K) \geq n\},$$

here τ is any "genus", i.e. any mapping $\tau : B(X) \to \{0, 1, 2, \dots\} \cup \{\infty\}$ that satisfies 1–4 above. We define

$$c_n = \sup_{K \in \Sigma_n} \inf\{a(y) : y \in K\}. \tag{2.5}$$

THEOREM 2.1. *The sequence $\{c_n\}$ is nonincreasing and*

$$\lim_{n \to \infty} c_n = 0.$$

If k and n are positive integers and

$$c_n = c_{n+k-1} \tag{2.6}$$

then the set E_n of eigenfunction of (1.1) on the level set

$$\{y \in S : a(y) = c_n\}$$

satisfies

$$\tau(E_n) \geq k, \tag{2.7}$$

in particular, E_n is non-empty.

The largest value of k for which (2.6) holds, with $n = 1$ or $c_{n-1} < c_n$, will be called the *multiplicity* of the critical value c_n and c_n will be called a *simple* critical value if its multiplicity is one.

REMARK: If τ is assumed only to satisfy 1,3,4 above but not the subadditive property 2 then the theorem remains valid with the exception that (2.7) must be weakened to

$$\tau(E_n) \geq 1.$$

Let

$$\Sigma = \{K \in B(W_0^{1,2}(\Omega)) : K \subseteq S\},$$

for $K \in \Sigma$ we define $\tau^c(K)$, the *co-genus* of K (cf. Heinz [**10**]), by

$$\tau^c(K) = \inf\{n \geq 0 : K \cap K_1 \neq \phi \text{ for every } K_1 \in \Sigma_{n+1}\}.$$

We have then, as a generalization of (1.6),

$$c_n = \inf_{K \in \Sigma_n^c} \sup\{a(x) : y \in K\},$$

where

$$\Sigma_n^c = \{K \in \Sigma : \tau^c(K) \leq n-1\}.$$

Next we observe that if M is a subspace of $W_0^{1,2}(\Omega)$ with co-dim $M = n-1$ then $\tau^c(M \cap S) = n-1$. We thus have the following estimates for the c_n,

$$c'_n \leq c_n \leq c''_n, \tag{2.8}$$

where

$$c'_n = \max_{\dim M \geq n} \min\{a(y) : y \in M, \quad b(y) = 1\} \tag{2.9}$$

and

$$c''_n = \min_{\text{co-dim}\, M \leq n-1} \max\{a(y) : y \in M, \quad b(y) = 1\}. \tag{2.10}$$

3. Sublinear and Superlinear Problems.

The classification of nonlinearities to be described here appears to be due originally to Pimbley [**19,20**]; the definition of these classes that we shall use, however, is not quite so restrictive as his. It will be convenient to write $f(x,y)$ in (1.1) in the form

$$f(x,y) = yF(y^2,x), \tag{3.1}$$

as we can clearly do if (1.2) is assumed. We shall then call the problem (1.1) *sublinear* if for all $x \in \Omega$

$$F(t_1,x) \geq F(t_2,x) > 0 \text{ for all } 0 < t_1 \leq t_2, \tag{3.2}$$

and we shall call it *superlinear* if for all $x \in \Omega$

$$0 < F(t_1,x) \leq F(t_2,x) \text{ for all } 0 < t_1 \leq t_2. \tag{3.3}$$

We observe that (3.3) obviously implies (2.2) while (3.2) just as obviously does not; the sublinear case of (1.1) is often studied without the assumption of (2.2), even the assumption of positivity of F when t is large is sometimes dropped.

We put

$$G(t,x) = \int_0^t F(s,x)\,ds, \quad s \geq 0, x \in \Omega, \tag{3.4}$$

and then for $y, v \in W_0^{1,2}(\Omega)$ (assuming of L the hypothesis that preceded Theorem 2.1) we have

$$\int_\Omega \left[G(v^2,x) - G(y^2,x)\right]\,dx \leq \int_\Omega (v^2 - y^2)F(y^2,x)\,dx, \tag{3.5}$$

if (3.2) holds and

$$\int_\Omega \left[G(v^2,x) - G(y^2,x)\right]\,dx \geq \int_\Omega (v^2 - y^2)F(y^2,x)\,dx, \tag{3.6}$$

if (3.3) holds.

This classification extends to quasi-linear problems of variational type, for example the equation of the classical elastica,

$$-y''/\sqrt{1-y'^2} = \lambda p(x)y, \quad |y'| < 1,$$

is sublinear and the equation

$$-y''/\sqrt{(1+y'^2)^3} = \lambda p(x)y^3,$$

is superlinear.

4. Nonlinear Sturm-Liouville Problems.

We consider the one-dimensional case

$$-y'' = \lambda f(x,y), \quad 0 < x < 1, \quad y(0) = y(1) = 0, \tag{4.1}$$

of (1.1), assuming f to have the properties indicated in sections 1 and 2. We assume τ to be a genus having properties 1–4 of section 2 and we assume that the $\{c_n\}$ and τ^c are defined in terms of τ. We have then the following general result.

PROPOSITION 4.1. *The critical values c_n are all simple, i.e.*

$$c_1 > c_2 > \cdots > c_n > c_{n+1} > \cdots.$$

PROOF: Let E_n denote the set of eigenfunctions on the critical level $\{y \in S : a(y) = c_n\}$. Uniqueness of the zero solution to the initial value problem for the ordinary differential equation in (4.1) implies that $y'(0) \neq 0$ for $y \in E_n$. Thus $y \to \operatorname{sgn}(y'(0))$ is an odd map of E_n to S^0, the existence of which implies that $\tau(E_n) = 1$, hence by (2.7) c_n is simple.

The special structure of the one-dimensional problem makes possible an additional upper estimate for the $\{c_n\}$. Let

$$0 < a_1 < a_2 < \cdots < a_{n-1} < 1, \tag{4.2}$$

then, as observed in [**11**], the conditions

$$y(a_1) = y(a_2) = \cdots = y(a_{n-1}) = 0,$$

determine a subspace M of $W_0^{1,2}(0,1)$ of co-dimension $n-1$ and thus we have the following result.

PROPOSITION 4.2. *Let*

$$C_n = \min\max\{a(y) : y \in S, \quad y(a_1) = y(a_2) = \cdots = y(a_{n-1}) = 0\}, \tag{4.3}$$

where the minimum is taken over all $\{a_i\}_{i=1,\ldots,n-1}$ satisfying (4.2), then

$$c_n' \le c_n \le c_n'' \le C_n. \tag{4.4}$$

5. Sublinear Sturm-Liouville Problem.

We consider now the eigenvalue problem

$$-y'' = \lambda y F(y^2, x), \quad 0 < x < 1, \quad y(0) = y(1) = 0, \tag{5.1}$$

with F satisfying (3.2). We shall assume also that F has a continuous partial derivative with respect to its first argument. We note that if G is defined by (3.4) then the functional a can be expressed as

$$a(y) = \tfrac{1}{2} \int_0^1 G(y^2, x)\, dx, \quad y \in W_0^{1,2}(0,1). \tag{5.2}$$

Two of the basic tools to be used are the comparisons of (5.1) with the "first variation"

$$-w'' = \sigma w \Phi(y^2, x), \quad 0 < x < 1, \quad w(0) = w(1) = 0, \tag{5.3}$$

and with the linear equation

$$-u'' = \mu u F(y^2, x), \quad 0 < x < 1, \quad u(0) = u(1) = 0; \tag{5.4}$$

in (5.3),

$$\Phi(y^2, x) = \frac{\partial}{\partial y} y F(y^2, x). \tag{5.5}$$

Below, when we refer to (5.3) or (5.4), it is to be understood that it is the function y under discussion that appears as argument in coefficient Φ of F.

We recall that $y \in S$ means $y \in W_0^{1,2}(0,1)$ and

$$\int_0^1 y'^2 \, dx = 1.$$

PROPOSITION 5.1. *Suppose that $y \in S$ is such that*

$$\int_0^1 y' u_i' \, dx = 0, \quad i = n+1, \ldots, \tag{5.6}$$

where u_i denotes the i-th eigenfunction of (5.4) (numbered in order of increasing eigenvalues and normalized by the requirement that it belong to S and that $u_i'(0) > 0$). Then

$$c_n \le a(y). \tag{5.7}$$

Equality holds in (5.7) only if y is an $(n-1)$-node eigenfunction of (5.1).

PROOF: From the theory of the Rayleigh quotient we have the inequality

$$\int_0^1 y'^2 \, dx \le \mu_n \int_0^1 y^2 F(y^2, x) \, dx, \tag{5.8}$$

if (5.6) holds and we have

$$\int_0^1 v'^2 \, dx \ge \mu_n \int_0^1 v^2 F(y^2, x) \, dx, \tag{5.9}$$

if v is orthogonal to the first $n-1$ eigenfunctions of (5.4), (here μ_n is the n-th eigenvalue in increasing order of (5.4)). If, in addition to (5.9), v satisfies

$$\int_0^1 v'^2\, dx = 1, \tag{5.10}$$

then since $y \in S$, it follows from (5.8) and (5.9) that the right-hand side in (3.5) is negative and thus

$$\int_0^1 \left[G(v^2,x) - G(y^2,x)\right]\, dx \le 0,$$

or

$$a(v) \le a(y),$$

for all such v. In view of the requirements on v it follows that

$$c_n \le c_n'' \le a(y);$$

cf. (2.10). In order to have equality in (5.7) the equality must hold in (5.8); that equality and (5.6) imply that y is an $(n-1)$-node eigenfunction of (5.1).

COROLLARY 5.1. *The problem (5.1) has exactly one positive eigenfunction on S.*

PROOF: A positive eigenfunction satisfies (5.6) with $n = 1$ and the inequality in (5.9) is strict unless v is proportional to y. Thus

$$a(y) > a(v),$$

for a positive v satisfying (5.10) unless $v = y$, i.e. a positive eigenfunction on S is to within a sign the unique maximizer on S of the functional a. The existence proof is standard.

REMARK 5.1: Corollary 5.1 remains valid for the higher dimensional problem (1.1) in which L is a second-order elliptic operator.

We will say that the problem (5.1) is *strictly sublinear* if F satisfies the stronger condition

$$F(t_1,x) > F(t_2,x) > 0 \text{ for } 0 < t_1 < t_2. \tag{5.12}$$

LEMMA 5.1. *Suppose that the problem (5.1) is strictly sublinear. Let*

$$0 = a_0 < a_1 < a_2 < \cdots < a_{n-1} < a_n = 1,$$

and let $V = V(a_1, \ldots, a_{n-1})$ denote the set of functions $v \in S$ such that

$$v(a_1) = v(a_2) = \cdots = v(a_{n-1}) = 0.$$

There exists a unique $\lambda > 0$ and a unique nonnegative $y \in V$ such that for each $i = 0, 1, \ldots, n-1$, $y \in C^2(a_i, a_{i+1})$ and satisfies

$$-y'' = \lambda y F(y^2, x), \quad x \in (a_i, a_{i+1}), \tag{5.13}$$

and for those intervals (a_i, a_{i+1}) in which y vanishes identically the least eigenvalue $\mu(a_i, a_{i+1})$ of the problem

$$-u'' = \mu u F(0, x), \quad x \in (a_i, a_{i+1}), \quad u(a_i) = u(a_{i+1}) = 0,$$

satisfies

$$\mu(a_i, a_{i+1}) \geq \lambda. \tag{5.14}$$

If y is as indicated and v is any other function in V then

$$a(v) < a(y), \tag{5.15}$$

unless $|v| = y$. Finally, y depends continuously on $(a_1, \ldots, a_{n-1})$ in $W_0^{1,2}(0,1)$.

PROOF: Suppose that y is as indicated. In view of (5.13) we have for $v \in V$ and $i = 0, 1, \ldots, n-1$,

$$\lambda^{-1} \int_{a_i}^{a_{i+1}} v'^2 \, dx \geq \int_{a_i}^{a_{i+1}} v^2 F(y^2, x) \, dx,$$

while

$$\lambda^{-1} \int_{a_i}^{a_{i+1}} y'^2 \, dx = \int_{a_i}^{a_{i+1}} y^2 F(y^2, x) \, dx.$$

Taking into account that both y and v belong to S and using (3.5) we obtain (5.15); we have used the fact that the problem is strictly sublinear to conclude that the inequality is strict unless $|v| = y$.

It follows from the inequality (5.15) that there can be at most one y satisfying the above conditions. Obviously there exists a maximizer y of the functional a on V. Standard arguments from the calculus of variations show that y must satisfy the equation (5.13) for some $\lambda > 0$ on each of the intervals (a_i, a_{i+1}). If the parameter λ in (5.13) did not take the same value on each of the intervals (a_i, a_{i+1}) where y does not vanish identically or if the other conditions where not met then by a repartitioning of the "energy"

$$\int_0^1 y'^2\, dx = 1,$$

among the intervals (a_i, a_{i+1}) the functional a could be increased.

Next we show that for $n = 1, 2, \ldots,$

$$c_n = c_n'' = C_n, \tag{5.16}$$

and that there is an $(n-1)$-node eigenfunction y of (5.1) with $y \in S$ and

$$a(y) = c_n.$$

To this end let K denote the 2^n-hedron in $\mathbb{R}^n$

$$K = \{\xi \in \mathbb{R}^n : |\xi_1| + \cdots + |\xi_n| = 1\},$$

and for $\xi \in K$ put

$$a_i(\xi) = \sum_{j=1}^{i} |\xi_j|, \quad i = 1, \ldots, n-1.$$

Next we define the map $\pi : K \to S$ by

$$(\pi(\xi))(x) = \operatorname{sgn} \xi_{i+1} y(x), \quad x \in (a_i, a_{i+1}),$$

where y is the unique maximizer of the functional a in $V(a_1, \ldots, a_{n-1})$. Let

$$K' = \pi(K),$$

then since $\tau(K) = n$ and π is odd we have $\tau(K') \geq n$. It follows that

$$\min\{a(v) : v \in K'\} \leq c_n,$$

(cf. (2.5)), on the other hand, by construction, we have

$$C_n \leq \min\{a(v) : v \in K'\},$$

(cf. (4.3)). In view of (4.4) the above two inequalities imply (5.16). According to the general theory (5.1) has an eigenfunction $y \in K'$ and satisfying

$a(y) = c_n$; this y must in fact, by virtue of the construction, have exactly $n-1$ nodes.

Finally, let y be the $(n-1)$-node eigenfunction obtained above, let λ be the corresponding eigenvalue of (5.1) and let $\sigma_1 < \sigma_2 < \ldots$, denote the eigenvalues of the corresponding problem (5.3). Then

$$\sigma_{n-1} \leq \lambda \leq \sigma_n. \tag{5.17}$$

The nodal characterization of y implies that λ is the n-th eigenvalue of (5.4); this implies the second inequality in (5.17) by Sturm comparison since (3.2) implies

$$\Phi(y^2, x) \leq F(y^2, x).$$

On the other hand y lies on an $(n-1)$-dimensional "face" of the set K' and the functional a achieves a minimum on K' at y. In other words there are $(n-1)$ independent directions tangent to S at y in which variation of y results in a positive incrementation of a, (these correspond to variation of the nodes of y). Upon inspection of the second order development of a at y,

$$a(y+h) - a(y) = \frac{\lambda^{-1}}{2} \int_0^1 \left(h'^2 - \lambda h^2 \Phi(y^2, x)\right) dx$$
$$+ \text{ higher order terms in } h$$

we see that (5.3) must have $n-1$ eigenvalues $\leq \lambda$.

We summarize these results in the following theorem.

THEOREM 5.1. *For each $n = 1, 2, \ldots$, the sublinear problem (5.1) has an $(n-1)$-node eigenfunction $y \in S$ that satisfies*

$$a(y) = c_n; \tag{5.18}$$

and (cf. (2.5), (2.10), (4.3)),

$$c_n = c_n'' = C_n.$$

For any $(n-1)$-node eigenfunction of (5.1) on S that satisfies (5.18) there holds

$$\sigma_{n-1} \leq \lambda \leq \sigma_n, \tag{5.19}$$

where the σ_i denote the eigenvalues of the corresponding problem (5.3). Finally any $(n-1)$-node eigenfunction y of (5.1) on S satisfies

$$a(y) \geq c_n.$$

The assumption of strict nonlinearity can obviously be relaxed to (3.2) in Theorem 5.1 because of the stability of the c_n and other quantities involved.

A most interesting example of the sublinear problem can be found in the well-known paper of Kolodner [**13**]. Other references which deal with the variational theory in particular are Coffman [**5**,**6**], Heinz [**11**], and Hempel [**12**]. The construction of the set K' was suggested by Rabinowitz [**21**].

6. Superlinear Sturm-Liouville Problem.

We now discuss the problem (5.1) in the case where F satisfies (3.3) which condition we shall assume for the remainder of this section without further mention. As with oscillation theory there is a symmetry or duality between the sublinear and the superlinear problem. Because the superlinear case has been treated in detail elsewhere (see [**7**]) we shall only quote results here.

PROPOSITION 6.1. *Suppose that $y \in S$ is such that*

$$\int_0^1 y' u_i' \, dx = 0; \quad i = 1, \ldots, n-1, \tag{6.1}$$

where u_i denotes the i-th eigenfunction of (5.4). Then

$$c_n \geq a(y). \tag{6.2}$$

Equality holds in (6.2) only if y is an $(n-1)$-node eigenfunction of (5.1).

The superlinear analogue of Theorem 5.1 is the following assertion.

THEOREM 6.1. *For each $n = 1, 2, \ldots,$ the superlinear problem (5.1) has an $(n-1)$-node eigenfunction $y \in S$ that satisfies*

$$a(y) = c_n; \tag{6.3}$$

and (cf. (2.5), (2.9)),

$$c_n' = c_n.$$

For any $(n-1)$-node eigenfunction of (5.1) on S that satisfies (6.3) there holds

$$\sigma_n \leq \lambda \leq \sigma_{n+1}, \tag{6.4}$$

where the σ_i denote the eigenvalues of the corresponding problem (5.3). Finally any $(n-1)$-node eigenfunction y of (5.1) on S satisfies

$$a(y) \leq c_n.$$

REMARKS:

(1) For the superlinear case of (1.1) where L is a second order elliptic operator the last assertion of Theorem 6.1 generalizes as follows: *if $y \in S$ is an eigenfunction satisfying $a(y) \geq c_n$ then the set*

$$\{x \in \Omega : y(x) \neq 0\}$$

has at most n components; this can be regarded as a generalization of the Courant nodal line theorem [9].

(2) The second inequality in (6.4) does not obviously admit as elementary a proof as does its dual, the first inequality in (5.19). For the former result we have had to make use of the results of section 9.

(3) For the sublinear problem of section 5 there is the third characterization (4.3) of the critical values. There is a dual of this for the superlinear problem but we have omitted it because it is neither as natural nor as interesting as (4.3).

7. The Nehari Problem.

By this we mean the problem

$$-y'' = y(p(x) + F(y^2, x)), \quad 0 < x < 1, \quad y(0) = y(1) = 0, \tag{7.1}$$

without a parameter and where $p(x)$ is nonnegative and F is assumed in place of (3.3) to satisfy the stronger condition

$$t_1^{-\epsilon} F(t_1, x) \geq t_2^{-\epsilon} F(t_2, x) > 0, \quad \text{for } t_1 \geq t_2 > 0, \tag{7.2}$$

for all $x \in (0,1)$; (actually this condition can be relaxed somewhat in the case of second order ordinary differential equations but strict superlinearity at least must be assumed). Nehari [**17,18**], attacked the problem (7.1) in the presence of (7.2) through the study of the following variational problem

$$H(y) = \text{extremal}, \quad \Gamma(y) = 0, \tag{7.3}$$

where

$$H(y) = \int_0^1 (y'^2 - G(y^2, x))\, dx, \tag{7.4}$$

and

$$\Gamma(y) = \int_0 (y'^2 - y^2 F(y^2, x))\, dx. \tag{7.5}$$

The problem (7.1) can be shown to have an $(n-1)$-node solution provided the problem

$$u'' + \mu p(x) u = 0, \quad u(0) = u(1),$$

has its n-th eigenvalue $\mu_n > 1$. The problem can be treated by methods very similar to those used to treat the problem discussed in section 6. If one introduces a parameter and considers the eigenvalue problem

$$-y'' = \lambda y(p(x) + F(y^2, x)), \quad 0 < x < 1, \quad y(0) = y(1) = 0,$$

then the set of eigenvalues corresponding to $(n-1)$-node eigenfunctions fills the interval $(0, \mu_n)$; the eigenfunctions obtained by the variational method cannot however be shown to form continuous branches.

8. Inverse Power Method.

Under the assumption (2.2) and with a defined by (1.3) there holds the inequality

$$a(v) - a(y) \geq \int_\Omega (v - y) f(x, y)\, dx.$$

It follows that if $y \in S$ and v satisfies

$$Lv = \alpha f(x, y), \quad \text{in } \Omega, \quad v = 0 \text{ on } \partial\Omega,$$

where α is chosen so that

$$\int_\Omega v\, Lv\, dx = \int_\Omega y\, Ly\, dx = 1, \tag{8.1}$$

then

$$a(v) \geq a(y),$$

with equality only if y is an eigenfunction of (1.1). This provides a basis for an iterative procedure for obtaining the dominant eigenfunction to the problem (1.1), cf. [**3,4**]. Although the sequence of iterates cannot be shown to converge in general, it can be shown that its set of limit points is a connected set of eigenfunction of (1.1) [**2**]. In those instances where the dominant eigenfunction on S is unique (cf. Corollary 5.1) the convergence of the full sequence follows.

Perhaps the first person to apply this method systematically to nonlinear problems was Nehari [**17,18**], who used this approach (with the normalization $\Gamma(v) = 0$ in place of (8.1), Γ given by (7.5)) to the variational problem (7.3) to obtain a dominant solution to (7.1). Here, even in the possible presence of nonuniqueness, solutions can sometimes be shown to be isolated – for example when y occurs analytically in F for each x – and full convergence follows. The iteration method does not work for the sublinear analogue of Nehari's problem.

More important in this context the iterative mapping $y \to v$ described above is odd and continuous and compact from S to S. As such it can play the essential role of a "push-up" operator for the Lyusternik-Schnirelman theory. The use of this map where applicable (the convexity of the functional a is essential) results in a considerable simplification of the theory.

9. Morse Index.

In this section we discuss the correlation of the generalized Morse index of a critical point with the order of the critical level on which it lies. Some very general results of this type have been obtained by Bahri and Lions [**1**]. The results here can be regarded as generalizing the spectral inequalities (5.19) and (6.4) above, (in fact, the second inequality of (6.4) was proved from the results below and we have been unable to find a substantially more elementary proof). The correlation is only partial; we can assert that on a given critical level there will be critical points whose index satisfies certain inequalities but we cannot assert that the Morse index of *every* critical point on that level satisfies those inequalities. The Morse index to which we have referred is characterized as follows: Let y be an eigenfunction of (1.1) with corresponding eigenvalue λ, let H_y denote the second gradient of $a - \lambda^{-1}b$ and let P_y denote the orthogonal projection onto the orthogonal complement of $sp\{y\}$; by the *Morse index of* y we refer to the number of nonnegative eigenvalues of $P_y H_y P_y$.

We now outline very briefly how the problem is attacked. We consider first a finite dimensional problem (on a finite dimensional sphere S) for an even C^∞ functional $a(y)$ without degenerate critical points. Let n, with $1 < n \leq \dim S + 1$, be given, and let

$$K = \{y \in S : a(v) \geq c_n\},$$

where c_n refers as usual to the n-th critical level determined by the max-min principle (2.5) (based on some genus τ). Note that we must have $\tau(K) = n$, (equality because the critical points are isolated and therefore the critical values are simple). Let y be a critical point on the level $a(v) = c_n$, if the Morse index k of the nondegenerate critical point y (the number of positive eigenvalues of the second gradient of a relative to S) is strictly less than $n - 1$ then by standard techniques [**16**], one can construct an odd map $\pi : K \to K$ such that π agrees with the identity except near y and $-y$ and for some neighborhood V of y, $V \cap \pi(K)$ has dimension k. In other words, we can "pinch" a neighborhood in K of y to a k-dimensional set. Suppose now that τ has the following property in addition to 1–4 of section 2.

5. If $\dim K_2 < \tau(K_1)$ then $\tau(K_1 \cup K_2) = \tau(K_1)$.

If this is the case then we can remove a balanced neighborhood of $(y, -y)$ from $\pi(K)$ without decreasing its genus (note that since π is odd $\tau(\pi(K)) \geq n$). It now follows that there must be at least one other critical point on the level $a(v) = c_n$, otherwise the set K' that resulted from the above deletion could be pushed up into $\{v \in S : a(v) > c_n\}$, contradicting the fact that $\tau(K') \geq n$. All critical points on the level $a(v) = c_n$ whose Morse indices are $< n - 1$ can be removed in this way and we could still conclude that one critical point would have to remain. In other words, assuming that τ satisfies 5 above, *there must be at least one critical point with Morse index* $\geq n - 1$ *on the level* $\{v \in S : a(v) = c_n\}$.

The Krasnosel'skii genus can in fact be shown to have property 5 [**7**], and thus the above assertion is valid when the critical values are determined by that genus.

For a finite dimensional problem, say on a sphere S of dimension N, critical values of a can also be determined by the principle

$$(9.1) \qquad e_n = \min_{K \in \Sigma_n} \max\{a(v) : v \in K\}, \quad n = 1, 2, \ldots, N+1,$$

where Σ_n denotes the class of closed balanced subsets of S of genus $\geq n$. From the linear case one would expect to find that

$$(9.2) \qquad c_n = e_{N-n+2}, \quad n = 1, 2, \ldots, N+1.$$

In any case, it is clear that, provided one can assume 5 of τ, then on each level $a(v) = e_n$ there is a critical point with Morse index $\leq n-1$, (we are still assuming nondegenerate critical points).

What is required to give both (9.2) and the existence on each critical level $a(v) = c_n$ of a critical point with Morse index $= n-1$ is the following property of the genus.

6. If K_1 and K_2 are closed balanced subsets of an N-sphere S and $\tau(K_1) + \tau(K_2) \geq N+2$ then $K_1 \cap K_2 \neq 0$.

The Krasnosel'skii genus fails to have property 6; this follows, as shown in [**7**], from an example of Yang [**23**]. Moreover, using the same example, one can construct a C^∞ functional $a(v)$ on a 7-dimensional sphere for which (9.2) fails. It turns out that one can however define a genus with the properties 1–6 and this also is essentially due to Yang [**22**].

We conclude now with the statement of a result for the isoperimetric problem associated with (1.1). Here of course there is no assumption of nondegeneracy.

PROPOSITION 9.1. *Let τ in (2.5) be a genus satisfying properties 1–6. For each $n = 1, 2, \ldots,$ there is an eigenfunction $y \in S$ of (1.1) that satisfies*

$$a(y) = c_n,$$

and such that $P_y H_y P_y$ has at least $n-1$ nonnegative eigenvalues and at most $n-1$ positive eigenvalues.

REMARK: It should be noted that for the problems discussed in sections 5 and 6, i.e. the sublinear and superlinear ordinary differential equations, because of the alternate characterizations (2.10) and (2.9) respectively, the characteristic values are independent of the choice of genus as long as properties 1 and 4 are satisfied.

Department of Mathematics, Carnegie-Mellon University, Pittsburgh, PA 15213

This research was supported by NSF Grant DMS 86-02954

REFERENCES

1. A. Bahri and P. L. Lions, *Remarque sur la théorie variationelle des points critiques et applications*, C. R. Acad. Sci. Paris **301, Ser. I** (1985), 145–147.
2. C. V. Coffman, *A minimum-maximum principle for a class of nonlinear integral equations*, J. d'Analyse Math. **22** (1969), 391–419.
3. C. V. Coffman, *Spectral theory of monotone Hammerstein operators*, Pacific J. Math. **36** (1971), 303–322.
4. C. V. Coffman, *Lyusternik-Schnirelman theory and eigenvalue problems for monotone potential operators*, J. Functional Analaysis **14** (1973), 203–238.
5. C. V. Coffman, *On variational principles for sublinear boundary value problems*, J. Diff. Equ. **17** (1975), 46–60.
6. C. V. Coffman, *The nonhomogeneous classical elastica*, Technical report, Department of Mathematics, Carnegie-Mellon University (1976).
7. C. V. Coffman, *Lyusternik-Schnirelman theory: complementary principles and the Morse index*, Nonlinear Anal., to appear.
8. E. R. Faddell and P. Rabinowitz, *Bifurcation for odd potential operators and an alternative topological degree*, J. Functional Analysis **26** (1977), 48–67.
9. P. R. Garabedian, "Partial Differential Equations," Wiley, New York, 1964.
10. H.-P. Heinz, *Un principe de maximum-minimum pour les valeurs critiques d'une fonctionelle non-lineaire*, C. R. Acad. Sci. Paris **275, Ser. A** (1972), 1317–1318.
11. H.-P. Heinz, *Nodal properties and variational characterizations of solutions to nonlinear Sturm-Liouville problems*, J. Diff. Equ. **69** (1986), 299–333.
12. J. Hempel, *Multiple solutions for a class of nonlinear boundary value problems*, Indiana Univ. Math. J. **20** (1971), 983–996.
13. I. I. Kolodner, *Heavy rotating string-a nonlinear eigenvalue problems*, Comm. Pure Appl. Math. **8** (1955), 395–408.
14. M. A. Krasnosel'skii, *An estimate of the number of critical points of functionals*, Uspehi Math. Nauk **7** (1952), 711–716.
15. M. A. Krasnosel'skii, "Topological Methods in the Theory of Nonlinear Integral Equations," Macmillan, New York, 1964.
16. J. Milnor, "Morse Theory," Princeton University Press, Princeton, 1963.
17. Z. Nehari, *On a class of nonlinear second-order differential equations*, Trans. Amer. Math. Soc. **95** (1960), 101–123.
18. Z. Nehari, *Characteristic values associated with a class of nonlinear second-order differential equations*, Acta Math. **105** (1961), 141–175.
19. G. H. Pimbley, *A sublinear Sturm-Liouville problem*, J. Math. and Mech. **11** (1960), 121–138.
20. G. H. Pimbley, *A superlinear Sturm-Liouville problem*, Trans. Amer. Math. Soc. **103** (1962), 229–248.
21. P. H. Rabinowitz, *Variational methods of nonlinear elliptic eigenvalue problems*, Indiana Univ. math. J. **23** (1973/74), 729–754.
22. C.-T. Yang, *On theorems of Borsuk-Ulam, Kakutani-Yamabe-Yujobô and Dyson, I*, Ann. of Math. **60** (1954), 262–282.
23. C.-T. Yang, *On theorems of Borsuk-Ulam, Kakutani-Yamabe-Yujobô and Dyson, II*, Ann. of Math. **62** (1955), 271–283.

Harnack-type Inequalities for some Degenerate Parabolic Equations

E. DI BENEDETTO

1. Introduction.

We give here a brief account on some recent results [**14**] concerning an intrinsic version of the Harnack inequality for nonnegative solutions of degenerate parabolic equations of the type

$$
\begin{aligned}
&u_t - \operatorname{div}(|Du|^{p-2}Du) = 0, \quad p > 2 \text{ in } \mathcal{D}'(\Omega_T) \\
&u \in C(0,T; L^2(\Omega)) \cap L^p(0,T : W^{1,p}(\Omega)),
\end{aligned} \tag{1.1}
$$

and of the type

$$
\begin{aligned}
&u_t - \Delta u^m = 0, \quad m > 1 \text{ in } \mathcal{D}'(\Omega_T) \\
&u \in C(0,T; L^2(\Omega)); u^m \in L^2(0,T;; W^{1,2}(\Omega)),
\end{aligned} \tag{1.2}
$$

where Ω is an open set in $\mathbb{R}^N$, $0 < T < \infty$, $\Omega_T \equiv \Omega \times (0,T]$ and D denotes the gradient with respect to the space variables only.

The notion of a weak solution in the specified classes is standard (see [**6**] and [**7**]).

A more complete version of the paper, including detailed proofs, will be published elsewhere.

2. Equations of the type (1.1).

The classical work of Moser [**10**] about the Harnack inequality for solutions of parabolic equations has been brought to full generality by Trudinger

[13] and Aronson-Serrin [2]. The generalization process reveals a gap between the elliptic and parabolic theories. To be specific, in the elliptic case, the principal part is allowed to have any growth with respect to the gradient of the solution, so that, for example, solutions of

$$\operatorname{div}(|Du|^{p-2}Du) = 0, \quad p > 1, \quad u \geq 0$$

do satisfy the Harnock inequality, whereas nonnegative solutions of the corresponding parabolic equation (1.1) in general do not (see [7] for a counterexample).

On the other hand, N. Trudinger [13] observed that if $u \geq 0$ solves

$$\frac{\partial}{\partial t}u^{p-1} - \operatorname{div}(|Du|^{p-2}Du) = 0, \quad p > 1, \tag{2.1}$$

then it satisfies the Harnack inequality. In view of this one might think of (1.1), loosely speaking, as (2.1) provided the time scale is changed in another one intrinsic to the solution, and conjecture that nonnegative solutions of (1.1) will satisfy the Harnack inequality with respect to such an intrinsic time scale.

This will be made rigorous in the following theorem.

THEOREM 1. *Let $u \geq 0$ be a local weak solution of (1.1) in Ω_T. Let $(x_0, t_0) \in \Omega_T$, let $B_R(x_0)$ be the ball of center x_0 and radius R, and assume $u(x_0, t_0) > 0$.*

There exist two constants C_0, C_1 depending only upon N, p such that

$$u(x_0, t_0) \leq C_0 \inf_{x \in B_R(x_0)} u(x, t_0 + \theta), \tag{2.2}$$

where

$$\theta = \frac{C_1 R^p}{[u(x_0, t_0)]^{p-2}}, \tag{2.3}$$

provided the box $B_{2R}(x_0) \times (t_0 - \theta, t_0 + \theta)$ is all contained in Ω_T.

The constants $C_0(N, p), C_1(N, P) \to \infty$ as $p \to \infty$ but they are "stable" as $p \searrow 2$, i.e. $C_i(N, p) \to C_i(N, 2) < \infty$, $i = 0, 1$.

Therefore letting $p \searrow 2$ in (2.2) and (2.3) we recover the classical Harnack inequality for nonnegative solutions of the heat equation proved by Hadamard [8], in the form given by Krylov-Safonov [9].

In Theorem 1, the value of θ is intrinsically linked to the solution u. The next result holds for all $\theta > 0$.

THEOREM 2. *Let $Q_{2R}(\theta) \equiv B_{2R}(x_0) \times \{t_0 - \theta, t_0 + \theta\}$. For all $R > 0$, $\theta > 0$ such that $Q_{2R}(\theta) \subset \Omega_T$*

$$(2.4)\qquad u(x_0,t_0) \le B\left\{\left(\frac{R^p}{\theta}\right)^{\frac{1}{p-2}} + \left(\frac{\theta}{R^p}\right)^{\frac{N}{p}}\left[\inf_{x\in B_R(x_0)} u(x,t_0+\theta)\right]^{\frac{\kappa}{p}}\right\}$$

where $\kappa = N(p-2)+p$ and $B = B(N,p)$ is a constant dependent only upon N and p.

Since $\theta > 0$ is arbitrary, Theorem 2 seems to be markedly different than Theorem 1. We prove that in fact they are equivalent statements.

We record the following immediate consequence of Theorems 1 and 2.

COROLLARY 1. *For all $R, \theta > 0$ such that $Q_{2R}(\theta) \subset \Omega_T$,*

$$(2.5)\qquad \fint u(x,t_0)dx \le B\left\{\left(\frac{R^p}{\theta}\right)^{\frac{1}{p-2}} + \left(\frac{\theta}{R^p}\right)^{\frac{N}{p}}[u(x,t_0+\theta)]^{\frac{\kappa}{p}}\right\},$$

where $B = B(N,p)$ is as in Theorem 2 and the left-hand side denotes the integral average of $u(x,t_0)$ in $B_R(x_0)$.

3. Equation of the type (1.2).

The following inequality has been established by Aronson and Caffarelli [**1**]. Let $u \ge 0$ be a continuous weak solution of (1.2) in the whole $\mathbb{R}^N \times (0,\infty)$. Then $\forall t_0 > 0$, $\forall T > 0$, $R > 0$,

$$(3.1)\qquad \int_{B_R(x_0)} u(x,t_0)dx \le C\left\{T^{-\frac{1}{m-1}}R^{\frac{\lambda}{m-1}} + T^{\frac{N}{2}}[u(x_0,t_0+T)]^{\frac{\lambda}{2}}\right\}$$

where $\lambda = N(m-1)+2$.

The proof in [**1**] is based on the global properties of such solutions.

As a by-product of our approach we find that inequalities like (3.1) are a purely local fact.

Precisely, we have

THEOREM 3. *Let $u \geq 0$ be any local weak solution of (1.2) in Ω_T. Let $(x_0, t_0) \in \Omega_T$ and assume $u(x_0, t_0) > 0$. There exist two constants C_0, C_1 depending only upon N, m, such that*

$$u(x_0, t_0) \leq C_0 \inf_{x \in B_R(x_0)} u(x, t_0 + \theta)$$

where

$$\theta = \frac{C_1 R^2}{u(x_0, t_0)^{m-1}},$$

provided the box $B_{2R}(x_0) \times (t_0 - \theta, t_0 + \theta)$ is all contained in Ω_T.

THEOREM 4. *For all $R, \theta > 0$ such that $Q_{2R}(\theta) \subset \Omega_T$,*

$$u(x_0, t_0) \leq D \left\{ \left(\frac{R^2}{\theta} \right)^{\frac{1}{m-1}} + \left(\frac{\theta}{R^2} \right)^{\frac{N}{2}} \left[\inf_{x \in B_R(x_0)} u(x, t_0 + \theta) \right]^{\frac{\lambda}{2}} \right\}$$

where $\lambda = N(m-1) + 2$ and D is a constant depending only on N, m.

Further comments and generalizations are in [**14**].

Department of Mathematics, Northwestern University, Evanston IL 60201

Research at MSRI supported in part by NSF Grant 812079-05.

REFERENCES

1. D.G. Aronson and L.A. Caffarelli, *The initial trace of a solution of the porous media equation*, Trans. AMS **280, No. 1** (1983), 351–366.
2. D.G. Aronson and J. Serrin, *Local behavior of solutions of quasilinear parabolic equations*, Arch. Rat. Mech. Anal. **25** (1967), 81–123.
3. G.I. Barenblatt, *On some unsteady motions of a liquid or a gas in a porous medium*, Prikl. Math. Meh. **16** (1952), 67–78.
4. M. Bertsch and L.A. Peletier, *A positivity property of solutions of nonlinear diffusion equations*, Jour. of Diff. Equ. **53, No. 1** (1984), 30–47.
5. B.E. Dahlberg and C.E. Kenig, *Nonlinear filtration*.
6. E. Di Benedetto and A. Friedman, *Hölder estimates for nonlinear degenerate parabolic systems*, J. Reine Angew. Math. **357** (1985), 1–22.
7. E. Di Benedetto, *On the local behavior of solutions of degenerate parabolic equations with measurable coefficients*, Ann. Sc. Norm. Sup. Pisa, Ser. IV **13, No. 3** (1986), 487–535.
8. J. Hadamard, *Extension à l'équation de la chaleur d'un théorème de A. Harnack*, Rend. Circ. Mat. Palermo **3, Ser. 2** (1954), 337–346.
9. N.V. Krylov and M.V. Safonov, *A certain property of solutions of parabolic equations with measurable coefficients*, Math. USSR Izvestija **16, No. 1** (1981), 151–164.
10. J. Moser, *A Harnack inequality for parabolic differential equations*, comm. Pure Appl. Math. **17** (1964), 101–134.
11. J. Serrin, *Local behavior of solutions of quasilinear elliptic equations*, Acta Math. **111** (1964), 302–347.
12. N.S. Trudinger, *On Harnack type inequalities and their application to quasilinear elliptic partial differential equations*, Comm. Pure Appl. Math **20** (1967), 721–747.
13. N.S. Trudinger, *Pointwise estimates and quasilinear parabolic equations*, Comm. Pure Appl. Math. **21** (1968), 205–226.
14. E. Di Benedetto, *Intrinsic Harnack-type inequalities for solutions of certain degenerate parabolic equations*, Archive for Rat. Mech. Anal. (to appear).

The Inverse Power Method for Semilinear Elliptic Equations

ALEXANDER EYDELAND AND JOEL SPRUCK

In this note we will discuss an iterative procedure that is very useful for both the numerical and theoretical study of nonlinear eigenvalue problems

$$(1)\qquad \begin{aligned} Lu \equiv \partial_i(a_{ij}\partial_j u) - a(x)u &= -\lambda\, f(x,u) && \text{in } D \\ u &= 0 && \text{on } \partial D \end{aligned}$$

where D is a bounded domain in R^n and L is self-adjoint and uniformly elliptic with $a_{ij} \in C^0(\bar{D})$ and $a(x) \geq 0$, $a \in L^\infty$.

The method is a generalization of the famous inverse power method for computing the eigenvalues of a symmetric matrix. Let $B(u,v) = \int_D (a_{ij}\partial_j u \partial_i v + a(x)uv)dx$ be the natural bilinear form associated to the operator L, where u, v belong to the Sobolev space $\overset{o}{H}_1(D)$, and set

$$(2)\qquad F(x,u) = \int_0^u f(x,t)dt.$$

Define $A_R = \{u \in \overset{o}{H}_1(D) : \|u\| = R\}$ where $\|u\|^2 = B(u,u)$. We construct a mapping $T : A_R \to A_R$ as follows: given $v \in S_R$, $u = T_v$ is defined by

$$(3)\qquad u = \frac{Rw}{\|w\|}$$

wherer w is the solution of the Dirichlet problem

$$(4)\qquad \begin{aligned} Lw &= -f(x,v) && \text{in} \quad D \\ w &= 0 && \text{on} \quad \partial D. \end{aligned}$$

The algorithm can now be described as follows.

(5) ALGORITHM I: Given $u_o \in A_R$, $u_{j+1} = Tu_j \qquad j = 0,1,2 \ldots$.

In order that the algorithm be well-defined we need $f(x,t)$ to satisfy the condition

$$v \in \overset{o}{H}_1, \qquad f(x,v) \equiv 0 \Longrightarrow v \equiv 0. \tag{6}$$

We now assume further that $f(x,t)$ satisfies the usual Sobolev growth conditions

$$\begin{aligned} &\lim_{t\to\infty} f(x,t)t^{-\frac{n+2}{n-2}} = 0 && n \geq 3 \\ &\lim_{t\to\infty} f(x,t)e^{-t^{\alpha}} = 0 && n = 2, \text{some } \alpha \in (0,2) \end{aligned} \tag{7}$$

uniformly for $x \in \bar{D}$. Now define

$$\Phi(u) = \int_D F(x,u)dx.$$

Associated with (1) we have the natural

(8) VARIATIONAL PROBLEM: $\Phi(u) \to \max,\ u \in A_R$.

Our simplest result connects the Algorithm (5) and the Variational Problem (8) and asserts a global convergence property.

THEOREM 1. *Let $f(x,t)$ satisfy (6)(7) and be nondecreasing in t. Let $u_o \in A_R$ be such that $\Phi(u_o) > 0$. Then the sequence $\{u_j\}$ defined by Algorithm I has the properties*

- *i)* $\Phi(u_{j+1}) \geq \Phi(u_j) \qquad j = 0,1, \ldots$.
- *ii) Any subsequence of the $\{u_j\}$ has a subsequence which converges strongly in $\overset{o}{H}_1(D)$ to a solution u of (1) with $\lambda > 0$.*
- *iii) If $u_0 \in L^\infty(D)$, $\|u_j\|_{L^\infty(D)}$ is uniformly bounded and the convergence in ii) is a strong convergence in $W^{2,p}(D)$ for any p.*

One interesting feature of Theorem 1 is that it is possible to produce solutions of (1) which change sign. Of course, if $u_0 \geq 0$ and $f(x,0) \geq 0$ then Algorithm I always produces a positive solution of (1).

In numerical experiments the entire sequence $\{u_j\}$ converges. Whether this is actually the case is a deep question which we intend to pursue in later work. However, we have the following result which shows that the possible set of limit points of the iteration is either very simple or very complicated.

PROPOSITION 2. *Let S_R be the limit points of Algorithm I for a fixed u_0. Under the hypotheses of Theorem 1, either*

i) S_R contains only one point, or

ii) S_R contains infinitely many points, none of which are isolated.

The hypothesis that $f(x,t)$ be nondecreasing in t in Theorem 1 is used only to deduce that $\Phi(u)$ is a convex functional and Theorem 1 is valid under this weaker hypothesis. In order to further weaken this assumption we will assume that

$$f_t(x,t) \geq -K \qquad x \in \bar{D}. \tag{9}$$

A typical example of such an f is given by

$$f(t) = a|t|^{p-1}t - b|t|^{q-1}t$$

where $1 \leq q < p < \frac{n+2}{n-2}$ and $a, b > 0$.

We construct $T^\sigma : A_R \to A_R$ as follows: given $v \in A_R$, $u = T^\sigma v$ is defined by

$$u = R\frac{(w+\sigma v)}{\|w+\sigma v\|} \qquad w + \sigma v \neq 0 \tag{10}$$

where as before, w satisfies (4). We will show in Lemma 2.2 of Section 2 that for σ large, positive T^σ is well-defined. We can now describe

(11) ALGORITHM II: Given $u_0 \in A_R$, $u_{j+1} = T^\sigma u_j$, $j = 0, 1, \ldots$.

THEOREM 3. *Let $f(x,t)$ satisfy (6)(7)(9). Let $u_0 \in A_R$ be such that $\Phi(u_0) > 0$. (Recall ($\Phi(u) = \int_D F(x,u)dx$). Then for σ large (depending on K) the sequence $\{u_j\}$ defined by Algorithm II has properties*

i) $\Phi(u_{j+1}) \geq \Phi(u_j)$.

ii) Any subsequence of the $\{u_j\}$ has a subsequence which converges weakly in $\overset{o}{H}_1(D)$ to a solution $u \not\equiv 0$ of

$$\begin{aligned} -(1-\lambda\sigma)Lu &= \lambda f(x,u) \qquad \text{in} \qquad D \\ u &= 0 \qquad \text{on} \qquad \partial D \end{aligned} \tag{12}$$

with $\lambda > 0$ and $(1-\lambda\sigma) \neq 0$. If $1-\lambda\sigma > 0$ then the convergence is strong.

iii) If $L = \Delta$, $f = f(t)$, and D is star-shaped, then $1-\lambda\sigma > 0$ always holds.

Both Algorithms I and II fit into a general theory of iterative procedures developed by Eydeland [2]. Concerning Algorithm I, there is very little literature. We mention a paper of Georg [4] who discusses the behavior of Algorithm I when u_0 starts in a small neighborhood of a strict relative maximum of the Variational Problem (8). In [4] there are also references to the plasma physics literature where Algorithm I has been used. Also, a very special case of part of Theorem 1 is implicit in the paper of Coffman [1].

We will discuss one further generalization that illustrates the flexibility of Algorithm I. Consider a diagonal system of nonlinear eigenvalue type:

$$(13)\qquad \begin{aligned} Lu^i &= \lambda f^i(x,\vec{u}) \quad \text{in} \quad D \quad i=1,\dots,N \\ u^i &= 0 \quad \text{on} \quad \partial D \end{aligned}$$

where $\vec{u}=(u^1,\dots,u^N)$ and L is given as in (1). We assume that the system (13) is of variational type, namely

$$(14)\qquad f^i(x,\vec{u}) = \frac{\partial F}{\partial u_i}(x,\vec{u}).$$

Writing $\vec{f}=(f^1,\dots,f^N)$ we can as before define

(15) ALGORITHM I'.: Given $\vec{u}_0 \in S_R, \vec{u}_{j+1} = Tu_j$, j =0,1,2 where T is defined by vectorizing (3), (4), and

$$\|\vec{u}\|^2 = \sum_{i=1}^{N} B(u^i,u^i).$$

The conditions (6), (7) are replaced by

$$(16)\qquad \vec{v} \in (\overset{o}{H}_1(D))^N, \vec{f}(x,v) \equiv 0 \Longrightarrow \vec{v} \equiv 0$$

$$(17)\qquad \begin{aligned} &\lim_{t\to\infty} |\vec{f}(x,t)| t^{-\frac{n+2}{n-2}} = 0 \qquad n \geq 3 \\ &\lim_{t\to\infty} |\vec{f}(x,t)| e^{-t^\alpha} = 0 \qquad n=2, \qquad \text{some } \alpha < 2 \end{aligned}$$

uniformly in $\bar{D}$.

THEOREM 4. *Let $\vec{f}(x,t)$ satisfy (14), (16), (17) and suppose that $F(x,\vec{u})$ is convex in $\vec{u}$. Then the sequence $\{\vec{u}_j\}$ defined by Algorithm I′ has the properties*

i) $\Phi(\vec{u}_{j+1}) \geq \Phi(\vec{u}_j) \qquad j = 0,1,\ldots$.

ii) Any subsequence of the $\{\vec{u}_j\}$ has a subsequence which converges strongly in $(\overset{o}{H}_1(D))^N$ to a solution $\vec{u}$ of (13) with $\lambda > 0$.

iii) If $\vec{u}_0 \in L^\infty(D)$, $\|\vec{u}_j\|_{L^\infty(D)}$ is uniformly bounded and the convergence in ii) is strong convergence in $(W^{2,p}(D))^N$ for any p.

The proof of Theorem 4 will be identical to that of Theorem 1 and will not be given.

Section 1. Analysis of Algorithm I.

Recall that the sequence $\{u_j\}$ given in Algorithm I is inductively defined by iterating the mapping $T: A_R \to A_R$ given by (3), (4). The mapping T enjoys the following variational property.

LEMMA 1.1. *Suppose $\Phi(u) = \int_D F(x,u)dx$ is convex. Let $v \in A_R$ and $u = Tv$. Then*

$$\Phi(u) \geq \Phi(v)$$

with equality if and only if $u = v$.

PROOF: Note that u satisfies

$$(18) \qquad \begin{aligned} -Lu &= \lambda f(x,v) && \text{in} \quad D \\ u &= 0 && \text{on} \quad \partial D \end{aligned}$$

where $\lambda = R/\|w\|$ (see (4)). From (18) we find

$$(19) \qquad \begin{aligned} B(u,u) &= \lambda \int_D uf(x,v)dx \\ B(u,v) &= \lambda \int_D vf(x,v)dx. \end{aligned}$$

Since $B(u,v) \leq \frac{1}{2}(B(u,u) + B(v,v)) = B(u,u)$, (19) implies that

$$\int_D (u-v)f(x,v)dx \geq 0$$

with equality if and only if $u = v$. Now the convexity of Φ implies that (here $u, v \in \overset{o}{H}_1(D)$ are arbitrary)

$$\Phi(u) \geq \Phi(v) + \int_D (u - v) f(x, v) \geq \Phi(v) \tag{20}$$

and the lemma follows.

COROLLARY 1.2.

$$\|u_{j+1} - u_j\|^2 \leq 2\lambda_{j+1}(\Phi(u_{j+1}) - \Phi(u_j))$$

PROOF: Let $u = u_{j+1}$, $v = u_j$, $\lambda = \lambda_{j+1}$ be as in Lemma 1.1. Then since $\|u\|^2 = \|v\|^2 = R^2$ we have

$$\|u - v\|^2 = 2B(u, u - v) = \lambda \int_D (u - v) f(x, v) dx$$

and the result follows from (20).

COROLLARY 1.3. *Let $\Phi(u_0) > 0$. Then*

$$0 < \lambda_{j+1} \leq R^2 / \Phi(u_0).$$

PROOF: Observe that from (19)

$$\lambda_{j+1} = \frac{B(u_{j+1}, u_j)}{\int_D u_j f(x, u_j) dx} \leq \frac{R^2}{\int_D u_j f(x, u_j) dx}.$$

Using (20) with $u = 0$ and $v = u_j$, we find

$$\int_D u_j f(x, u_j) dx \geq \Phi(u_j) \geq \Phi(u_0)$$

and the upper bound for λ_{j+1} follows.

LEMMA 1.4. *Let $u_0 \in L^\infty(D)$, $u_0 \in A_R$ be such that $\Phi(u_0) > 0$. Then $\|u_j\|_{L^\infty(D)} \leq K$ for a universal constant K depending only on $\|u_0\|_{L^\infty(D)}$, the operator L and the nonlinearity $f(x, t)$.*

PROOF: We consider only the case $n > 2$. Let $u = u_{j+1}$, $v = u_j$, $\lambda = \lambda_{j+1}$. Writing (18) in the weak form, we have

$$\int_D (a_{ij} u_i \varsigma_j + a u \varsigma) = \lambda \int_D \varsigma f(x, v) \qquad \forall \varsigma \in \overset{o}{H}_1(D).$$

Choose $\varsigma = (u_+)^\beta$, $\quad \beta \geq 1$ where $u_+ = \max(u, 0)$. Then

$$\lambda_0 \int_D \beta (u_+)^{\beta-1} |\nabla u_+|^2 \leq \lambda \int_D |f(x,v)| u_+^\beta. \tag{21}$$

Let $\varepsilon > 0$ be a small constant to be chosen. Then by (7) there is a large constant $M = M(\varepsilon)$ so that

$$|f(x,v)| \leq \varepsilon |v|^{\frac{n+2}{n-2}} \qquad \text{for } |v| \geq M. \tag{22}$$

It follows from (22) that

$$\int_{D \cap \{|v|>M\}} \left| \frac{f(x,v)}{v} \right|^{\frac{n}{2}} \leq \varepsilon^{n/2} \int_D |v|^{\frac{2n}{n-2}} \leq C \varepsilon^{n/2}.$$

(We have used the Sobolev embedding theorem). Therefore, (21) implies (C is a generic constant under control)

$$\int_D \left| \nabla u_+^{\frac{\beta+1}{2}} \right|^2 \leq C \varepsilon \beta \| v u_+^\beta \|_{L^{\frac{n}{n-2}}} + C(\varepsilon) \beta \int_D u_+^\beta. \tag{23}$$

Now with $p = \frac{n}{n-2}(\beta + 1)$,

$$\| v u_+^\beta \|_{L^{\frac{n}{n-2}}} \leq \|v\|_{L^p} \|u_+\|_{L^p}^{\beta+1} \leq \frac{1}{\beta+1} \|v\|_{L^p}^{\beta+1} + \frac{\beta}{\beta+1} \|u_+\|_{L^p}^{\beta+1}$$

$$\int_D \left| \nabla u_+^{\frac{\beta+1}{2}} \right|^2 \geq C \|u_+\|_{L^p}^{\beta+1}$$

$$\int_D u_+^\beta \leq C^{\frac{2\beta+n}{n(\beta+1)}} \|u_+\|_{L^p}^\beta.$$

Substitution into (23) gives

$$\|u_+\|_{L^p}^{\beta+1} \leq C\varepsilon\beta \left(\frac{1}{\beta+1} \|v\|_{L^p}^{\beta+1} + \frac{\beta}{\beta+1} \|u_+\|_{L^p}^{\beta+1} \right) + C(\varepsilon) C^{\frac{2\beta+n}{n(\beta+1)}} \beta \|u_+\|_{L^p}^\beta.$$

Using Young's inequality and simplifying gives

$$\|u_+\|_{L^p} \leq (C\varepsilon\beta)^{\frac{1}{\beta+1}} \|v\|_{L^p} + (C\varepsilon\beta^2)^{\frac{1}{\beta+1}} \|u_+\|_{L^p} + \beta C(\varepsilon).$$

Assume $u_0 \in L_p(D)$ with $p > \frac{2n}{n-2}$. This fixes $\beta > 1$ and we choose ε so small that $(C\varepsilon\beta)^{\frac{1}{\beta+1}} = \frac{1}{4}$. Then

$$\|u_+\|_{L^p} \leq \rho \|v\|_{L^p} + C(\beta), \rho = \frac{1}{3} \beta^{-\beta+1} < \frac{1}{3}.$$

Replacing u by $-u$ we have the same estimate for $u_- = \max(-u, 0)$, and so

$$\|u\|_{L^p} \le \sigma\|v\|_{L^p} + C(\beta) \qquad \sigma = \frac{2}{3}\beta^{-\beta+1} < \frac{2}{3}. \tag{24}$$

Iterating (24) gives

$$\|u_{j+1}\|_{L^p} \le \sigma^{j+1}\|u_0\|_{L^p} + C(\beta)\frac{\sigma}{1-\sigma} \qquad j = 0, 1, \dots \ . \tag{25}$$

To finish the proof we go back to (18) and use standard elliptic estimates. With $r = \frac{n-2}{4}p > \frac{n}{2}$, we have

$$\begin{aligned} \|u_{j+1}\|_{L^\infty} &\le C\|f(x, u_j)\|_{L^r} \\ &\le C(\delta) + \delta C\|u_j^{\frac{n+2}{n-2}}\|_{L^r} \\ &\le C(\delta) + \delta C\|u_j^{\frac{4}{n-2}}\|_{L^r}\|u_j\|_{L^\infty} \qquad \text{any } \delta > 0 \\ &\le C + C\delta\|u_j\|_{L^\infty} \qquad \text{by (25)} \end{aligned} \tag{26}$$

For $\theta = C\delta < 1$, (26) gives

$$\|u_{j+1}\|_{L^\infty} \le \theta^{j+1}\|u_0\|_{L^\infty} + \frac{C\theta}{1-\theta}$$

and the lemma is proven.

We can now prove Theorem 1.

PROOF OF THEOREM 1: Since $\|u_j\| = R$ we can choose subsequences $\{u_{j_i+1}\}$, $\{u_{j_i}\}$ converging weakly in $\overset{o}{H}_1(D)$ to u, v respectively. By Corollary 1.3 we may assume that $\lambda_{j_i} \to \lambda$. But $u_{j_i+1} - u_{j_i}$ converges weakly to $u - v$ so from Corollary 1.2 and lower semicontinuity,

$$\|u - v\| \le \lim\ \inf \|u_{j_i+1} - u_{j_i}\| = 0.$$

Hence $u = v$ and u is a solution of (1). Clearly $\|u\| \le R$. We show that $u \ne 0$; in fact $\Phi(u) = \lim \Phi(u_j)$. For

$$\Phi(u) - \Phi(u_j) = \int_D (F(x, u) - F(x, u_j)) = \int_D f(x, u_j^*)(u - u_j)$$

by the mean value theorem. Using (7) we estimate

$$\begin{aligned} |\Phi(u) - \Phi(u_j)| &\le \int_D |u - u_j| \left(C(\varepsilon) + \varepsilon|u_j^*|^{\frac{n+2}{n-2}}\right) \\ &\le C(\varepsilon)\int_D |u - u_j| + \varepsilon\|u - u_j\|_{L^{\frac{2n}{n-2}}}\|u_j^*\|^{\frac{n-2}{n2}}_{L^{\frac{2n}{n-2}}} \\ &\le C(\varepsilon)\int_D |u - u_j| + C\varepsilon. \end{aligned} \tag{27}$$

It follows that $\Phi(u_{j_i}) \to \Phi(u)$.

To show strong convergence, observe that since $\|u\| \leq R$, $\|u_{j_i+1}\| = R$ we have

$$\frac{1}{2}\|u - u_{j_i+1}\|^2 \leq B(u_{j_i+1}, u_{j_i+1} - u)$$

$$= \lambda_{j_i+1} \int_D (u_{j_i+1} - u) f(x, u_{j_i}).$$

Arguing as in (27), we see that $\|u - u_{j_i+1}\| \to 0$.

Finally, by Lemma 1.4 the entire sequence u_j is uniformly bounded in L^∞ and so by elliptic regularity theory, it is also uniformly bounded in $W^{2,p}(D)$ for any p. This proves part iii) of the theorem and concludes the proof.

REMARK: If a_{ij}, $a \in C^\alpha(\bar{D})$ and $f(x,t)$ is C^α (uniformly in $\bar{D}$, locally in t) the convergence may be taken in $C^{2,\beta}(\bar{D}) \quad \beta < \alpha$.

We will utilize part ii) of Theorem 1 and Corollary 1.2 to prove Proposition 2.

PROOF OF PROPOSITION 2: It suffices to show that if S_R contains one isolated solution u, then the entire sequence converges to u.

Since u is isolated we can find disjoint neighborhoods N_1 of u and N_2 containing all other elements of S_R. Let $2\varepsilon = \text{dist}(N_1, N_2)$. By Corollary 1.2 there is an integer j_0 so that for $j \geq j_0$, $\|u_{j+1} - u_j\| \leq \varepsilon$. We claim that there is an integer j_1 so that for $j \geq j_1$, u_j belongs to one of the neighborhoods N_1, N_2. Otherwise, we can find a subsequence u_{j_i} lying in the complement of $N_1 \cup N_2$. But by Theorem 1 the distance from u_{j_i} to S_R tends to zero, a contradiction. Since $u \in S_R$ we can find a $u_j \in N_1$, with $j > \max(j_0, j_1)$. Then $\text{dist}(u_{j+1}, N_2) \geq \varepsilon$ so that we must have $u_{j+1} \in N_1$. By induction $u_k \in N_1$ for $k \geq j$ and the proposition is proven.

Section 2. Analysis of Algorithm II.

Recall that $\Phi(u) = \int_D F(x,u)dx$. Define

$$\Phi_\sigma(u) = \Phi(u) + \sigma/2 \quad B(u,u). \tag{28}$$

LEMMA 2.1. *$\Phi_\sigma(u)$ is convex for $\sigma > \frac{K}{\lambda_0 \lambda_1}$ where λ_1 is the first Dirichlet eigenvalue of $-\Delta$ and*

$$\lambda_0 = \inf_{\substack{|\xi|=1 \\ x \in \bar{D}}} a_{ij}(x)\xi_1\xi_j > 0.$$

PROOF: We rewrite Φ_σ in the form

$$\Phi_\sigma(u) = \int_D (F(x,u) + K\frac{u^2}{2}dx + \left(\sigma/2 \quad B(u,u) - \frac{K}{2}\int_D u^2 dx\right)$$
$$\equiv \Phi_1(u) + \Phi_2(u).$$

Since $\frac{d^2}{dt^2}(F(x,t) + K\frac{t^2}{2}) = f_t(x,t) + K \geq 0$, Φ_1 is convex. Now

$$\frac{1}{2}(\Phi_2(u) + \Phi_2(v)) - \Phi_2\left(\frac{u+v}{2}\right) = \Phi_2\left(\frac{u-v}{2}\right)$$

so that Φ_2 is convex whenever $\Phi_2(u) \geq 0$ for $u \in \overset{o}{H}_1(D)$. But

$$\Phi_2(u) \geq \frac{\sigma\lambda_0}{2}\int_D |\nabla u|^2 dx - \frac{K}{2}\int_D u^2 dx$$
$$\geq \left(\frac{\sigma\lambda_0\lambda_1 - K}{2}\right)\int u^2 dx$$

so Φ_2 is convex for $\sigma\lambda_0\lambda_1 > K$ as required.

LEMMA 2.2. *For σ chosen as in Lemma 2.1 $u = T^\sigma(v)$ is well-defined if $\Phi_\sigma(v) > 0$.*

PROOF: Suppose $w + \sigma v = 0$ (see (10)). Then

$$\sigma L v = f(x,v). \tag{29}$$

Multiplying (29) by v and integrating by parts gives

$$\sigma B(v,v) + \int_D v f(x,v)dx = 0. \tag{30}$$

On the other hand, the convexity of $\Phi_\sigma(u)$ implies

$$0 = \Phi_\sigma(0) \geq \Phi_\sigma(v) - \Phi'_\sigma(v)v$$

or

$$\sigma B(v,v) + \int_D v f(x,v)dx \geq \Phi_\sigma(v) > 0$$

contradicting (30).

LEMMA 2.3. *Let $v \in A_R$ with $\Phi_\sigma(v) > 0$ and $u = T^\sigma(v)$. Then*

$$\Phi_\sigma(u) \geq \Phi_\sigma(v)$$

with equality if and only if $u = v$.

PROOF: Note that u satisfies

$$\begin{aligned} -Lu &= \lambda(f(x,v) - \sigma Lv) \quad &\text{in} \quad D \\ u &= 0 \quad &\text{on} \quad \partial D \end{aligned} \tag{31}$$

where $\lambda = \frac{R}{\|w+\sigma u\|}$. By Lemma 2.2 this is well-defined. From (31) we see that

$$\begin{aligned} B(u,u) &= \lambda\left(\int_D uf(x,v)dx + \sigma B(u,v)\right) \\ B(u,v) &= \lambda\left(\int_D vf(x,v)dx + \sigma B(v,v)\right). \end{aligned} \tag{32}$$

Since $B(u,v) \leq B(u,u)$, (32) implies that

$$\sigma(B(u,v) - B(v,v)) + \int_D (u-v)f(x,v)dx \geq 0. \tag{33}$$

Now the convexity of Φ_σ implies that

$$\begin{aligned} \Phi_\sigma(u) &\geq \Phi_\sigma(v) + \Phi_\sigma'(v)(u-v) \\ &= \Phi_\sigma(v) + \sigma(B(u,v) - B(v,v)) + \int_D (u-v)f(x,v)dx \\ &\geq \Phi_\sigma(v) \end{aligned} \tag{34}$$

by (33) and the lemma follows.

REMARK: Since $u, v \in A_R$, Lemma 2.3 says that $\Phi(u) \geq \Phi(v)$.

COROLLARY 2.4. *Let $\{u_j\}$ be defined by Algorithm II. Then*

$$\begin{aligned} \|u_{j+1} - u_j\|^2 &\leq 2\lambda_{j+1}(\Phi_\sigma(u_{j+1}) - \Phi_\sigma(u_j) \\ &= 2\lambda_{j+1}(\Phi(u_{j+1}) - \Phi(u_j)). \end{aligned}$$

PROOF: As in Corollary (1.2) with $u = u_{j+1}$, $v = u_j$, $\lambda = \lambda_{j+1}$,

$$\begin{aligned} \|u-v\|^2 = 2B(u,u-v) &= 2\lambda\left(\int_D (u-v)f(x,v) + \sigma(B(u,v) - B(v,v))\right) \\ &\leq 2\lambda(\Phi_\sigma(u) - \Phi_\sigma(v)) \end{aligned}$$

by (34).

COROLLARY 2.5. $0 < \lambda_{j+1} \le R^2/\Phi_\sigma(u_0)$.

PROOF: Observe that from (32)

$$\lambda_{j+1} = \frac{B(u_{j+1}, u_j)}{\int_D u_j f(x, u_j)dx + \sigma B(u_j, u_j)}.$$

Using (34) with $u = 0$, $v = u_j$ we have

$$\int_D u_j f(x, u_j)dx + \sigma B(u_j, u_j) \ge \Phi_\sigma(u_j).$$

Since $B(u_{j+1}, u_j) \le R^2$ and $\Phi_\sigma(u_j) \ge \Phi_\sigma(u_0)$, the Corollary follows.

PROOF OF THEOREM 3: We have shown in Lemma 2.3 that $\Phi_\sigma(u_{j+1}) \ge \Phi_\sigma(u_j)$ so part i) is finished. Exactly as in the proof of Theorem 1 using Corollaries 2.4 and 2.5 we see that we can choose a subsequence (of the given subsequence) converging weakly to a solution u of (12) with $\lambda > 0$. Moreover $\Phi(u) = \lim_{j\to\infty} \Phi(u_j) > 0$ so that $u \ne 0$. Hence $1 - \lambda\sigma \ne 0$. To complete the proof of part ii) we show strong convergence if $1 - \lambda\sigma > 0$. We may assume that $1 - \lambda_{j_i+1}\sigma \ge \theta > 0$ for j_i large. As in the proof of Theorem 1,

$$\frac{1}{2}\|u - u_{j_i+1}\|^2 \le B(u_{j_i+1}, u_{j_i+1} - u). \tag{35}$$

But,

$$B(u_{j_i+1}, u_{j_i+1} - u) = \lambda_{j_i+1}\left(\int_D (u_{j_i+1} - u) f(x, u_{j_i}) + \sigma B(u_{j_i}, u_{j_i+1} - u)\right). \tag{36}$$

From Corollary 2.4

$$\begin{aligned} B(u_{j_i}, u_{j_i+1} - u) &= B(u_{j_i+1}, u_{j_i+1} - u) + B(u_{j_i} - u_{j_i+1}, u_{j_i+1} - u) \\ &\le B(u_{j_i+1}, u_{j_i+1} - u) + C(\Phi(u_{j_i+1}) - \Phi(u_{j_i}))^{\frac{1}{2}}. \end{aligned} \tag{37}$$

Combining (36),(37) (using $1 - \lambda_{j_i+1}\sigma > \theta$) gives

$$\begin{aligned} \theta B(u_{j_i+1}, u_{j_i+1} - u) \le \lambda_{j_i+1} &\int (u_{j_i+1} - u) f(x, u_{j_i}) \\ &+ C(\Phi(u_{j_i+1}) - \Phi(u_{j_i}))^{\frac{1}{2}}. \end{aligned} \tag{38}$$

Exactly as in the proof of Theorem 1, the right-hand side of (38) tends to zero. Thus (35) and (38) show the strong convergence.

To prove part iii) we use the well-known Pohazaev identity. Namely, u is a solution of

$$\begin{aligned} \Delta u + \frac{\lambda}{1-\lambda\sigma} f(u) &= 0 \quad \text{in} \quad D \\ u &= 0 \quad \text{on} \quad \partial D. \end{aligned} \tag{39}$$

Multiplying (39) by $x \cdot \nabla u$ (D is star-shaped about the origin) and integrating by parts gives

$$\left(1 - \frac{n}{2}\right) \int_D |\nabla u|^2 dx + \frac{\lambda n}{1-\lambda\sigma} \int_D f(u) dx = \frac{1}{2} \int_{\partial D} (x \cdot \nu) u_\nu^2 d\sigma. \tag{40}$$

Therefore, (40) gives that

$$\frac{\lambda n}{1-\lambda\sigma} \int_D f(u) dx > \frac{n-2}{2} \int_D |\nabla u|^2 dx > 0.$$

Since $\int_D f(u) dx > 0$, we must have $1 - \lambda\sigma > 0$, concluding the proof.

Department of Mathematics and Statistics
University of Massachusetts, Amherst, MA 01003

Research at MSRI supported in part by NSF Grant DMS-812079-05.

REFERENCES

1. C. Coffman, *Uniqueness of the Ground State Solution for $\Delta u - u + u^3 = 0$ and a Variational Characterization of Other Solutions*, Arch. Rat. Mech. Anal. **46** (1972), 81–95.
2. A. Eydeland, *A Method of Solving Nonlinear Variational Problems by Nonlinear Transformation of the Objective Functional, Part I*, Numerische Math. **43** (1984), 59–82.
3. A. Eydeland and B. Turkington, *On the Computation of Nonlinear Planetary Waves*, preprint.
4. K. Georg, *On the Convergence of an Inverse Iteration Method for Nonlinear Elliptic Eigenvalue Problems*, Number. Math. **32** (1979), 69–74.

Radial Symmetry of the Ground States for a Class of Quasilinear Elliptic Equations

BRUNO FRANCHI AND ERMANNO LANCONELLI

1. Introduction.

Let us consider the semilinear equation in $\mathbb{R}^n$

$$\Delta u + f(u) = 0. \tag{1}$$

In [**3**], Gidas, Ni and Nirenberg proved that if $f(u) = -u + g(u)$ near $u = 0$, with, roughly speaking, $g(u) = O(u^{1+\alpha})$ for given $\alpha > 0$, then the ground states of (1.a) are necessarily spherically symmetric about some point. Throughout this note, a *ground state* will be a positive, twice differentiable solution of the given equation in $\mathbb{R}^n$ vanishing at the infinity.

The aim of this note is to prove the radial symmetry of the ground states of the quasilinear equation

$$\operatorname{div}\left(\frac{Du}{\sqrt{1+|Du|^2}}\right) + f(u) = 0 \tag{2}$$

under the same assumptions on the function f.

A basic tool of the proof is an L^∞-estimate of Du that we can obtain, for example, from the results in [**5**]. We note (see Remark 2.2 for the details) that the same technique may be applied to more general quasilinear equations in $\mathbb{R}^n$ if an analogous estimate is known.

Finally, both existence and uniqueness of the radial ground states for a general class of quasilinear elliptic equations in $\mathbb{R}^n$ (containing (2)) are proved in [**1**]; these results are announced in [**2**].

2. Symmetry Results.

In this section, we shall prove the following result.

THEOREM 2.1. *Let u be a positive twice differentiable solution of*

$$Lu = \operatorname{div}\left(\frac{Du}{\sqrt{1+|Du|^2}}\right) + f(u) = 0 \quad in\ \mathbb{R}^n, n \geq 2$$

with $u(x) \to 0$ as $|x| \to \infty$. Assume that $f(u) = -u + g(u)$ near zero, where g is a $C^{1+\alpha}$-function such that $g(0) = g'(0) = 0$. Then u is spherically symmetric about some point $\mathbb{R}^n$.

The proof will be carried out following the technique in [**3**]. We put $A(p) = (1+p^2)^{-\frac{1}{2}}$ and $E(p) = (pA(p))'$. Note that $E(p) > 0$ for every $p \geq 0$.

Step 1. For every $\varepsilon \in (0,1)$ we have $u(x) = 0(\exp(-\varepsilon|x|))$ as x goes to infinity.

In fact, put $v_\lambda(x) = \lambda \exp(-\varepsilon|x|))$, where $\lambda > 0$ will be fixed in the sequel. We get

$$\begin{aligned}
&Lv_\lambda/E(|Dv_\lambda|) \\
&\quad = v_\lambda\left(\varepsilon^2 - \varepsilon(n-1)r^{-1}(A/E)(|Dv_\lambda|) - 1 + g(v_\lambda)/(E(|Dv_\lambda|)v_\lambda)\right) \\
&\quad \leq v_\lambda\left(\varepsilon^2 - 1 + g(v_\lambda)/E(|Dv_\lambda|)v_\lambda)\right).
\end{aligned}$$

Now put

$$c_1 = \left(\sup_{0\leq t\leq 1} |g'(t)t^{-\alpha}|\right) \Big/ \left(\inf_{0\leq t\leq 1} E(t)\right).$$

Since $|Dv_\lambda| = \varepsilon|v_\lambda|$, we have

$$Lv_\lambda/E(|Dv_\lambda|) \leq v_\lambda(\varepsilon^2 - 1 + c_1 v_\lambda^\alpha), \tag{3}$$

where $v_\lambda \leq 1$.

Let us choose $r^* > 0$ such that

$$u(x) < ((1-\varepsilon^2)c_1^{-1})^{1/\alpha} \text{ if } |x| \geq r^*. \tag{4}$$

Without loss of generality, we may suppose $((1-\varepsilon^2)c_1^{-1}) < 1$. Moreover, let us choose λ such that

$$\lambda \exp(-\varepsilon r^*) = ((1-\varepsilon^2)c_1^{-1})^{1/\alpha}. \tag{5}$$

Then, if $|x| \geq r^*$,

$$v_\lambda(x) \leq \lambda \exp(-\varepsilon r^*) = ((1-\varepsilon^2)c_1^{-1}) < 1,$$

so that, by (3) and (5),

$$Lv_\lambda/E(|Dv_\lambda|) \leq v_\lambda(\varepsilon^2 - 1 + 1 - \varepsilon^2) = 0$$

if $|x| \geq r^*$. Moreover, if $|x| = r^*$ by (4) we have

$$v_\lambda(x) = \lambda \exp(-\varepsilon r^*) = ((1-\varepsilon^2)c_1^{-1})^{1/\alpha} > u(x).$$

Thus the assertion follows via standard comparison principles (for the sake of simplicity, we assume that $f'(t) < 0$ if $0 \leq t < 1$).

Step 2. We have $u(x), Du(x) = O(|x|^{(1-n)/2}\exp(-|x|))$ as x goes to infinity.

By the results in [5] we have $C^* = \sup|Du| < +\infty$. In fact, it is enough to apply Theorem 4 of [5] to the balls $B(x, \frac{1}{2}) = \Omega'$ and $B(x,1) = \Omega$, keeping in mind that, by our hypotheses on f, outside of a given compact subset of $\mathbb{R}^n$ we may suppose $f'(u) \leq 0$.

Now, by standard Hölder estimates for C^2 solutions of quasilinear elliptic equations, we get

$$[Du]_{\sigma, B(x,1)} \leq c_2 \quad \forall x \in \mathbb{R}^n,$$

where c_2 and σ depend only on $\sup u$ and C^*. Thus u is a solution of the linear elliptic equation

$$\sum_{i,j} a_{ij}(x)\partial_{ij}u + c(x)u = 0, \tag{6}$$

where

$$a_{ij}(x) = \delta_{ij}A(|Du(x)|) + \big(A'(|Du(x)|)/|Du(x)|\big)\partial_i u(x)\partial_j u(x)$$

and

$$c(x) = f(u(x)/\int_0^1 ds f'(su(x)).$$

Due to our hypotheses on u and f, the coefficients a_{ij} and c are Hölder continuous functions. Then, by classical Schauder estimates, for every $\varepsilon < 1$ we have

$$\sup_{B(x,\frac{1}{2})} |Du| + \sup_{B(x,\frac{1}{2})} |D^2 u| \leq c_3 \sup_{B(x,1)} u \leq c_4 \exp(-\varepsilon|x|),$$

where c_3, c_4 depend only on $\sup u$, C^* and ε.

Since the main equation can be written in the form

$$(7)\quad \Delta u - u = \sum_{i,j}(A'/A)(|Du|)\partial_i u\, \partial_j u\, \partial_{ij} u\, |Du|^{-1} - \big[u(A(|Du|) - 1) + g(u)\big]/A(|Du|),$$

we get

$$\Delta u - u = h = O(\exp(-(\alpha+1)\varepsilon|x|) \quad \forall \varepsilon < 1.$$

If we choose ε such that $(\alpha+1)\varepsilon > 1$, the assertion follows by the representation formula

$$(8)\qquad u(x) = \int G(x-y)h(y)\,dy,$$

where G is the Green function for $-\Delta + 1$ in $\mathbb{R}^n$.

Step 3. The radial symmetry of u follows exactly in the same way as in the proof of Theorem 2 in [**3**] keeping in mind that:

(i) denoting by x^λ the symmetric point of x with respect to a fixed hyperplane and putting $u^\lambda(x) = u(x^\lambda)$, we have

$$(Lu^\lambda)(x) = (Lu)(x^\lambda);$$

(ii) by the formula (7), we can write

$$Lu = \Delta u - u + \sum_j b_j\, \partial_j u + cu$$

where b_j, $c = O(\exp(-\beta|x|))$ as x goes to infinity, for a suitable $\beta > 0$, $j = 1, \ldots, n$;

(iii) by the representation formula (8), we can apply Proposition 4.2 in [**3**] to the solution u.

This completes the proof. □

REMARK 2.2: Let us consider, instead of (2), the more general quasilinear equation in $\mathbb{R}^n$

$$(9)\qquad \operatorname{div}(A(|Du|)Du) + f(u) = 0,$$

where A is continuously differentiable in $[0, \infty)$ and f satisfies the assumptions of Theorem 2.1. It is not difficult to see that the above arguments can be applied to ground states u of (9) if we suppose $A(0) = 1$, $A' \leq 0$, and $E(p) = (pA(p))' > 0$ in $[0, \infty)$ when Du is bounded on all of $\mathbb{R}^n$. Radial symmetry of the ground states therefore follows, for example, for the generalized mean curvature operator

$$Lu = \operatorname{div}\big((1 + |Du|^2)^{-s/2} Du\big) + f(u)$$

for $0 \leq s \leq 1$. In fact, the boundedness of the gradient holds in the case $0 \leq s < 1$ by the results in [**4**], and the case $s = 1$ has been established earlier.

REMARK 2.3: Geometrically, the equation (2) describes an n-manifold V whose mean curvature depends only on the distance from the hyperplane $x_{n+1} = 0$ and that is tangent at infinity to the same hyperplane. Here, we restrict ourselves to the case where V is the graph of a given function. The general problem (V a connected immersed regular n-manifold) has been recently studied by Atkinson, Peletier and Serrin (see these Proceedings). Also in this case we can prove that V is rotationally symmetric with respect to some axis normal to $x_{n+1} = 0$, if asymptotically V has no vertical tangent hyperplanes. The proof can be carried out by re-examining the proof of Theorem 2.1 and by using the technique of moving hyperplanes together with a suitable version of the strong maximum principle for quasilinear equations as in the paper by J. Serrin [**6**].

Dipartimento di Matematica, Piazza di Porta S. Donato, 5, I-40126 Bologna, ITALY

The authors are supported by the G.N.A.F.A. of C.N.R. and M.P.I, Italy

Research at MSRI supported in part by NSF Grant DMS 812079-05.

REFERENCES

1. B. Franchi, E. Lanconelli and J. Serrin, *Existence and uniqueness of nonnegative solutions of quasilinear equations in* $\mathbb{R}^n$, to appear.
2. B. Franchi, E. Lanconelli and J. Serrin, *Existence and uniqueness of ground state solutions of quasilinear elliptic equations*, in "Nonlinear Diffusion Equations and their Equilibrium States," J. Serrin, ed., to appear.
3. B. Gidas, W.-M. Ni and L. Nirenberg, *Symmetry of positive solutions of nonlinear elliptic equations in* $\mathbb{R}^n$, in "Mathematical Analysis and Applications, Advances in Math," Academic Press, 1981, pp. 369–402.
4. O. A. Ladyzhenskaya and N. N. Ural'tseva, "Linear and Quasilinear Elliptic Equations," Academic Press, 1968.
5. O. A. Ladyzhenskaya and N. N. Ural'tseva, *Local estimates for gradient of solutions of non-uniformly elliptic and parabolic equations*, Comm. Pure Appl. Math. **23** (1970), 677–703.
6. J. Serrin, *A symmetry problem in potential theory*, Arch. Rat. Mech. Anal. **43** (1971), 304–318.

Existence and Uniqueness of Ground State Solutions of Quasilinear Elliptic Equations

BRUNO FRANCHI, ERMANNO LANCONELLI AND JAMES SERRIN

1. In this note we study the existence and the uniqueness of radially symmetric ground states for the quasilinear elliptic equation

$$\operatorname{div}(A(|Du|)Du) + f(u) = 0. \tag{1.1}$$

Here, by a ground state we mean a nonnegative classical solution of (1.1) on $\mathbb{R}^n$, vanishing at infinity but not identically zero. Our results improve those stated in [**4**]. Complete proofs will appear elsewhere [**5**].

We note that in the semilinear case ($A \equiv 1$) if $f(u) = -u + g(u)$ near $u = 0$, with, roughly speaking, $g(u) = 0(u^{1+\alpha})$ for a given $\alpha > 0$, then positive ground states are necessarily spherically symmetric [**6**]. An analogous result holds for some classes of quasilinear equations including the mean curvature operator [**3**]. Thus, taking into account the radial symmetry of equation (1.1), it is natural to restrict ourselves to the case of radial ground states. Putting $u = u(r)$, with $r = |x|$, the problem is then reduced to the following one:

$$(*) \qquad \begin{aligned} &(A(|u'|)u')' + \frac{n-1}{r}A(|u'|)u' + f(u) = 0, \quad r > 0 \\ &u \geq 0 \text{ if } r \geq 0, \quad u'(0) = 0, \quad u(r) \to 0 \text{ as } r \to \infty. \end{aligned}$$

We always suppose that A and f are continuous functions on $(0, \infty)$ such that $f(u) \to 0 = f(0)$ as $u \to 0$, $pA(p)$ is strictly increasing and $pA(p) \to 0$ as $p \to 0$.

Further hypotheses will be introduced in the sequel. In any case, however, all of our results apply in particular to the following important cases:

(a)	$A(p) \equiv 1$	(Laplace operator);
(b)	$A(p) = p^{m-2}, \quad 1 < m < 2$	(degenerate Laplace operator);
(c)	$A(p) = (1 + p^2)^{-1/2}$	(mean curvature operator)

or more generally

$$(d) \qquad A(p) = (1+p^2)^{-s/2}, \quad 0 < s \le 1.$$

We note that, for the results in Sections 2 and 3, the condition $m \le 2$ in (b) can be avoided.

Since we are dealing with problem $(*)$, we may consider n as a real parameter greater than 1. The results in the case $n = 1$ are slightly different, but simpler; for the sake of brevity, we shall omit this case. In the semilinear case ($A \equiv 1$) the existence of radial ground states was studied by several authors (Strauss, Berestycki and Lions, Atkinson and Peletier; see [**11**,**2**,**1**] and the references therein). The uniqueness of radial ground states is further treated by H. G. Kaper and Man Kam Kwong (cf. their article [**7**] in these proceedings), who have improved the results of Peletier and Serrin.

In Section 2 we will deal with general properties of solutions; Section 3 contains our existence results, and finally, Section 4 is devoted to the uniqueness theorems.

2. In the sequel we shall say that a continuously differentiable function u is a classical solution of the equation in $(*)$ if $u'(0) = 0$, $w = A(|u'|)u'$ is continuously differentiable in $(0, \infty)$ and

$$w' + \frac{n-1}{r} w + f(u) = 0, \quad r > 0,$$

where we put $w = 0$ if $u' = 0$. In order to state our results we will use the notation

$$H(p) = \int_0^p \rho d(\rho A(\rho)), \quad H(\infty) = \lim_{p \to \infty} H(p);$$

$$F(u) = \int_0^u f(t)\, dt.$$

If u is a classical solution, we get

$$(2.1) \qquad H(p_1) - H(p_0) + (n-1) \int_{r_0}^{r_1} A u'^2 \tfrac{dr}{r} = F(u_0) - F(u_1),$$

where $r_0, r_1 \ge 0$, $u_i = u(r_i)$, $p_i = |u'(r_i)|$, $i = 0, 1$. It follows without great difficulty that if u is a classical solution of $(*)$ then

$$(2.2) \qquad u'(r) \to 0 \text{ as } r \to 0;$$

$$(2.3) \qquad u'(r) \le 0 \text{ for } r \ge 0$$

hence if $u(R) = 0$ for a given $R > 0$ then $u \equiv 0$ on $[R, \infty)$;

$$(2.4) \qquad f(u(0)) \geq 0;$$

$$(2.5) \qquad (n-1)\int_0^\infty A u'^2 \frac{dr}{r} = F(u(0)).$$

In particular, if problem $(*)$ has a solution, the following necessary condition is verified:

(2.6) There exists a positive number δ such that $F(\delta) > 0$.

Putting

$$(2.7) \qquad \beta = \inf\{u > 0; F(u) > 0\}$$

we see from (2.5) that $u(0) > \beta$. Moreover, another necessary condition is

$$(2.8) \qquad \max_{[0,\beta]} |F| < H(\infty).$$

Condition (2.8) is satisfied for any f in case of the (degenerate) Laplace operator, since in this case $H(\infty) = \infty$. On the other hand, (2.8) is quite restrictive for the mean curvature operator, since then $H(\infty) = 1$.

Some stronger regularity properties of the functions f and A give us a more precise description of the solutions of $(*)$. In fact, let us suppose

(R1) *f is Lipschitz continuous on* $(0, \infty)$,

(R2) *$E(p)$ is bounded away from zero on any interval* $(0, \varepsilon)$,

where $E(p)$ denotes the right derivative of the function $p \to pA(p)$. Then

$$(2.3') \qquad u'(r) < 0 \text{ for } r > 0 \text{ as long as } u(r) > 0;$$

$$(2.4') \qquad f(u(0)) > 0.$$

We note that condition (R2) is not satisfied by the degenerate Laplace operator if $m > 2$.

If f is Lipschitz continuous up to zero, then the ground states are strictly positive. In other cases, compactly supported ground states are allowed. In fact, if $\beta > 0$, a *sufficient* condition for the existence of compactly supported ground states is that

$$(2.10) \qquad \int_0 \frac{1}{H^{-1}(|F(s)|)}\, ds < \infty.$$

(for the case $A \equiv 1$ see [9]). This condition is also *necessary* if $p^2 A(p)/H(p)$ is bounded near zero (as in the special cases (a)–(d) noted above).

3. By a shooting method, we can prove the following existence theorems.

THEOREM 3.1. *Suppose $\beta > 0$ and assume that*

(i) $F(u) < 0$ *for* $0 < u < \beta$, $\quad f(\beta) > 0$;
(ii) $F(\gamma) + \sup_{[0,\beta]} |F| < H(\infty)$,

where $\gamma = \sup\{t > \beta; f(u) > 0$ *for* $\beta < u < t\}$. *Then, problem* $(*)$ *has at least one solution.*

If $H(\infty) = \infty$ (as in the case of the degenerate Laplace operator) condition (ii) is automatically satisfied if $\gamma < \infty$. When $\gamma = \infty$ we can prove a stronger result.

THEOREM 3.2. *Suppose the hypotheses of Theorem 3.1 hold with the exception that condition (ii) is replaced by*

(iii) $H(\infty) = \infty, \quad \gamma = \infty, \quad \liminf_{u \to \infty} H^{-1}(F(u))/u = 0.$

Then, problem $(*)$ *has at least one solution.*

4. In this section we state our main uniqueness theorems and sketch the main steps of the proofs.

THEOREM 4.1. *Let f satisfy condition (R1) and let A be a continuously differentiable function on $(0,\infty)$ satisfying (R2). Moreover, let us assume $n \geq 3/2$ and*

(T1) $\beta > 0$;
(T2) $A(p)p^{2-m} \to 1$ *as* $p \to 0$, *for a suitable* $m \in (1,2]$;
(T3) $t \to (A(t^{1/2}))^{-2}$ *is concave;*
(T4) f *is a decreasing function on* (β,η), *where* $\eta \in (\beta,\infty]$.

Then $(*)$ *has at most one solution u such that* $u(0) \leq \eta$.

THEOREM 4.2. *Suppose the hypotheses of Theorem 4.1 hold, with the exception that condition (T4) is replaced by*

(T4′) *the function* $u \to f(u)(u-\beta)^{1-m}$ *is positive and decreasing for* $\beta < u < \eta$.

Assume also that

(T5) E/A *is decreasing and* $p^{2-m}E$ *is increasing on* $(0,\infty)$.

Then, $(*)$ *has at most one solution* u *such that* $u(0) \le \eta$.

We note that if $\eta = \infty$ or if $f(u) \le 0$ for $u \ge \eta$ then the condition $u(0) \le \eta$ is unnecessary, since $f(u(0)) > 0$ as we noted in Section 2.

Obviously (T4′) is weaker than (T4); on the other hand (T5) is satisfied by the degenerate Laplace operator but not satisfied, for example, by the mean curvature operator. Finally, we note that if $m = 2$ condition (T4′) is the same as condition (S) of [**9**,**10**].

The proofs of Theorems 4.1 and 4.2 are accomplished as follows. Suppose for contradiction that $(*)$ has two different solutions u and v with $u(0) \le \eta$ and $v(0) \le \eta$. Then $u' < 0$ ($v' < 0$) as long as $u > 0$ ($v > 0$). Let $r = r(u)$ and $s = s(u)$ be the inverse functions of $u(r)$ and $v(r)$, defined respectively on the intervals $(0, u(0)]$ and $(0, v(0)]$. We show successively that

(i) there exists $\bar{u} > 0$ such that $r(\bar{u}) = s(\bar{u})$;
(ii) if $r(\bar{u}) = s(\bar{u})$ for some $\bar{u} > 0$, then $\bar{u} > \beta$;
(iii) if $r(\bar{u}) = s(\bar{u})$ for some $\bar{u} > 0$, then $\bar{u} \le \beta$.

Thus we get a contradiction and the uniqueness is proved.

This procedure follows the scheme introduced by Peletier and Serrin for the semilinear case. We start with an important, and rather unusual, identity. Denote by $l = l(u,p)$ the *positive* function defined implicitly by the equation

$$H(l) = H(p) + F(u) \tag{3.1}$$

on the domain

$$\mathsf{P} = \{(u,p); \quad p > 0, \quad 0 < H(p) + F(u) < H(\infty)\}.$$

Moreover we put

$$K(u,p) = 2(n-1)A(l)\{p^2 A(p) - l^2 A(l)\}. \tag{3.2}$$

The following lemma can be proved by direct calculation.

LEMMA 4.3. *Suppose that condition (T1) holds. If u is a solution of $(*)$ such that $0 < u(r) \le \beta$ on (r_0, r_1) then*

$$(r^{n-1} la(l))^2 \Big|_{r_0}^{r_1} = -\int_{r_0}^{r_1} r^{2n-3} K(u,p)\,dr, \tag{3.3}$$

where $p = p(r) = |u'(r)|$ and $l = l(r) = l(u(r), p(r))$.

Now, since H is an increasing function, the condition $F(u) \le 0$ implies $l \le p$ and hence $K(u,p) \ge 0$. Thus from (3.3) it follows that

$$\text{limit } \lambda = \lim_{r\to\infty} r^{n-1} lA(l) \ \text{ exists } (\ge 0). \tag{3.4}$$

Moreover if (T1) and (T2) are satisfied then one can show that

$$\begin{aligned} &\lambda = 0 \ \text{ when } n \le m \\ &\lambda = ((n-m)/(m-1))^{m-1} \lim_{r\to\infty} r^{n-m}(u(r))^{m-1} \ \text{ when } n > m. \end{aligned} \tag{3.5}$$

A further consequence of (3.3) is the following monotone separation lemma.

LEMMA 4.4. *Suppose that hypothesis (T1) is satisfied. If $r - s > 0$ on some interval $(0, u)$, then $(r-s)' < 0$ on $(0, u)$.*

Step (i) of the proof now follows directly from Lemma 4.4.

In order to prove step (ii), we assume for contradiction that $\bar{u} \le \beta$. By (3.3) and (3.4) we get

$$\begin{aligned} &R^{2(n-1)}\{L^2 A^2(L) - M^2 A^2(M)\} - (\lambda^2 - \mu^2) \\ &\qquad = \int_0^{\bar{u}} \left(r^{2n-3} \frac{K(u,p)}{p} - s^{2n-3} \frac{K(u,q)}{q} \right) du, \end{aligned} \tag{3.6}$$

where $R = r(\bar{u})$, $L = l(\bar{u}, |u'(R)|)$, $p = p(u) = |u'(r(u))|$, λ is defined by (3.4) and M, q and μ are the corresponding quantities for v. Now, since we can prove (again using the identity (3.3)) that r and s intersect only at a finite number of points, we may assume $r - s > 0$ on $(0, \bar{u})$ and hence $r' - s' < 0$ on $(0, \bar{u}]$.

Therefore $L < M$, while also $\lambda \ge \mu$ by (3.5) since $u \ge v$ on (R, ∞). Hence the left-hand side of (3.6) is strictly negative. On the other hand, the right-hand side of (3.6) is nonnegative. To see this, we note first that $r > s$ and hence $p < q$. Next, it can be shown with the help of the hypothesis (T3) that

$$\frac{\partial}{\partial t}(K(u,t)/t) \le 0.$$

Consequently $K(u,p)/p \geq K(u,q)/q$ and the result follows. We thus obtain a contradiction, proving step (ii).

Step (iii) when hypothesis (T4) holds is a consequence of the comparison principle for second order quasilinear equations. For the remaining case (T4′) and (T5) the proof can be carried out by appropriately adapting the argument of Peletier and Serrin [**9**, Lemma 10].

Dipartimento di Matematica, Università di Bologna

Dipartimento di Matematica, Università di Bologna

Department of Mathematics, University of Minnesota

The second author is partially supported by grants from the Italian government.

REFERENCES

1. F. V. Atkinson and L. A. Peletier, *Ground states of $-\Delta u = f(u)$ and the Emden-Fowler equation*, Arch. Rational Mech. Anal. **93** (1986), 103–127.
2. H. Berestycki and P. L. Lions, *Nonlinear scalar field equations I*, Arch. Rational Mech. Anal. **82** (1983), 313–345.
3. B. Franchi and E. Lanconelli, *Radial symmetry of ground states for a class of quasilinear elliptic equations*, in "Proceedings of the MSRI Microprogram in Diffusion Equations and Their Equilibrium States," Springer-Verlag, 1988.
4. B. Franchi, E. Lanconelli and J. Serrin, *Esistenze e unicità degli stati fondamentali per equazioni ellittiche quasilineari*, Atti Accad. Naz. Lincei (8) **79** (1985), 121–126.
5. B. Franchi, E. Lanconelli and J. Serrin, *Existence and uniqueness of nonnegative solutions of quasilinear equations in $\mathbb{R}^n$*, to appear.
6. B. Gidas, W. M. Ni and L. Nirenberg, *Symmetry of positive solutions of nonlinear elliptic equations in $\mathbb{R}^n$*, Advances in Math. Studies **7A** (1981), 369–402.
7. H. G. Kaper and M. K. Kwong, *Uniqueness of nonnegative solutions of semilinear elliptic equations*, in "Proceedings of the MSRI Microprogram in Diffusion Equations and Their Equilibrium States," Springer-Verlag, 1988.
8. K. McLeod and J. Serrin, *Uniqueness of positive radial solutions of $\Delta u + f(u) = 0$ in $\mathbb{R}^n$*, Arch. Rational Mech. Anal. **99** (1987), 115–145.
9. L. A. Peletier and J. Serrin, *Uniqueness of positive solutions of semilinear equations in $\mathbb{R}^n$*, Arch. Rational Mech. Anal. **81** (1983), 181–197.
10. L. A. Peletier and J. Serrin, *Uniqueness of non-negative solutions of semilinear equations in $\mathbb{R}^n$*, J. Diff. Eq. **61** (1986), 380–397.
11. W. A. Strauss, *Existence of solitary waves in higher dimensions*, Comm. Math. Phys. **55** (1977), 149–162.

Blow-up of solutions of nonlinear parabolic equations

AVNER FRIEDMAN

§1. Sufficient conditions for blow-up.

Consider the first initial-boundary value problem

$$\frac{\partial u}{\partial t} - Lu = F(u, \nabla_x u) \quad \text{in} \quad \Omega \times (0, \infty), \tag{1.1}$$

$$u = 0 \quad \text{on} \quad \partial\Omega \times (0, \infty), \tag{1.2}$$

$$u(x, 0) = \phi(x) \quad \text{on} \quad \Omega \tag{1.3}$$

where Ω is a bounded domain in $\mathbb{R}^N$ with C^2 boundary,

$$Lu = \sum_{i,j=1}^{N} \frac{\partial}{\partial x_j}(a_{ij}(x) u_{x_j}) + \sum_{i=1}^{N} b_i(x) u_i + c(x) u,$$

(a_{ij}) is uniformly positive definite matrix, the coefficients of L are sufficiently smooth in $\bar{\Omega} \times [0, \infty)$, say in C^1, ϕ is Holder continuous in $\bar{\Omega}$, and $F(u, w)$ is Lipschitz continuous in $\mathbb{R}^{N+1}$.

By standard parabolic theory, there exists a unique local solution of (1.1)-(1.3); the solution can be continued step-by-step in time provided one can establish the following a priori estimate:

For any $T > 0$, if the solution exists in $\Omega_T \equiv \Omega \times (0, T)$ then

$$|u|_{L^\infty(\Omega_T)} + |\nabla_x u|_{L^\infty(\Omega_T)} \le C_T \tag{1.4}$$

where C_T is a positive constant.

Consider the case where

$$\begin{gathered} \phi \ge 0, \; F(u, \nabla_x u) = f(u) \ge 0, f'(u) > 0 \quad \text{if} \quad u > 0, \\ f \quad \text{convex}, \quad \int^{\infty} \frac{du}{f(u)} < \infty \end{gathered} \tag{1.5}$$

THEOREM 1.1. *If (1.5) holds and if $\int_\Omega \phi$ is sufficiently large then there exists a $T > 0$ such that the solution of (1.1)-(1.3) exists in Ω_T but it does not exist in $\Omega_{T+\epsilon}$ for any $\epsilon > 0$.*

PROOF: By the maximum principle we deduce that $u \geq 0$. Denote by $\psi(x)$ the principal eigenfunction of L^*, the adjoint of L, i.e.,

$$\begin{aligned} L^*\psi + \lambda\psi = 0 \quad &\text{in} \quad \Omega, \\ \psi > 0 \quad &\text{in} \quad \Omega, \\ \psi = 0 \quad &\text{on} \quad \partial\Omega \end{aligned}$$

and normalize ψ by $\int_\Omega \psi = 1$. Consider the function

$$g(t) = \int_\Omega u(x,t)\psi(x)dx.$$

Then

$$g'(t) = \int_\Omega u_t\psi = \int_\Omega (Lu + f(u))\psi = \int_\Omega uL^*\psi + \int f(u)\psi$$

Since $L^*\psi = -\lambda\psi$ and since $f(u)$ is convex, we get

$$\begin{aligned} g'(t) &\geq -\lambda \int_\Omega u\psi + f\Big(\int_\Omega u\psi\Big) \\ &= -\lambda g(t) + f\big(g(t)\big) \end{aligned}$$

from which the assertion of the theorem easily follows.

The above well-known method can be extended to other equations. Consider

$$(1.6) \qquad \begin{aligned} &F(u, \nabla_x u) = \sum_{i=1}^N \frac{\partial h_i(u)}{\partial x_i} + f(u) \quad \text{where} \\ &\big|h_i(u)\big| \leq C\big(|u|^q + 1\big), f(u) \geq \mu|u|^p - C \quad (\mu > 0, C > 0), \\ &f \geq 0, f'(u) > 0 \quad \text{if} \quad u > 0, \ f \quad \text{convex.} \end{aligned}$$

THEOREM 1.2. *If $p > q > 1$ then the assertion of Theorem 1.1 is again valid.*

In the special case $L = \Delta, N = 1, g_1(u) = u^2$ this theorem was proved by Friedman and Lacey [**14**] by constructing a subsolution which increases to $+\infty$ in finite time.

PROOF: (Due to Xinfu Chen) by the maximum principle, $u \geq 0$. Consider

$$g(t) = \int_\Omega u(x,t)\phi^{\alpha+1}(x,t)dx \qquad (\alpha > 0)$$

where ψ is now normalized by $\int \psi^{\alpha+1} = 1$. Then

$$g'(t) = \int_\Omega Lu \cdot \psi^{\alpha+1} + \sum \int_\Omega \frac{\partial h_i(u)}{\partial x_i}\psi^{\alpha+1} + \int_\Omega f(u)\psi^{\alpha+1}$$
$$\geq \int_\Omega uL^*\psi^{\alpha+1} - \sum(\alpha+1)\int_\Omega h_i(u)\psi^\alpha\psi_{x_i} + \int_\Omega f(u)\psi^{\alpha+1}.$$

Notice that

$$L^*\psi^{\alpha+1} \geq (\alpha+1)\psi^\alpha L^*\psi - C_0\psi^{\alpha+1},$$
$$\sum_i \left|\int h_i(u)\psi^\alpha\psi_{x_i}\right| \leq C_0 \int u^q\psi^\alpha|\nabla\psi| + C_0$$
$$\leq \delta \int (u^q\psi^\alpha)^{p/q} + C_1 \qquad \left((\alpha+1)\delta = \frac{\mu}{2}\right.$$

where C_j are constants. Hence, with

$$\alpha = \frac{q}{p-q}$$

we get

$$g'(t) \geq -C_2 g(t) + \frac{\mu}{2}\int_\Omega u^p\psi^{\alpha+1} - C_3$$
$$\geq -C_2 g(t) - \frac{\mu}{2}(g(t))^p - C_3$$

and the assertion easily follows.

Another approach to establishing non-existence of global solutions was introduced by Levine [**25**]. This method consists of establishing concavity of $G(t)^{-\alpha}$ for some $\alpha > 0$, where

$$G(t) = \int_0^t ||u||^2 + (T-t)||\phi||^2 + \beta(t+\tau)^2;$$

β, T, τ are suitable parameters and $||u||$ is the $L^2(\Omega)$ norm (or related norm) of $u(\cdot, t)$.

Consider the special case

$$u_t - \Delta u = \delta f(u), \qquad f(0) > 0 \tag{1.7}$$

with $f'(u) > 0$ if $u > 0$, and denote by δ_* the g.l.b. of the spectrum of $\Delta u + \delta f(u)$ in Ω under zero boundary conditions; thus for $\delta < \delta_*$ there is a positive solution of (1.7) with $u = 0$ on $\partial\Omega$, but for $\delta > \delta_*$ no such solution exists.

THEOREM 1.3. *Bellout [5]. If $\delta > \delta_*$ and if f/f' is concave, then for any $\delta > \delta_*$ the following holds: for any $\phi \geq 0$ there does not exist a global solution of (1.7), (1.2), (1.3).*

Examples:

$$f(u) = e^u, \tag{1.8}$$

$$f(u) = (u+\mu)^p \qquad (p > 1, \mu > 0). \tag{1.9}$$

The proof is based on constructing a subsolution that increases to $+\infty$ in finite time.

Lacey [22] had previously proved a result similar to Theorem 1.3, requiring, however, that δ_* is in the spectrum of $\Delta u + \delta f(u)$ (but not that f/f' is concave). This assumption on δ_* is not easily verifiable; in case (1.8) this is only known if $N \leq 10$ and in case (1.9) if $N \leq 6$. Lacey's proof is based on extending the method of averages used in Theorem 1.1.

DEFINITION 1.1: If a solution of (1.1)-(1.3) exists in Ω_T but not in $\Omega_{T+\epsilon}$, for any $\epsilon > 0$, then we call T the *blow-up time.*

We shall henceforth consider only the case

$$|F(u, \nabla_x u)| \leq A(u)(|\nabla_x u| + 1), \tag{1.10}$$

where $A(u)$ is a continuous function of u.

THEOREM 1.4. *If T is a finite blow-up time for (1.1)-(1.3), (1.10), then*

$$\sup_{x\in\Omega} |u(x,t)| \to \infty \quad \text{as} \quad t \to T. \tag{1.11}$$

PROOF: Suppose for contradiction that

$$\sup_{x\in\Omega} |u(x,t_n)| \leq C, \qquad t_n \to T$$

and compare $u(x,t)$ for $t \geq t_n$ with the solution $U(t)$ of

$$U_t = F(U,0), \qquad U(t_n) = C.$$

Clearly $U(t) \leq C + 1$ if $t_{n_0} \leq t \leq T + \epsilon$ if n_0 is sufficiently large; therefore also $u(x,t) \leq C + 1$ if $t_{n_0} \leq t < T$. Similarly we get a lower bound $u(x,t) \geq -C - 1$. Recalling (1.10) we can now use parabolic estimate in order to bound $\nabla_x u$ and higher derivatives of u in Ω_T and thus extend the solution beyond T; this is a contradiction.

DEFINITION 1.2: A point $x_0 \in \Omega$ is called a *blow-up point* if there exist sequences $x_n \in \Omega$, $t_n \in (0,T)$ such that

$$x_n \to x_0, t_n \to T, \big|u(x_n, t_n)\big| \to \infty \quad \text{as} \quad n \to \infty.$$

We denote by S the set of all blow-up points.

Clearly S is a closed set and, by Theorem 1.4, S is nonempty.

It was proved by Baras and Cohen [**1**] (under mild conditions on f) that the nonnegative solution of

$$u_t - \Delta u = f(u) \tag{1.12}$$

satisfying (1.2), (1.3) cannot be extended beyond the blow-up time. More precisely, if $f_n(u)$ are bounded truncations of $f(u)$ such that $f_n(r) \to f(r)$ as $n \to \infty$ (for each r) then the corresponding solutions u_n satisfy:

$$u_n(x,t) \to \begin{cases} u(x,t) & \text{if} \quad t < T \\ +\infty & \text{if} \quad t > T. \end{cases}$$

It is generally difficult to estimate precisely the blow-up time, except in special cases. Consider

$$\begin{aligned} u_t - \epsilon\Delta u = f(u) \quad & \text{if} \quad x \in \Omega, t > 0, \\ u = 0 \quad & \text{if} \quad x \in \partial\Omega, t > 0, \\ u(x,0) = \phi(x) \quad & \text{if} \quad x \in \Omega \end{aligned} \tag{1.13}$$

and let

$$\phi \geq 0, \quad \phi(x_0) = \max_{x \in \bar{\Omega}} \phi(x), \; x_0 \in \Omega.$$

THEOREM 1.5. *(Friedman-Lacey [**13**]). Denote by T_ϵ the blow-up time for (1.13) and by T_0 the blow-up time for*

$$v'(t) = f\big(v(t)\big), v(0) = \phi(x_0).$$

If f is as in (1.5) and if

$$-c_0\big(1 + o(1)\big)|x - x_0|^{2\alpha} \leq \phi(x) - \phi(x_0) \leq -c_1\big(1 + o(1)\big)|x - x_0|^{2\alpha}$$

for $|x - x_0|$ *small, where* $c_0 > 0, c_1 > 0, \alpha > 0$*, then*

$$c\epsilon^\alpha + o(\epsilon^\alpha) \le T_\epsilon - T_0 \le C\epsilon^\alpha + o(\epsilon^\alpha)$$

for some positive constants c, C.

The constants c, C can be computed and, in particular, in case (1.8) with $\alpha = 2$, we get $C = 4c$ where

$$c = \frac{1}{4} e^{-2\phi(x_0)} \big|\Delta\phi(x_0)\big|.$$

§2. The blow-up set.

We begin with the 1-dimensional case and take for simplicity $Lu = u_{xx}$. Thus we consider

$$u_t - u_{xx} = f(u) \quad \text{if} \quad 0 < x < \beta,\ t > 0, \tag{2.1}$$

$$u(0,t) = u(\beta,t) = 0 \quad \text{if} \quad t > 0, \tag{2.2}$$

$$u(x,0) = \phi(x) \quad \text{if} \quad 0 < x < \beta. \tag{2.3}$$

Assume that

$$\phi \ge 0, \tag{2.4}$$

and that

$$\begin{gathered} f(u) > 0 \quad \text{if} \quad u > 0, f'(u) > 0 \quad \text{if} \quad u > 0, \\ f'(u) \to \infty \quad \text{if} \quad u \to \infty, \quad \text{and} \quad f^\theta(u) \quad \text{is convex} \\ \int^\infty \frac{du}{f^\theta(u)} < \infty \quad \text{for some} \quad 0 < \theta < 1. \end{gathered} \tag{2.5}$$

Suppose T is the blow-up time and let

$$R = \big\{\alpha_1 < x < \alpha_2,\ T - \delta < t < T\big\}$$

be any rectangle in Ω_T.

THEOREM 2.1. *If $u_x < 0$ in R then u does not blow-up at any point of the interval $\alpha_1 < x \le \alpha_2$.*

*The proof is essentially due to Friedman and McLeod [**15**].*

PROOF: Suppose the assertion is not true. Then there is a sequence (x_n, t_n) in R such that $u(x_n, t_n) \to \infty, x_n \to x_0, t_n \to T$ and $x_0 \in (\alpha_1, \alpha_2]$. Since $u_x < 0$ in R,

$$u(x, t_n) \to \infty \quad \text{uniformly in} \quad x \in \gamma_n \equiv \{\alpha_1 < x < x_0\} \quad \text{as} \quad n \to \infty.$$

By comparing u with the solution v of the heat equation for $x \in \gamma_n, t_n < t < T$ having the same values as u on the parabolic boundary, we deduce that for any $\alpha_1 < \alpha' < \alpha'' < x_0$,

$$u(x,t) \to \infty \quad \text{uniformly if} \quad \alpha' \le x \le \alpha'', t \to T. \tag{2.6}$$

Consequently, for any large M there exists a small δ' such that

$$u > M \quad \text{in} \quad R' \equiv \{\alpha' < x < \alpha'', T - \delta' < t < T\}; \tag{2.7}$$

M will be chosen later on.

Consider the function

$$J = u_x + \epsilon h(x) F(u) \quad \text{in} \quad R' \; (0 < \epsilon < 1) \tag{2.8}$$

where (h, λ) is a solution of

$$\begin{aligned} &h > 0, h'' + \lambda h = 0 \quad \text{if} \quad \alpha' < x < \alpha'', \\ &h(\alpha') = h(\alpha'') = 0. \end{aligned}$$

Then

$$J_t - J_{xx} = f'(u)u_x - \epsilon h F'(u) f(u) - \epsilon h F''(u) u_x^2 - 2\epsilon h' F'(u) u_x - \epsilon h'' F(u).$$

Choosing $F = f^\theta$ and substituting u_x from (2.8), we find that for some coefficient $b(x,t)$,

$$\mathcal{L}J \equiv J_t - J_{xx} + bJ \le \epsilon h[f'F - F'f] + \epsilon\lambda h F - 2\epsilon^2 h' h F' F,$$

or

$$\mathcal{L}J \le -\epsilon h f^\theta \left\{(1-\theta)f'(u) - \lambda - 2\epsilon\theta h' f^{\theta-1}(u) f'(u)\right\}.$$

Since $f'(u) \to \infty$ if $u \to \infty$, the expression in braces is positive if M in (2.7) is chosen large enough, i.e., $\mathcal{L}J < 0$ in R'. Next, if ϵ is small enough then $J \le 0$ on the parabolic boundary of R'. Consequently, by the maximum principle, $J < 0$ in R', i.e.,

$$\frac{u_x(x,t)}{f^\theta(u(x,t))} \le -\epsilon h(x).$$

Integrating with respect to $x, \alpha' < x < \alpha''$, we get

$$\int_{u(\alpha',t)}^{u(\alpha'',t)} \frac{du}{f^\theta(u)} \le -\epsilon \int_{\alpha'}^{\alpha''} h(x)dx.$$

As $t \to T$ the left-hand side converges to zero (by (2.6) and (2.5)), which is a contradiction.

THEOREM 2.2. *If $\phi'(x)$ changes sign just once and if (2.4), (2.5) hold, then the blow-up set consists of a single point.*

This result is due to Friedman and McLeod [**15**] in case $\phi'' + f(\phi) \ge 0$; the last restriction was removed by Caffarelli and Friedman [**8**].

PROOF: Let $t_0 \in (0,T), \delta$ small and set

$$R_0 = \left\{\beta - 2\delta < x < \beta - \delta,\ t_0 < t < T\right\}.$$

Consider the function

$$w(x,t) = u(x,t) - u\big(2(\beta-\delta) - x, t\big) \quad \text{in} \quad R_0.$$

Since $u_x(\beta, t_0) < 0$ we see that $w \le 0$ on the parabolic boundary of R_0. Also

$$w_t - w_{xx} = \tilde{c}w$$

for some coefficient $\tilde{c}(x,t)$. Hence, by the maximum principle, $w < 0$ in R_0 and $w_x(\beta-\delta,t) < 0$, i.e., $u_x(\beta-\delta,t) < 0$. By varying δ we conclude that $u_x(x,t) < 0$ if $\beta - \delta_0 < x < \beta, t_0 < t < T$. Applying Theorem 2.1 it follows that there is no blow-up in $\beta - \delta_0 < x < \beta$. Similarly there is no blow-up in $0 < x < \delta_1$ for some $\delta_1 < 0$.

The assumption on ϕ' implies that there is a curve $x = s(t)(0 \le t < T)$ such that

$$u_x(x,t) > 0 \quad \text{if} \quad 0 \le x < s(t), \ u_x(x,t) \le 0 \quad \text{if} \quad s(t) \le x \le \beta.$$

Suppose for simplicity that $s(t)$ is continuous. If $s(t) \to s_0$ as $t \to T$ then s_0 is the single blow-up point, by Theorem 2.1. Thus it suffices to show that $\lim s(t)$ exists as $t \to T$.

Suppose for contradiction that

$$s^- = \liminf_{t \to T} s(t) < \limsup_{t \to T} s(t) = s^+.$$

By Theorem 1.4, $u(s(t),t) \to \infty$ if $t \to T$. Consider a solution v of the heat equation in

$$R_n \equiv \left\{s^- + \frac{\epsilon}{2} < x < s^+ - \frac{\epsilon}{2}, \ \tau_n < t < T\right\}$$

where $v = \min\{u(s(t),t), t > \tau_n\}$ on $t = \tau_n$ and $v = 0$ on $x = s^- + \epsilon/2, x = s^+ - \epsilon/2$, and τ_n is such that the arc $\gamma_n : x = s(t)$ for $\tau_n \le t \le \sigma_n$ initiates at $x = s^- - \epsilon/2$, terminates at $x = s^+ - \epsilon/2$ and lies in $(s^- + \epsilon/2, s^+ - \epsilon/2)$ for $\tau_n < t < \sigma_n$. By comparison,

$$u(x,t) \ge v(x,t) \quad \text{if} \quad s^- + \frac{\epsilon}{2} < x < s^+ - \frac{\epsilon}{2}, \quad (x,t) \ \text{lies above} \ \gamma_n.$$

It follows that

$$u(x,t) \to \infty \quad \text{uniformly if} \quad s^- + \epsilon < x < s^+ - \epsilon, t \to T. \tag{2.9}$$

On the other hand, $u(s^- + \epsilon, t)$ remains bounded as $t \to T$. We now apply the preceeding reflection agrument about $x = s^- + \epsilon \quad (4\epsilon < s^+ - s^-)$ with initial time $t = t_0$ such that $s(t_0) > s^- + 4\epsilon)$ and deduce that

$$u(x,t) - u\big(2(s^- + \epsilon) - x, t\big) \le 0$$

if $s^- - \epsilon < x < s^- + \epsilon, t_0 < t < T$. It follows that $s(t) > s^- + \epsilon$ if $t_0 < t < T$, which is a contradiction.

THEOREM 2.3. *(Caffarelli and Friedman [8]). If ϕ' changes sign twice and if (2.4) (2.5) hold then the blow-up set consists of, at most, two points.*

The proof is based on extending the previous method, showing that the curves $x = s_i(t)$ where $x \to u(x,t)$ has local maximum or local minimum have limits as $t \to T$.

Theorem 2.1 extends to equations with variable coefficients; however, in that case it is generally difficult to ensure that the condition $u_x \leq 0$ is satisfied in some regions.

Theorems 2.1, 2.2 can be extended [**15**] to the boundary condition

$$\frac{\partial u}{\partial \nu} + \mu u = 0 \quad (\mu > 0).$$

Friedman and Lacey [**14**] studied the blow-up behavior of solutions of

$$u_t + uu_x - u_{xx} = f(u) \tag{2.10}$$

and, in particular, proved a single point blow-up provided ϕ' changes sign just once and

$$\phi(x) \geq \phi(\beta - x) \quad \text{if} \quad \frac{\beta}{2} < x < \beta.$$

Consider next the case where Ω is a bounded domain in $\mathbb{R}^N$

THEOREM 2.4. *(Friedman and McLeod [15]). Assume that Ω is strictly convex, that (2.4), (2.5) hold and that $f^{\theta-1}f' \leq C$ for θ as in (2.5) and C is some positive constant. Then the blow-up set of the solution of (1.12), (1.2), (1.3) is a compact subset of Ω.*

PROOF: Since u is a super solution of the heat equation, $\partial u/\partial \nu \leq -c$ on $\partial\Omega \times (t_0, T)$ for any $t_0 \in (0,T)$ where c is a positive constant. Suppose $0 \in \partial\Omega$ and Ω lies in $\{x_1 < 0\}$. Then $\partial u/\partial x_1 < -c/2$ if $x \in B_\delta \cap \partial\Omega$, $t_0 < t < T$ for some $\delta > 0$. We proceed as in Theorems 1.1, 1.2, showing first (by reflection) that $u_{x_1} \leq 0$ and then that

$$J \equiv u_{x_1} + \epsilon h(x) F(u) < 0 \quad \text{if} \quad x \in B'_\delta \cap \Omega, t_0 < t < T$$

for some $\delta' < \delta$; here we take $h(x) = (x_1 + \eta)^2$ for any $0 < \eta < \delta'$.

When Ω is a ball and ϕ radially symmetric and decreasing in r, the method of Theorems 1.1, 1.2 can be extended to show that $r = 0$ is the only blow-up point [**15**]. In fact, the relevant estimate here is

$$r^{N-1} u_r + \epsilon h(r) f^\theta(u) \leq 0$$

with $h(r) = r^N$. In the radial case Mueller and Weissler [**27**] also proved a single point blow-up by entirely different approach; the one-dimensional symmetric case was first considered by Weissler [**29**].

§3. Pointwise asymptotic behavior.

Set

$$G(u) = \int_u^\infty \frac{du}{f(u)} \tag{3.1}$$

THEOREM 3.1. *(Friedman and McLeod [**15**]). Assume that either $f(0) > 0$, and the assumptions of Theorem 2.4 hold, or $f(0) = 0$ and $f(u) > 0, f'(u) > 0$ and $f(u)$ is convex if $u > 0$, with $\int^\infty (1/f) < \infty$. If T is the blow-up time for a solution of (1.12), (1.2), (1.3), if*

$$f(\phi) + \Delta\phi \geq 0 \tag{3.2}$$

then

$$G\big(u(x,t)\big) \geq c(T-t) \quad \text{in} \quad \Omega_T \quad (c > 0). \tag{3.3}$$

This estimate implies:

$$\text{(3.4)} \quad \text{if} \quad f(u) = (u+\lambda)^p$$
$$\text{then} \quad u(x,t) \leq \frac{C}{(T-t)^{1/(p-1)}} \qquad (1 < p < \infty, \lambda \geq 0);$$

$$\text{if} \quad f(u) = e^u \quad \text{then} \quad u(x,t) \leq \log\frac{1}{T-t} + C \tag{3.5}$$

The proof is based on showing that for small $\delta > 0$, $J \equiv u_t + \delta f(u)$ is a nonnegative function; if $f(0) = 0$ then we consider J in $\Omega \times (0,T)$, whereas if $f(0) > 0$ then we consider J in $\Omega' \times (0,T)$ with $\bar{\Omega}' \subset \Omega$, $\partial\Omega'$ a small neighborhood of $\partial\Omega$, and use the fact that u does not blow-up in an neighborhood of $\Omega \backslash \Omega'$.

For $p < (N+2)/(N-2), \lambda = 0$, Giga and Kohn [**21**] established the estimate (3.4) provided Ω is a convex domain, without assuming the restriction

(3.2). Their proof is based on energy inequalities, exploiting also Theorem 2.4.

The estimate (3.3) allows one to consider scaled solutions near a blow-up point (x_0, T) in order to study the asymptotic behavior of u near the blow-up point. Setting, in case (3.4),

$$w(y,s) = (T-t)^{1/(p-1)}u(x_0 + x, t)$$

where

$$x = (T-t)^{\frac{1}{2}}y, T-t = e^{-s}$$

we get

$$w_s - \Delta w + \frac{1}{2}y \cdot \nabla w + \frac{w}{p-1} - w^p = 0; \tag{3.6}$$

The asymptotic behavior of $w(y,s)$ as $s \to \infty$ gives the asymptotic behavior of $u(x,t)$ near (x_0, T).

THEOREM 3.2. *(Giga and Kohn [20]). If $N \leq 2$ or $N \geq 3, p \leq (N+2)/(N-2)$ then for any constant $C > 0$ there holds:*

$$(T-t)^{-1/(p-1)}u(x,t) \to \gamma$$

uniformly if $|x - x_0| \leq C(T-t)^{\frac{1}{2}}, t \to T$ where either $\gamma = 0$ or $\gamma = \gamma_p \equiv \left(\frac{1}{p-1}\right)^{1/(p-1)}$.

Recently Giga and Kohn proved (oral communication) that actually $\gamma = \gamma_p$.

The proof of Theorem 3.2 depends upon showing that

(i) any limit point of the trajectory $s \to w(\cdot, s)$ is a stationary solution, i.e., it satisfies

$$-\Delta v + \frac{1}{2}y \cdot \nabla v + \frac{v}{p-1} - v^p = 0; \tag{3.7}$$

(ii) if $N \leq 2$ or if $N \geq 3, p \leq (N+2)/(N-2)$ then any nonnegative bounded solution of (3.7) is either $v \equiv 0$ or $v \equiv \gamma_p$.

The proof of Theorem 3.2 can be extended to the case (3.5); the only new difficulty is to show that for the stationary equation

$$-\Delta v + \frac{1}{2}y \cdot \nabla v + 1 - e^v = 0 \tag{3.8}$$

there are no solutions satisfying an appropriate asymptotic condition at ∞; in the radial case with $\phi' \leq 0$, this condition is

$$(3.9) \qquad 1 + \frac{1}{2} y \cdot \nabla v \to 0 \quad \text{if} \quad |y| \to \infty.$$

Bebernes and Troy [**4**] have proved that for $N = 1$ there are no radial solutions of (3.8), (3.9); for $N = 2$ the same assertion (for radial solutions) was proved by Bebernes, Bressan and Eberly [**2**]; see also [**10**] for another proof.

If $p > (N+2)/(N-2)$ then there exists an infinite sequence of positive radial solutions of (3.7) (Troy [**28**]); if $N > 2$ then there exists an infinite sequence of positive radial solutions of (3.8), (3.9) (Eberly and Troy [9]); see also Giga [**19**] for related results.

§4. Miscellaneous.

4.1. L^r estimates.

Consider (1.12), (1.2), (1.3) with $f(u) = u^p$ and Ω convex. Then [**15**]

$$(4.1) \qquad \liminf_{t \to T} ||u(\cdot, t)||_{L^q(\Omega)} = \infty \quad \text{if} \quad q > \frac{(p-1)N}{2};$$

on the other hand, in the radially symmetric case with $\phi'(r) \leq 0$,

$$(4.2) \qquad \limsup_{t \to T} ||u(\cdot, t)||_{L^q(\Omega)} < \infty \quad \text{if} \quad q < \frac{(p-1)N}{2}.$$

(If $N \leq 2$ or $N \geq 3, p \leq \frac{N+2}{N-2}$ then (4.1) holds also for $q = \frac{(p-1)N}{2}$, by the results of Giga and Kohn mentioned above.)

4.2. Nonlinear degenerate equations.

Consider

$$(4.3) \qquad \begin{aligned} &u_t = u^2(u_{xx} + u) \quad \text{if} \quad -a < x < a, t > 0, \\ &u(\pm a, t) = 0 \quad \text{if} \quad t > 0, \\ &u(x, 0) = \phi(x) \quad \text{if} \quad -a < x < a \end{aligned}$$

where $\phi(x) = \phi(-x), \phi'(x) \leq 0$ if $0 < x < a$; this problem arises in a plasma model. Friedman and McLeod [**16**] proved the following results:

> If $a < \pi/2$ then a global solution exists, whereas is $a > \pi/2$ then the solution blows-up in finite time T provided that, say, $\phi'' + \phi \geq 0$. In the latter case,
>
> $$u(x,t) \sim u(0,t)\cos x \quad \text{if} \quad 0 \leq x < \frac{\pi}{2}, t \to T,$$
>
> $$\frac{u(0,t)}{\sqrt{T-t}} \to \infty \quad \text{if} \quad t \to T,$$
>
> $$u\left(\frac{\pi}{2},t\right) \to \infty \quad \text{at a smaller rate than} \quad u(0,t), \quad \text{as} \quad t \to T,$$
>
> $$u(x,t) \quad \text{remains bounded for} \quad \frac{\pi}{2} < x < a,\ t \to \infty.$$

The precise asymptotic behavior of $u(0,t)$ and $u(\pi/2,t)$ is not known. Some of the above results extend to N-dimensions.

4.3. Systems.

Friedman and Giga [**11**] have established a single point blow-up for some parabolic systems.

$$u_t - u_{xx} = f(u,v),$$
$$v_t - v_{xx} = g(u,v)$$

in $-a < x < a, t > 0$ with

$$u(x,0) = \phi(x),\ v(x,0) = \psi(x),$$
$$u(\pm a,t) = v(\pm a,t) = 0,$$

where $\phi(x) = \phi(-x), \psi(x) = \psi(-x)$ and $\phi'(x) \leq 0,\ \psi'(x) \leq 0$ if $0 < x < a$.

Burnell, Lacey and Wake [**7**] established non-existence for some parabolic systems with third initial-boundary conditions.

Bellout [**5**] proved a single point blow-up for parabolic equations with nonlinear memory term:

$$u_t - u_{xx} = \int_0^t h(t,\tau) f\big(u(x,\tau)\big)\,d\tau + g(x)$$

if $-a < x < a, t > 0$ with

$$u(x,0) = \phi(x) \quad \text{if} \quad -a < x < a,$$
$$u(\pm a, t) = 0 \quad \text{if} \quad t > 0,$$

where $\phi(-x) = \phi(x), \phi'(x) \leq 0$ if $0 < x < a$.

Bebernes, Bressan and Lacey [3] considered blow-up for equations $u_t - \Delta u = f(u) + g(t)$ with g depending on u : $g(t) = \lambda \int u(x,t)dx$.

4.4. Extinction.

Consider

$$\text{(4.4)} \qquad \begin{aligned} &u_t - \Delta u + \lambda u^q = 0 \quad \text{if} \quad x \in \Omega,\ t > 0, \\ &u = 0 \quad \text{if} \quad x \in \partial\Omega,\ t > 0 \\ &u(x,0) = \phi(x) \quad \text{if} \quad x \in \Omega,\ \phi \leq 0 \end{aligned}$$

where $\lambda > 0$ and $0 < q < 1$. It is well-known that there exists a positive and smallest T such that $u(x,t) \equiv 0$ if $t \geq T$. A point $x_0 \in \Omega$ is called an *extinction point* if $u \not\equiv 0$ in any $\{t < T\}$- neighborhood of (x_0, T).

Friedman and Herrero [12] studied the extinction set and proved that it consists of a single point if $N = 1$ and $\phi'(x)$ changes sign just once. In the N-dimensional radial case with $\phi'(r) \leq 0$, they proved that 0 is the only extinction point. The asymptotic formula

$$(T-t)^{-1/(1-q)} u(r,t) \to \left(\frac{1}{1-q}\right)^{1/(1-q)}$$
$$\text{as} \quad r \leq C(T-t)^{\frac{1}{2}}, t \to T$$

is then valid; the proof is similar to that of [20]. Here one needs the fact (proved by A. Friedman, J. Friedman and B. McLeod [15]) that there exist no global radial solution in $C^2[0,\infty)$ of

$$\Delta w(r) - \frac{r}{2} w'(r) = \lambda w^q - \frac{w}{1-q}, w \geq 0,\ w' \leq 0.$$

4.5. Open problems.

(i) It is not known whether the blow-up set always has zero measure (not even if $N = 1$ and ϕ general). Such a result would of course follow if one could establish (4.2) for some $q > 0$.

(ii) It would be also interesting to study the behavior of $u(x,t)$ for $|x - x_0| > C(t)(T-t)^{\frac{1}{2}}$ where $C(t) \to \infty$ as $t \to T$; cf. Galaktianov and Pusashkov [**18**]. A. Friedman and S. Kamin recently proved (say, in the symmetric 1-dimensional case) that

$$|u_x| \geq cxu^p/(\log 1/(T-t))$$

when $f(u)$ is given by (3.4) with $\lambda = 0$. This implies that

$$(T-t)^{-\frac{1}{p-1}} u(x,t) \to 0 \quad \text{if} \quad \frac{|x|}{(T-t)^{\frac{1}{2}} \log 1/(T-t)} \to \infty \quad \text{as} \quad t \to T.$$

A similar result holds for general $f(u)$.

(iii) Establish the estimate (3.3) without imposing the restriction (3.2).

(iv) Extend the results of the preceeding sections, such as Theorem 2.2, to the case where ϕ changes sign.

(v) Study the blow-up set for the Cauchy problem.

(vi) Study the blow-up behavior in case the nonlinearity in the parabolic problem arisees in the boundary conditions (such as in Levine and Payne [**26**]).

Department of Mathematics, Purdue University, West Lafayette IN 47907

This work is partially supported by National Science Foundation Grants DMS-8420896 and DMS-8501397.

REFERENCES

1. P. Baras and L. Cohen, *Complete blow-up after $T_{\max}$ for the solution of a semilinear heat equation*, to appear.
2. J. Bebernes, A. Bressan and D. Eberly, *A description of blow-up for the solid fuel ignition model*, to appear.
3. J. Bebernes, A. Bressan and A. A. Lacey, *Total blow-up verses single point blow-up*, to appear.
4. J. Bebernes and W. C. Troy, *Nonexistence for the Kassoy problem*, SIAM J. Math. Anal., to appear.
5. H. Bellout, *A criterion for blow-up of solutions to semilinear heat equations*, SIAM J. Math. Anal., to appear.
6. H. Bellout, *Blow-up of solutions of parabolic equations with nonlinear memory*, to appear.
7. J. G. Burnell A. A. Lacey and G. C. Wake, *Steady states of the reaction-diffusion equations. Part I: Questions of existence and continuity of solution branches*, J. Austral. Math. Soc., Ser. B 24 (1983), 374-391.
8. L. A. Caffarelli and A. Friedman, *Blow-up of solutions of nonlinear heat equations*, J. Math. Anal. Appl., to appear.
9. D. Eberly and W. C.Troy, *Existence of logarithmic-type solutions to the Kassoy problem in dimension* $2 < N < 10$, to appear.
10. A. Friedman, J. Friedman and B. McLeod, *Concavity of solutions of nonlinear ordinary differential equations*, J. Math. Anal. Appl., to appear.
11. A. Friedman and Y. Giga, *A single point blow-up for solutions of semilinear parabolic systems*, Faculty of Sci., Univ. of Tokyo, to appear **Sec. 1A**.
12. A. Friedman and M. A. Herrero, *Extinction properties of semilinear heat equations with strong absorption*, J. Math. Anal. Appl., to appear.
13. A. Friedman and A. A. Lacey, *The blow-up time for solutions of nonlinear heat equations with small diffusion*, SIAM J. Math. Anal., to appear.
14. A. Friedman and A. A. Lacey, *Blow-up of solutions of semilinear parabolic equations*, J. Math. Anal. Appl., to appear.
15. A. Friedman and B. McLeod, *Blow-up of solutions of semilinear heat equations*, Indiana Univ. Math. J. **34** (1985), 425-447.
16. A. Friedman and B. McLeod, *Blow-up of nonlinear degenerate parabolic equations*, Archive Rat. Mech. Anal., to appear.
17. J. Fujita, *On the blowing up of the Cauchy problem for* $u_t = \Delta u_u^{1+\alpha}$, J. Faculty Science, Univ. of Tokyo **13** (1966), 109-124.
18. B. A. Galaktianov and C. A. Pusashkov, *Application of a new comparison theorem to the study of unbounded solutions of nonlinear parabolic equations*, Differ. Uravnen **22** (1986), 1165-1173.
19. Y. Giga, *On elliptic equations related to self-similar solutions for nonlinear heat equations*, Hiroshima Math. J..
20. Y. Giga and R. V. Kohn, *Asymptotically self-similar blow-up of semilinear heat equations*, Comm. Pure Appl. Math. **38** (19895), 297-319.
21. Y. Giga and R. V. Kohn, *Characterizing blow-up using similarity variables*, to appear.
22. A. A. Lacey, *Mathematical analysis of thermal runaway for spatially inhomogeneous reactions.*, SIAM J. Appl. Math. **43** (1983), 1350-1366.
23. A. A. Lacey, *The form of blow-up for nonlinear parabolic equations*, Proc. Royal Soc. Edinb. **98A** (1984), 183-202.
24. A. A. Lacey and D. Tzanetis, *Global existence and convergence to a singular steady state for a semilinear heat equation*, to appear.
25. H. A. Levine, *Some nonexistence and instability theorems of formally parabolic equations of the form* $P_{u_t} = -Au + F(u)$, Archive Rat. Mech. Anal. **51** (1973), 371-386.

26. H. A. Levine and L. E. Payne, *Nonexistence theorems for the heat equation with nonlinear boundary conditions and for the porous medium equation backward in time*, J. Diff. Eqs. **16** (1974), 319-334.
27. C. E. Mueller and F. B. Weissler, *Single point blow-up for a general semilinear heat equation*, Indiana Univ. Math. J. **34** (1984), Issue 4.
28. W. C. Troy, *The existence of bounded solutions of a semilinear heat equation*, to appear.
29. F. B. Weissler, *Single point blow-up of semilinear initial value problems*, J.Diff. Eqs. **55** (1984), 202-224.

Solutions of Diffusion Equations in Channel Domains

ROBERT GARDNER

0. Introduction.

A. The Problem. We will discuss the existence and the qualitative properties of solutions of the problem

$$\begin{aligned} \Delta u + f(u) &= 0 \quad \text{in} \quad \Omega \\ u &= 0 \quad \text{on} \quad \partial\Omega; \end{aligned} \tag{1}$$

here Ω is a channel of the form

$$\Omega = \{(x, y) : x \in \mathbb{R}^1, \quad 0 < y < \phi(x)\}.$$

The function ϕ is assumed to be smooth and positive and the support of $\phi'(x)$ is assumed to lie in an interval $[-R, R]$. Thus Ω is asymptotically flat; let ϕ_-, ϕ_+ denote the channel widths at $x = -\infty, \infty$.

We shall look for solutions which equilibriate at $x = \pm\infty$; such solutions of (1) should tend to solutions of the *endstate equation*,

$$\begin{aligned} u_{yy} + f(u) &= 0 \\ u(0) = u(L) &= 0, \end{aligned} \tag{2}$$

with $L = \phi_-, \phi_+$. It is therefore essential that we first determine the solution set of problem (2). We assume that $f(u)$ is such that the following hypothesis is valid:

(H) *There exists $L_0 > 0$ such that if $L < L_0$ then (2) admits only the zero solution, while if $L > L_0$ then (2) admits exactly three solutions, the zero solution together with two positive solutions, $u_\alpha(y), u_1(y)$, with*

$$0 < u_\alpha(y) < u_1(y).$$

The problem of verifying (H) for a particular $f(u)$ is quite delicate. This has been accomplished when $f(u)$ assumes the form

$$f(u) = u(1-u)(u-\alpha), \quad 0 < \alpha < \tfrac{1}{2}$$

(see [**9**]). These results have recently been extended to certain qualitative cubics (see [**8**]). In the following we shall attach superscripts of $\pm$ to the solutions $u_i(y)$ of (2), $i = \alpha, 1$, when $L = \phi_-$ or ϕ_+.

The full problem is to find a solution u of (1) which satisfies

(3) $$\lim_{x \to -\infty} u = 0, \quad \lim_{x \to +\infty} u = u_1^+(y).$$

B. The formation of propagating fronts and standing waves in reactive and diffusive media have been extensively studied in one space variable (see, e.g. [**2**]). The problem (1), (3) appears to be a natural extension of such equations to several space variables in that the problem retains a certain one-dimensional character. In the context of population genetics, we view solutions of (1), (3) as two-dimensional clines.

In one space variable, a typical and highly intuitive condition which enables the equation to support a cline (i.e., a stable, steady solution) is that travelling fronts near $x = \pm\infty$ propagate inwards, i.e. towards the origin (see, e.g. [**4**] and [**5**]). In a previous paper [**6**], it has been shown that the parabolic equation associated with (1) does support travelling waves when Ω is a uniform strip and $L > L_0$, where L_0 is as in (H). Moreover the sign of the velocity of the front is determined by L. Thus the ingredients for extending the results of [**4**] to two-dimensional domains are available.

C. Travelling Waves. The travelling waves obtained in [**6**] are solutions of the elliptic equation

(4) $$\begin{aligned} -\theta u_\xi &= u_{\xi\xi} + u_{yy} + f(u) \\ u(\xi, 0) &= u(\xi, L) = 0, \end{aligned}$$

and

(5) $$\lim_{\xi \to -\infty} u = 0, \quad \lim_{\xi \to +\infty} u = u_1(y).$$

The hypotheses are that $f(u)$ satisfies (H) and that $L > L_0$. In addition to existence of solutions u, θ of (4), (5) it was also shown that $u_\xi > 0$ and that the wave speed $\theta = \theta(L)$ depends on the channel width L in the following manner: there exists $L_1 > L_0$ such that

$$\begin{aligned} \theta(L) < 0 \quad & \text{if } L > L_1 \\ \theta(L) > 0 \quad & \text{if } L_0 \leq L < L_1. \end{aligned}$$

This should be contrasted to the case of one-dimensional fronts where the wave speed is always negative for the cubic f with $\alpha < \frac{1}{2}$.

The monotonicity of u together with conditions on the sign of $\theta(L)$ make such travelling fronts natural candidiates for upper and lower solutions of problem (1).

D. Results. The following theorems are proved below.

THEOREM 1. *Suppose that (H) holds and that*

$$\text{(6)} \qquad \begin{cases} \theta(\phi_+) < 0 \\ \text{and either } \theta(\phi_-) > 0 \text{ if } \phi_- \geq L_0 \text{ or } \phi_- < L_0. \end{cases}$$

Then there exists a solution $u(x,y)$ of problem (1), (3).

THEOREM 2. *Suppose that in addition to the hypotheses of Theorem 1, $\phi'(x) \geq 0$ for all x. If $u(x,y)$ is the solution obtained in Theorem 1, then the following are also true:*

(A) *$u_x > 0$ in Ω.*

(B) *u is linearly stable, i.e., the spectrum of the linearized problem*

$$\begin{aligned} \Delta w + f'(u)w = \lambda w \quad & \text{in } \Omega \\ w = 0 \quad & \text{on } \partial\Omega \end{aligned}$$

lies in the ray $\{\lambda : \lambda < -\rho\}$ for some $\rho > 0$.

(C) *u is unique in the class of functions $v(x,y)$ for which $v = 0$ on $\partial\Omega$ and which, for some sufficiently small $\lambda > 0$, satisfy*

$$(7_+) \qquad \liminf_{x \to +\infty} v(x,y) \geq u_1^+(y) - \lambda\rho_+(y)$$

$$(7_-) \qquad \limsup_{x \to -\infty} v(x,y) \leq \lambda\rho_-(y);$$

here, $\rho_\pm(y)$ are positive functions which vanish at $y = 0, \phi_\pm$.

(D) *Every solution of the parabolic equation associated with (1) whose initial data $v(x,y)$ satisfy (7) tends to the unique, monotone steady state as $t \to \infty$.*

E. Further Remarks. With minor modifications, the method of proof of Theorem 2.C yields the following result.

THEOREM 3. *The monotone travelling front solution $u(\xi, y), \theta$ of problem (4), (5) is stable in the C^0 sense and the wave speed, θ, is uniquely determined.*

This can be proved by using the adaptation of the Fife-McLeod comparison theorem for travelling fronts in one space variable (see § 2.C) to show that solutions of time-dependent problem can be sandwiched between translates of a (modified) travelling wave solution. In particular, this implies that θ is unique. The asymptotic stability and uniqueness of the travelling front is more delicate here due to the zero eigenvalue arising from translation invariance. Presumably this problem could be resolved by constructing a suitable Liapunov function along the lines of Fife and McLeod [**3**]. We will not pursue these questions further here.

The results described above can be used to obtain a large multiplicity of (linearly stable) stationary solutions in domains Ω where $\phi(x)$ has exactly $n+1$ local extrema, say at $x_0 < x_1 < \cdots < x_n$. Roughly, the conditions are that $x_{i+1} - x_i = O(\epsilon^{-1})$ where ϵ is a small parameter and that the variation of $\phi(x)$ is confined to 0(1) intervals as $\epsilon \to 0$. Also, if $\phi_j = \phi(x_j)$ then conditions analogous to (6) must be imposed for each pair ϕ_j, ϕ_{j+1} of local extrema. there then exist $2^{[n/2]}$ linearly stable steady states. For each odd j, a typical solution is either near zero or $u_1^j(y)$ for x near x_j, where $u_1^j(y)$ is the maximal solution of (2) with $L = \phi_j$. Linearized stability is proved using the results described here on monotone channels to approximate the transition layers of the oscillatory solutions. These results will appear elsewhere.

Acknowledgements. I would like to thank L.A. Peletier for several suggestions which substantially improved the construction of the upper solution $\bar{u}$ in Theorem 1. I would also like to thank Joel Spruck for a remark which simplified the proof of Theorem 2.B.

1. Proof of Theorem 1.

The proof employs the familiar theory of upper and lower solutions. Thus we find (weak, H^1) solutions of

$$\begin{aligned} \Delta\bar{u} + f(\bar{u}) &\leq 0, \quad \bar{u} \geq 0 \text{ on } \partial\Omega \\ \Delta\underline{u} + f(\underline{u}) &\geq 0, \quad \underline{u} \leq 0 \text{ on } \partial\Omega \end{aligned}$$

with $\underline{u} \leq \bar{u}$, where $\underline{u}$ and $\bar{u}$ both satisfy the endstate conditions (3). An exact solution u is obtained as the fixed point of the monotone iteration scheme

$$(8) \qquad \begin{aligned} (\Delta - h)u_j &= -f(u_{j-1}) - hu_{j-1}, \\ u_j &= 0 \text{ on } \partial\Omega, \end{aligned}$$

where $u_0 = \underline{u}$ and

$$h > \max_{0 \leq s \leq 1} |f'(s)|.$$

A. The lower solution. The travelling wave $u(\xi, y), \theta$ of (4) will be a lower solution when $x \geq R$ and $L = \phi_+$; however, it fails to satisfy $u \leq 0$ on $\partial\Omega$ when, e.g., $x \leq -R$. The existence theorem for travelling waves actually was proved for any linear boundary conditions; we shall perturb the boundary conditions for (2) and (4) in such a manner that the zero solution of (2) perturbs to a negative solution $u_0^+(y, \epsilon)$ while the solution $u_{1(y)}^+$ of the Dirichlet problem coincides exactly with the large, positive solution of the mixed boundary value problem. More precisely we first perturb the condition $u = 0$ at $y = 0, \phi_+$ to $u = -\epsilon$ at $y = 0, \phi_+$. Next, if $p = u_{1y}^+(0)$ we perturb this to the mixed problem

$$\begin{aligned} \epsilon u_y - pu &= p\epsilon \text{ at } y = 0 \\ \epsilon u_y + pu &= p\epsilon \text{ at } y = \phi_+ \end{aligned}$$

(see Figure 1). The existence results of [6] then provide a travelling wave $u_\epsilon(\xi, y), \theta_\epsilon$ with

$$\lim_{\xi \to -\infty} u_\epsilon = u_0(y, \epsilon), \qquad \lim_{\xi \to +\infty} u_\epsilon = u_1^+(y).$$

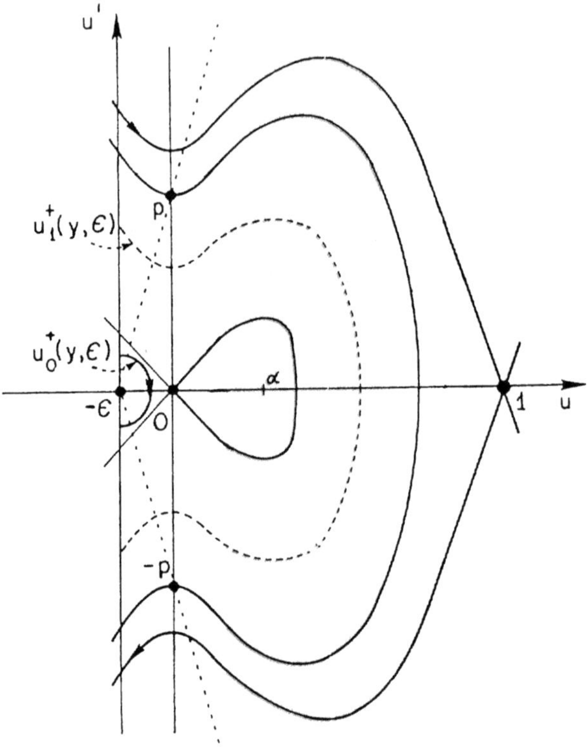

Figure 1

Moreover, if ϵ is small enough then θ_ϵ will be negative since $\theta(\phi_+) < 0$. From the above, there exists a translate to u_ϵ (again denoted by u_ϵ) such that $u_\epsilon(\xi, y) \leq 0$ for $\xi \leq R$. The subsolution $\underline{u}$ is now defined to be

$$\underline{u}(x,y) = \begin{cases} \min(u_\epsilon(x,y), 0), & x \geq R \\ 0, & x \leq R. \end{cases}$$

B. The upper solution. The construction of $\bar{u}$ is somewhat different. Let $\bar{\phi} = \max \phi(x)$; the difficulty is that $\bar{\phi}$ may be larger than ϕ_+. Let $\bar{\phi}(x)$ be a monotone increasing function such that $\bar{\phi}' \in C_0^\infty$, $\bar{\phi}(-\infty) = \phi_-$, $\bar{\phi}(+\infty) = \bar{\phi}$, and $\bar{\phi}(x) \geq \phi(x)$ for all x. Let $\bar{\Omega}$ be the channel with upper boundary $y = \bar{\phi}(x)$. We will obtain an exact solution $\bar{u}$ of (1), (3) in $\bar{\Omega}$; such a solution is clearly an upper solution of (1) in Ω. By translating $\underline{u}$ even further to the right we can easily arrange that $\underline{u} \leq \bar{u}$ in Ω.

In order to obtain an exact solution in $\bar{\Omega}$ we define a subsolution (again denoted by $\underline{u}$) in $\bar{\Omega}$ as in A., above. For a supersolution, we take the x-independent solution $\bar{u}_1(y)$ of (2) with $L = \bar{\phi}$. The iterates u_j in (8) with $u_0 = \underline{u}$ converge monotonically to a limit $u^*(x,y)$ in $\bar{\Omega}$. It easily follows that since $\bar{\phi}' \geq 0$ then $\bar{u}_x > 0$ in $\bar{\Omega}$ (see Theorem 2.A). Thus the limit u^* must converge to limits, $u^*_{\pm}(y)$ as x tends to $\pm\infty$. By some elementary arguments (see [**6**,Lemma 2.11]) the limits $u^*_{\pm}(y)$ must satisfy the endstate equation (2) with L alternatively equal to ϕ_- and $\bar{\phi}$. By construction, $u^*_+(y) = \bar{u}_1(y)$.

If $u^*_-(y) \equiv 0$ then u^* satisfies (3) near $x = -\infty$. In the event that $\phi_- < L_0$ this is necessarily the case. If, on the other hand $L \geq L_0$ then $\bar{u}_-(y)$ could equal $u^-_\alpha(y)$ or $u^-_1(y)$. In this case let $u_\epsilon, \theta_\epsilon$ be the travelling wave solution of (4), (5) with $L_\epsilon = \phi_- + \epsilon$ where ϵ is small and positive. The limiting state at $+\infty$, denoted by $u_1(y,\epsilon)$, in (5) will be uniformly larger than $u^-_1(y)$ since $L_\epsilon > \phi_-$. It follows that by suitably translating $u_\epsilon(\xi, y)$ to the left we will have that for some $\gamma \leq -R$,

$$u_\epsilon(\gamma, y) > u^*(\gamma, y), \quad 0 < y < \phi_-. \tag{9}$$

Now by assumption u^* tends to a positive limit $u^*_-(y)$ as x tends to $-\infty$; from (5) it follows that for some $\Gamma < \gamma$

$$u_\epsilon(\Gamma, y) < u^*(\Gamma, y), \quad 0 < y < \phi_-. \tag{10}$$

It follows from (9) and (10) that $\min(u_\epsilon, u^*)$ will be a supersolution which equals u_ϵ for $x \leq \Gamma$ and u^* for $x \geq \gamma$. Finally, we modify u^* near $x = +\infty$, since u^* does not satisfy (3) at $x = +\infty$. In particular, u^* tends to $\bar{u}_1(y)$ at $x = +\infty$ which is larger than $u^+_1(y)$. We will define a supersolution $\hat{u}$ in the region $x \geq R$ in Ω with the properties that

$$\hat{u}(R, y) \equiv 1 > u^*(R, y), \quad 0 < y < \phi_+ \tag{11}$$

$$\lim_{x \to +\infty} \hat{u}(x, y) = u^+_1(y) < \bar{u}_1(y). \tag{12}$$

To this end let $\psi(x)$ be a smooth, monotone decreasing function for $x \geq R$ and suppose in addition that $\psi' \in C_0^\infty$ and that

$$\psi(R) = 1, \quad \psi(\infty) = 0.$$

Consider the boundary value problem

$$\begin{aligned} \Delta\hat{u} + f(\hat{u}) = 0, &\quad x > R, \quad 0 < y < \phi_+ \\ \hat{u} = 1 &\text{ at } x = R \\ \hat{u}(x,y) = \psi(x) &\text{ for } x > R \text{ and } y = 0, \phi_+. \end{aligned} \tag{13}$$

Problem (13) has $u \equiv 1$ as a supersolution and $u(x,y) = u_1^+(y)$ as a subsolution. Using the iteration scheme (8) we obtain an exact solution $\hat{u}$ of (13). Moreover, by an argument similar to that of Theorem 2.A, below, it follows that $\hat{u}$ is monotone decreasing in x since $\psi' \leq 0$ for all x. Thus $\hat{u}$ has a limit at $x = +\infty$ which, since it is a solution of (2) with $L = \phi_+$ that lies above $u_1^+(y)$, must in fact equal $u^+(y)$. Thus $\hat{u}(x,y)$ satisfies (11) and (12).

The supersolution $\bar{u}$ of (1) in Ω is finally obtained by setting

$$\bar{u}(x,y) = \begin{cases} \min\{u_\epsilon(x,y), u^*(x,y)\}, & x \leq \gamma \\ u^*(x,y), & \gamma \leq x \leq R \\ \min\{u^*(x,y), \hat{u}(x,y)\}, & R \leq x. \end{cases}$$

Properties (9)–(12) ensure that $\bar{u}$ will be a supersolution in Ω which satisfies the limiting conditions (3) at $x = \pm\infty$. By suitably translating $\underline{u}$ we can also ensure that $\underline{u} < \bar{u}$ in Ω.

2. Proof of Theorem 2.

A. Monotonicity. We now suppose that the channel is monotone, i.e. that $\phi' \geq 0$. The subsolution $\underline{u} = u_0$ is monotone increasing in x since the travelling wave used in its construction has this property. We will show inductively that if u_{j-1} in (8) is monotone in x then $u_{jx} \geq 0$ also. If $w_j = \partial u_j / \partial x$ then

$$(\Delta - h)w_j = \left(-f'(u_{j-1}) - h\right)w_{j-1}.$$

By the induction hypothesis and our choice of h in (8) the right-hand side of the above is negative. The nonnegativity of w_j in Ω will follow from the maximum principle if it can be shown that $w_j \geq 0$ on $\partial\Omega$.

Since $u_j = 0$ on $\partial\Omega$ we clearly have that $w_j = 0$ on the horizontal portions of $\partial\Omega$. The sign of w_j on the curved portion of $\partial\Omega$ can be checked by noting that here,

$$\frac{\partial}{\partial x} = -\frac{\phi'}{1+\phi'^2}\frac{\partial}{\partial n} + \frac{1}{1+\phi'^2}\frac{\partial}{\partial \tau}$$

where $n = (-\phi', 1)$ is an outward normal to $\partial\Omega$ and $\tau = (1, \phi')$ is a tangent vector. Since $\partial u_j/\partial\tau = 0$ we have that

$$w_j = \frac{\partial u_j}{\partial x} = -\frac{\phi'}{1+\phi'^2}\frac{\partial u_j}{\partial n}.$$

By the strong maximum principle and the positivity of u_j we have that $\partial u_j/\partial n < 0$; it follows from the above that $w_j > 0$ on $\partial\Omega$ whenever $\phi'(x) > 0$.

B. Linearized Stability. We next prove that the monotone solution u obtained above is stable in the sense that the spectrum $\sigma(L)$ of the linear operator

$$Lw = \Delta w + f'(u)w$$

with domain $H^2(\Omega) \cap H_0^1(\Omega)$ is contained in the negative half-plane.

Since Ω is unbounded, L may possess both point and continuous spectrum. Thus we define the *normal points* N of $\sigma(L)$ to consist of isolated eigenvalues of finite multiplicity. The *essential spectrum*, $\sigma_e(L)$, of L is then defined to be $\sigma(L)/N$. Note that since L is symmetric, $\sigma(L)$ is real.

Intuitively σ_e should depend only on the end states, $u_0 \equiv 0$ and $u_1^+(y)$, and since theses solutions are both stable, σ_e should be negative. Although such results are available in one space variable (see e.g. Henry [**7**,Theorem 5.A.2]), we have not been able to locate an analogous theorem to the channel problem with distinct limits at each end. We therefore prove the following lemma:

LEMMA 1. *Let $\rho_j^\pm, \psi_j^\pm$ be the eigenvalues and eigenfunctions for the two point boundary value problems*

$$(14_\pm) \qquad \psi_{yy} + a_\pm(y)\psi = \rho\psi, \quad \psi(0) = \psi(\phi_\pm) = 0,$$

where $a_+(y) = f'(u_1^+(y))$ and $a_- \equiv f'(0)$, so that $\rho_j^\pm < \rho_1^\pm < 0$ for $j < 1$. Let $\rho^* = \max(\rho_1^-, \rho_1^+) < 0$; then

$$\sigma_e(L) \subset \{\lambda : \lambda \leq \rho^*\}.$$

The proof of Lemma 1 is provided in an appendix. It is now quite easy to prove stability. If the normal points N lie in $\{\lambda \le \rho^*\}$ then $\sigma(L)$ is negative and the result follows.

Suppose then that N contains a point λ with $\lambda > \rho^*$. There are at most finitely many such points so we may suppose that λ is the largest element of N. It follows that λ is simple and that the associated eigenfunction w, satisfying

$$\begin{aligned} &\Delta w + f'(u)w = \lambda w \\ &w = 0 \text{ on } \partial\Omega, \end{aligned} \tag{15}$$

is positive. If $z = u_x$, then z satisfies the equation

$$\Delta z + f'(u)z = 0. \tag{16}$$

By the argument in A., above, it follows that $z = 0$ on the horizontal portions of $\partial\Omega$ and that $z > 0$ on the curved portion

$$\Gamma = \{(x, \phi(x) : \phi'(x) > 0\}$$

of $\partial\Omega$.

Multiply (15) by z, (16) by w, integrate, and subtract to obtain

$$\int_\Gamma z\frac{\partial w}{\partial dn}\, ds = \int_\Gamma \left(z\frac{\partial w}{\partial n} - w\frac{\partial z}{\partial n}\right) ds = \lambda \int_\Omega wz.$$

By the strong maximum principle $\partial w/\partial n < 0$ on $\partial\Omega$; since $z > 0$ in Ω and on Γ it follows that $\lambda < 0$. This completes the proof.

C. Uniqueness. Let u_m denote the monotone (and hence, linearly stable) solution obtained from the iteration scheme (8). We assume that there exists another (possibly non-monotone) solution u_* distinct from u_m which satisfies (7) and obtain a contradiction. The proof proceeds along the following lines. We first use u_* to obtain another *monotone* solution $\hat{u} \not\equiv u_m$ which satisfies (3); in fact u_m and $\hat{u}$ both lie between suitable translates of the upper and lower solutions $\underline{u}, \bar{u}$ obtained in Theorem 1. Noting that the class V_m of functions v defined in Ω which have values between $\underline{u}$ and $\bar{u}$ and which are monotone increasing in x form a positively invariant set for the parabolic equation associated with (1)

$$u_t = \Delta u + f(u) \quad u_0 \text{ on } \partial\Omega, \tag{17}$$

we are able to use a Liapunov function associate with (17) to see that the maximal invariant of (17) in V_m consists of equilibria. The proof is completed by using the linearized stability result in B. to conclude that each equilibrium in V_m has the Conley index of an attractor. Since V_m is positively invariant it follows that there is exactly one equilibrium in V_m, yielding the desired contradiction. The central problem is therefore to produce a second monotone solution $\hat{u}$ obtained from u_* which also satisfies the boundary conditions (3).

LEMMA 2. *There exists a monotone solution $\hat{u}$ of (1) distinct from u_m, with either $\hat{u} > u_m$ or $\hat{u} < u_m$ in Ω.*

PROOF: Suppose $u_* \neq u_m$ is a second solution of (1) which also satisfies (7). Suppose that $u_*(x,y) > u_m(x,y)$ at some $(x,y) \in \Omega$; if $u_* \leq u_m$ then we can assume that $u_*(x,y) < u_m(x,y)$ at some point and the proof proceeds along similar lines. For $\tau \geq 0$ let

$$\Omega_\tau = \{(x,y) \in \Omega : 0 < y < \phi(x-\tau), \quad x \in \mathbb{R}^1\}.$$

Since $\phi(x)$ is monotone increasing, $\Omega_\tau \in \Omega$ for $\tau \geq 0$. Define

$$u_\tau(x,y) \begin{cases} u_*(x-\tau,y), & (x,y) \in \Omega_\tau \\ 0, & (x,y) \in \Omega - \Omega_\tau; \end{cases}$$

since $u_* \geq 0$ in Ω, u_τ is a subsolution of (1) for each $\tau \geq 0$. Next, consider the function

$$\underline{u}(x,y) = \operatorname{Max}\{\operatorname*{Max}_{\tau \geq 0} u_\tau(x,y), u_m(x,y)\}. \tag{18}$$

Since each u_τ is a subsolution and u_m is an exact solution is follows that $\underline{u}(x,y)$ is a subsolution of (1). By construction $\underline{u}$ is monotone increasing in x and $\underline{u} \geq u_m$ in Ω with strict inequality at at least one point.

We now employ the monotone iteration scheme (8) with initial iterate $u_0 = \underline{u}$ as defined in (18). The limit, which we denote by $\hat{u}$ is monotone increasing in x by Theorem 2(A). Also, since $\hat{u} \geq \underline{u}$ and $\underline{u} \geq u_m$ it follows from the strong maximum principle that $\hat{u} > u_m$ in Ω since we have strict inequality at at least one point.

Finally we remark that if $u_* \leq u_m$ in Ω then we replace max with min and $\tau \geq 0$ with $\tau \leq 0$ in (18) to obtain a supersolution $\bar{u} \leq u_m$. In this case we obtain a monotone solution $\hat{u} < u_m$ in Ω.

It must also be shown that $\hat{u}$ satisfies the limiting conditions (3). Since $\hat{u}$ is monotone in x there exist limits $u_\pm(y)$ at $\hat{u}$ at $x = \pm\infty$ which must be solutions of (2) with $L = \phi_\pm$.

LEMMA 3. *With notation as above,* $u_-(y) = 0$ *and* $u_+(y) = u_1^+(y)$.

PROOF: We assume that $\hat{u} > u_m$ in Ω; if $\hat{u} < u_m$ the proof proceeds along similar lines.

Since the original solutions u_m and u_* satisfy (7_+) it follows that $\underline{u}$ and hence, $\hat{u}$ satisfies (7_+). Thus $u_+(y) > u_\alpha^+(y)$; since $u_1^+(y)$ is the unique solution of (2) which is larger than $u_\alpha^+(y)$ it follows that $u_+(y) = u_1^+(y)$.

If $\phi_- < L_0$ then (2) has only the zero solution when $L = \phi_-$ and so $u_-(y) = 0$. If $\phi_- \geq L_0$ this may no longer be the case. The approach in this case is to construct an upper solution $\bar{u}(x, y, t)$ of the parabolic equation (17) which dominates $\hat{u}(x, y)$ for all $t \geq 0$. As $t \to +\infty$, $\bar{u}$ tends to a function $u^\sharp$ which decays to zero as $x \to -\infty$. The construction of $\bar{u}$ is similar to a method introduced by Fife and McLeod [**3**] in their stability analysis of travelling fronts in one space variable.

Let $\Omega^\sharp \supset \Omega$ be another monotone channel which coincides with Ω when $\phi(x) = \phi_-$, i.e. when $x \leq -R$, and whose width near $x = +\infty$ is $\phi^\sharp > \phi_+$. Let $u^\sharp(x, y)$ be a monotone solution of (1), (3) in $\Omega^\sharp$.

Next, let $\rho_1 < 0$ and $\psi_1(y) > 0$ be the principal eigenvalue and eigenfunction of (14_-) and define

$$\beta(x, y, t) = \epsilon(t)\psi_1(y)h(x - \varsigma(t))$$
$$\epsilon(t) = \epsilon_0 e^{-\mu t}.$$

Here ϵ_0 and μ are positive parameters, $\varsigma(t)$ is a positive, monotone increasing function to be determined, and $h(z)$ is a smooth monotone decreasing cutoff function,

$$h(z) = \begin{cases} 1, & z \leq \gamma \\ 0, & z \geq \gamma + 1. \end{cases} \tag{19}$$

The supersolution $\bar{u}$ is defined by

$$\bar{u}(x, y, t) = \beta(x, y, t) + u^\sharp(x + \varsigma(t), y).$$

We shall show that for appropriately chosen $\mu, \epsilon_0, \varsigma$, and $\gamma, \bar{u}$ satisfies the following inequalities:

$$
\begin{aligned}
&(20a) && N[\bar{u}] \stackrel{\text{def}}{=} \bar{u}_t - \Delta\bar{u} - f(\bar{u}) \geq 0 && \text{in} \quad \Omega \times \mathbb{R}_+ \\
&(20b) && \bar{u} \geq 0 && \text{on} \quad \partial\Omega \times \mathbb{R}_+ \\
&(20c) && \bar{u}(x,y,0) \geq \underline{u}(x,y) && \text{in} \quad \Omega.
\end{aligned}
$$

Note that since β and $u^\sharp$ are nonnegative and ς is positive (20b) is satisfied.

Next we verify (20a). To this end, note that

(21)
$$
\begin{aligned}
N[\bar{u}] &= -\mu\epsilon\psi_1 h - \varsigma'\epsilon\psi_1 h' + \varsigma' u_x^\sharp - \epsilon\psi_1'' h - \epsilon\psi_1 h'' - \Delta u^\sharp - f(\bar{u}) \\
&\geq -\mu\epsilon\psi_1 h + \varsigma' u_x^\sharp - \rho_1\epsilon\psi_1 h - \epsilon\psi_1 h'' + f'(0)\epsilon\psi_1 h + f(u^\sharp) \\
&\quad - f(u^\sharp + \epsilon\psi_1 h);
\end{aligned}
$$

we have used that $-\varsigma' h' \geq 0$ to eliminate one term. Since $u^\sharp(\tau, y) \to 0$ as $x \to -\infty$ given $\delta > 0$ there exists $T < -R - 1$ such that $u(\tau, y) \leq \delta$ for $\tau \leq T$. (Recall that $\phi(x) = \phi_-$ for all $x \leq -R$.) Next, choose $\gamma = T - 2\varsigma(0)$ where γ is as in (19). Since $\varsigma(t)$ is monotone increasing we have that

$$
x - \varsigma(t) = \tau - 2\varsigma(t) \leq T - 2\varsigma(0) \text{ for } \tau \leq T.
$$

Thus for $\tau \leq T$ and $t \geq 0$, $h(x - \varsigma(t)) = h(\tau - 2\varsigma(t)) \equiv 1$. Finally, note that for $\tau \leq T$ and some constant $K > 0$ depending only on $f(u)$,

$$
\left| f'(0)\epsilon\psi_1 + f'(u^\sharp) - f(u^\sharp + \epsilon\delta_1) \right| \leq K(\epsilon + \delta)\epsilon\psi_1.
$$

We now use these estimates in (21) to obtain

$$
N[\bar{u}] > \left[-\mu - K(\epsilon + \delta) - \rho_1 \right] \epsilon\psi_1
$$

for $\tau \leq T$ and $t \geq 0$; here we have used that $\varsigma' u_x^\sharp > 0$. Thus (20a) is valid whenever $\tau \leq T$ provided that we choose

$$
\mu = \frac{|\rho_1|}{3}, \quad \delta = \frac{|\rho_1|}{(3K)}, \quad \epsilon_0 \leq \frac{|\rho_1|}{(3K)}.
$$

Next, we estimate $N[\bar{u}]$ when $\tau \geq T$. First note that whenever $\tau - 2\varsigma > \gamma + 1$ then $h(x - \varsigma) = 0$ so that $\bar{u} = u^\sharp$ satisfies $N[\bar{u}] = \varsigma' u_x^\sharp > 0$. Now consider an interval $T \leq \tau \leq \hat{T}$ where $\hat{T} > T + 1$ is given. By choosing T

sufficiently negative we can arrange that $\hat{T} < -R$ so that $\phi^\sharp(\tau) = \phi_-$ for $\tau \leq \hat{T}$. Since $u^\sharp_x > 0$ in $\Omega^\sharp$ it follows from the strong maximum principle that $u^\sharp_{xy} \neq 0$ at $y = 0, \phi_-$ and for $\tau \leq \hat{T}$. It follows that there exists $d > 0$ depending only on δ, T, and $\hat{T}$ such that

$$\text{(22)} \qquad \frac{u^\sharp_x(\tau,y)}{\psi_1(y)} > d, \quad (\tau,y) \in [T,\hat{T}] \times [0,\phi_-],$$

since $\psi_1(y)$ is positive for $0 < y < \phi_-$. From (21) and (22) there exists $K > 0$ such that

$$N[\bar{u}] \geq [d\varsigma' - K\epsilon]\psi_1(y)$$

for (τ, y) as in (22). Choose $\varsigma' = K\epsilon/d$, i.e.

$$\varsigma(t) = \varsigma(0) + \frac{K\epsilon_0}{\mu d}(1 - e^{-\mu t});$$

if $\varsigma(0) > 0$ then $\varsigma(t)$ is a monotone increasing positive function, as required, and (20a) holds on the interval $T \leq \tau \leq \hat{T}$ for all $t \geq 0$.

It remains to verify (20a) when $\tau \geq \hat{T}$. We will show for ϵ_0 sufficiently small that $h(x - \varsigma) = h(\tau - 2\varsigma) = 0$ for all $\tau \geq 0$; by a previous remark, $\bar{u} = u^\sharp$ is an upper solution here. In order that $h(\tau - 2\varsigma) = 0$ we require that $\tau - 2\varsigma \geq \gamma + 1$. Recalling that $\gamma = T - 2\varsigma(0)$ this becomes

$$\tau - 2\varsigma \geq T + 1 - 2\varsigma(0).$$

Thus for $\tau \geq \hat{T}$ our condition is that

$$\hat{T} - (T + 1) \geq 2 \max_{t \geq 0}(\varsigma(t) - \varsigma(0)) = 2K\epsilon_0/(\mu d).$$

Since $\hat{T} > T + 1$ is fixed and d depends only on $T, \hat{T}$, and δ this condition is fulfilled if

$$\epsilon_0 \leq \frac{\mu d}{2K}[\hat{T} - (T + 1)].$$

This verifies (20a) in $\Omega \times \mathbb{R}_+$ for sufficiently small ϵ_0.

We finally show that (20c) holds. By a previous remark $\hat{u}$, which is larger than $\underline{u}$, tends to $u_1^+(y)$ at $x = \pm\infty$; hence $\underline{u}(x,y) \leq u_1^+(y)$ by monotonicity in x. On the other hand $\phi^\sharp > \phi_+$ implies that the endstate $u_1^\sharp(y)$ of $u^\sharp$ is pointwise larger than $u_1^+(y)$. Thus if ρ is sufficiently large and positive we have that for all x,

$$\text{(23)} \qquad \underline{u}(x,y) < u_1^+(y) < u^\sharp(\rho,y), \quad 0 < y < \phi_+.$$

Fix $\Gamma \leq -R$; Γ will be specified later. It follows from (23) that for $\varsigma(0)$ sufficiently large and positive (depending only on Γ) that

$$\bar{u}(x,y,0) \geq u^\sharp(x+\varsigma(0),y) \geq \underline{u}(x,y), \quad x \geq \Gamma,$$

so that (20c) holds for $x \geq \Gamma$.

Now consider u_τ and u_m in (18); since each of these functions satisfies (7_-) as $x \to -\infty$ for $\tau \geq 0$ it follows that $\underline{u}$ satisfies (7_-) also. Thus there exists $\Gamma \leq -R$ such that

$$\underline{u}(x,y) \leq 2\lambda\rho_-(\lambda) \text{ for } x \leq \Gamma, \tag{24}$$

(see (7_-)). Clearly we have that

$$\bar{u}(x,y,0) \geq \max\{u^\sharp(x+\varsigma(0),y), \beta(x,y,0)\} \tag{25}$$

since $u^\sharp$ and β are nonnegative. We will consider two subintervals of $x \leq \Gamma$: $x \leq T - \varsigma(0)$ and $T - \varsigma(0) < x < \Gamma$. (If $\Gamma < T - \varsigma(0)$ we need only consider points of the first type.) Since $u^\sharp > 0$ in Ω it follows that $u^\sharp_y \neq 0$ at any point on $\partial\Omega$ where $u^\sharp = 0$. It follows for λ sufficiently small, for $T - \varsigma(0) \leq x \leq \Omega$, and $0 < y \leq \phi_-$ that

$$2\lambda\rho_-(y) \leq u^\sharp(x+\varsigma(0),y).$$

Thus from (24), (25) and the above we have that (20c) holds whenever $T - \varsigma(0) \leq x \leq \Gamma$.

Finally suppose that $x \leq T - \varsigma(0)$, so that

$$x - \varsigma(0) \leq T - 2\varsigma(0) = \gamma;$$

for such x we have that $h(x - \varsigma(0)) = 1$ (see (19)), so that $\beta(x,y,0) = \epsilon_0\psi_1(y)$. For sufficiently small λ depending on ϵ_0 we have that

$$2\lambda\rho_-(y) \leq \epsilon_0\psi_1(y), \quad 0 \leq y \leq \phi_-.$$

Thus from (24), (25) and the above we have that

$$\bar{u}(x,y,0) \geq \epsilon_0\psi_1(y) \geq 2\lambda\rho_-(y) \geq \underline{u}(x,y)$$

for $x \leq T - \varsigma(0)$. This completes verification of (20c).

It can now be proved that $u_-(y) = \lim_{x\to-\infty} \hat{u}(x,y) = 0$. We will show that the iterates $u_k(x,y)$ of the monotone iteration scheme in (8) with $u_0 = \underline{u}$ all satisfy $u_k(x,y) \le \bar{u}(x,y,t)$ for $t \ge 0$. This follows for $k = 0$ from (20) and standard comparison theorems for parabolic equations.

Next, consider the case $k = 1$. Let $U^1(x,y,t)$ be the solution of

$$U_t^1 = (\Delta - h)U^1 + f(u_0) + hu_0$$
$$U^1 = 0 \text{ on } \partial\Omega \tag{26}$$
$$U^1(x,y,0) = u_0(x,y).$$

It is routine to show that $U^1(x,y,t)$ tends to the unique steady state solution of (26), $u_1(x,y)$, as $t \to +\infty$. Moreover, from (20) and (26) we have that

$$(\bar{u} - U^1)_t \ge (\Delta - h)(\bar{u} - U^1) + (f'(*) + h)(\bar{u} - u_0)$$
$$\bar{u}(x,y,0) = U^1(x,y,0) \ge 0;$$

here $f'(*)$ is evaluated at some value between u_0 and $\bar{u}$. Since $\bar{u} - u_0 \ge 0$ it follows from the above that $\bar{u} - U^1 \ge 0$ for all t; hence $\bar{u} \ge u_1(x,y)$ for all t. The proof now follows for all k by an induction argument.

Since $\hat{u} = \lim u_k$ it follows that $u_k \le \bar{u}$ for all t, so that

$$\hat{u}(x,y) \le \lim_{t\to\infty} \bar{u}(x,y,t) = u^\sharp(x + \bar{\varsigma}, y),$$

where $\bar{\varsigma} = \varsigma(0) + K\epsilon_0/(d\mu)$. Since $u^\sharp$ tends to zero as $x \to -\infty$, this completes the proof of Lemma 3.

We now conclude the proof of Theorem 2.C. From Lemmas 2 and 3 we have distinct monotone solutions $\hat{u}$ and u_m which satisfy (3); without loss of generality, $\hat{u} > u_m$ in Ω. By Theorem 2.B, u_m and $\hat{u}$ are both linearly stable. Let $\psi_m(x,y)$, ρ_m and $\hat{\psi}$, $\hat{\rho}$ be the principal eigenfunction and eigenvalue of the linearization about u_m and $\hat{u}$, so that $\psi_m, \hat{\psi} > 0$ and $\rho_m, \hat{\rho} < 0$. It follows that if

$$\bar{u} = \hat{u} + \epsilon\hat{\psi}$$
$$\underline{u} = u_m - \epsilon\psi_m$$

then $\bar{u}$ (resp. $\underline{u}$) is a strict upper solution (resp. lower solution) of (1), (3) provided that $\epsilon > 0$ is sufficiently small. Also $\underline{u} \le u_m$, $\hat{u} \le \bar{u}$ in Ω.

Let

$$V_m = \left\{ v \in C^{1+\alpha}(\Omega) : v = 0 \text{ on } \partial\Omega, v_x \geq 0, \text{ and } \underline{u} \leq v \leq \bar{u} \text{ in } \Omega \right\}.$$

It follows from arguments similar to those of Theorem 2.B and standard comparison theorems for parabolic equations that V_m is a positively invariant set for the parabolic equation (17). Since functions in V_m have a uniform decay rate as $x = \pm\infty$, and (17) is smoothing, the time t image of V_m, denoted by $V_m(t)$, is a compact subset of $C^{1+\alpha}(\Omega)$ and so, the Conley index h of the maximal invariant set $S(V_m)$ in $V_m(t)$ defined. Since $V_m(t)$ is positively invariant, $h = \bar{1}$ (the index of an attractor is the pointed one point space).

Next we show that every solution $u(x, y, t)$ of (17) in V_m tends to a steady state. To this end let

$$J(t) = \int_\Omega \left[-|\nabla u|^2 + F(u) - \alpha(x, y) \right] \, dx \, dy;$$

here $F'(u) = f(u)$, $F(0) = 0$, and $\alpha(x, y)$ is a function which tends to zero as $x \to -\infty$ and to

$$-\left[u_{1y}^+\right]^2 + F(u_1^+)$$

as $x - +\infty$. It is not hard to show that $\underline{u}$ and $\bar{u}$ decay exponentially fast to their limits at $x = \pm\infty$; (this uses the stability of the endstates $u_0 = 0$ and u_1^+). It follows that for $t > 0$ solutions u of (17) in V_m will obey similar decay rates at $x = \pm$; thus the integral in J will converge for $t > 0$. We omit the technical details of this argument.

Using the decay rates as $x = \pm\infty$ it follows that $J' = \|u_t\|_{L^2(\Omega)}$; (the decay rate ensures that boundary terms involving u_x drop out after integrating by parts). Thus J is a Liapunov function on V_m. By standard theorems on dynamical systems it follows that every solution u of (17) in V_m tends to the set of steady states in V_m.

Since every steady state in V_m is monotone in x each steady state in V_m is linearly stable and hence, isolated, by Theorem 2.B. It immediately follows that the set of steady states if finite, since it lies in the compact set $S(V_m)$. Moreover, since each rest point in V_m is an attractor we have that $S(V_m)$ consists entirely of the set of steady states, say $S(V_m) = \{S_1, \ldots, S_2\}$ with $N \geq 2$. However, by the sum formula for Conley's index

$$h(S(V_m)) = \bigvee_{i=1}^{N} h(S_i) = \bigvee_{i=1}^{N} \bar{1}$$

providing the desired contradiction.

D. Asymptotic Stability. For brevity's sake we will only describe the main ideas; the details are all similar to previous arguments. First, the comparison argument of Lemma 3 is used to show that solutions of (17) with initial data as in (7) have values between sub- and supersolutions of (17), say $\underline{u}$ and $\bar{u}$. As $t \to +\infty$, $\underline{u}$ and $\bar{u}$ tend to steady sub- and supersolutions $u_\sharp$ and $u^\sharp$ which satisfy the limiting conditions (3) at $x = \pm\infty$. (Here, it may be necessary to modify $u^\sharp, u_\sharp$ in Lemma 3 by the construction appearing in (11)–(13) to ensure that (3) is satisfied exactly.) From this it follows that any point $u_0(x, y)$ in the w-limit set of a solution of (17) has values between $u_\sharp$ and $u^\sharp$, and hence, u_0 lies in a compact subset of $C^\alpha(\Omega)$. The Liapunov function J can now be used to conclude that either u_0 is a rest point or that u_0 lies on an orbit connecting two distinct rest points. Since there is only one rest point, which is an attractor, it follows that the entire w-limit set is the singleton consisting of the unique steady state.

Appendix.

PROOF OF LEMMA 1: The first step is to map Ω in a nice way onto the flat strip

$$\Omega_1 = \{(\xi, \eta) : \xi \in \mathbf{R}^1, \quad 0 \leq \eta \leq 1\}.$$

LEMMA A.1:

(i) There exists a conformal mapping of Ω onto Ω_1 given by two harmonic functions $\xi = \xi(x, y)$, $\eta = \eta(x, y)$. If $u(x, y) = U(\xi, \eta)$ then

$$\Delta_{(x,y)} u = |\nabla \xi|^2 \Delta_{(\xi,\eta)} U$$

where $|\nabla \xi|^2$ is evaluated at $x = x(\xi, \eta)$, $y = y(\xi, \eta)$.

(ii) There exist functions $p(x, y), q(x, y)$ such that

$$\xi = \phi_+^{-1} x + p(x, y)$$
$$\eta = \phi_-^{-1} y + q(x, y)$$
$$|\nabla^\alpha p|, |\nabla^\alpha q| \leq C e^{-kx} \text{ for } x \leq a, |\alpha| \leq n.$$

Here C depends only on n and a, and k is a positive constant. A similar decay rate holds as $x \to -\infty$ wherein ϕ_+^{-1} is replaced with ϕ_-^{-1}.

The proof of Lemma A.1 can be found in [**1**, Theorem A1.2]. The decay rate is only proved there for p; the decay rate for q then follows from Cauchy-Riemann equations.

The resolvent equation for the operator

$$Lw = \Delta w + f'(u)w, \quad w = 0 \text{ on } \partial\Omega$$

is $Lw - \lambda w = g \in L^2(\Omega)$. The associated equation under the change of variables $(x, y) \to (\xi, \eta)$ is

$$(A.1) \qquad |\nabla \xi|^2 \Delta W + (f'(U) - \lambda)w = G \in L^2(\Omega_1);$$

here, an upper case variable is the image in Ω_1 of a lower case variable defined in Ω under this transformation. By conformality and (ii) of Lemma A.1, $|\nabla \xi|^2$ is unformly bounded away from zero in Ω_1 and tends to constant limits at $\xi = \pm\infty$.

Define

$$a_\lambda(\xi, \eta) = |\nabla \xi|^{-2}(f'(U) - \lambda)$$
$$G_1 = |\nabla \xi|^{-2} G;$$

then the resolvent equation becomes

$$A_\lambda W \stackrel{\text{def}}{=} \Delta W + a_\lambda W = G_1 \in L^2(\Omega_1).$$

Next, define

$$\hat{a}_\lambda(\xi, \eta) = \begin{cases} a_\lambda(+\infty, \eta), & \xi > 0 \\ a_\lambda(-\infty, \eta), & \xi > 0 \end{cases}$$

and consider the equation

$$(A.2) \qquad \hat{A}_\lambda W \stackrel{\text{def}}{=} \Delta W + \hat{a}_\lambda W = G_1, \quad W = 0 \text{ on } \partial\Omega_1.$$

Since $a_\lambda - \hat{a}_\lambda \to 0$ as $|\xi|$ tends to infinity, it follows from a standard theorem that if (A.2) is solvable for all $G_1 \in L^2(\Omega_1)$ then either zero is in the resolvent set of A_λ or zero is a normal point of A_λ, since $\hat{A}_\lambda$ and A_λ essentially differ by a compact perturbation (see [**7**, Theorem A.1, p. 136]). Thus if (A.2) is solvable for all G_1 then λ is either a normal point of the original operator L or in the resolvent set of L.

We finally show that (A.2) is solvable for suitable λ. To this end consider the eigenvalue problems

$$(A.3_{\pm}) \qquad \psi_{\eta\eta} + \phi_{\pm}^2 b^{\pm}(\eta)\psi = \rho\psi, \quad \psi(0) = \psi(1) = 0,$$

where $b^{\pm}(\eta)$ are the limits of $f'(U(\xi,\eta))$ at $\xi = \pm\infty$. Since the endstates of U are stable the eigenvalues $\{\rho_j^{\pm}\}_{j=1}^{\infty}$ of (A.3$_{\pm}$) are negative. Let

$$R = \max_{j\geq 1}\{\rho_j^+, \rho_j^-\} < 0,$$

and let $\psi_j^{\pm}$ be eigenfunctions associated with $\rho_j^{\pm}$.

LEMMA A.2: Define

$$\hat{J}(W) = \int_{\Omega}\int_{1} \left[-|\nabla W|^2 + \hat{a}_{\lambda}W\right] \, dx\, dy.$$

If $\lambda > R$ then

$$\sup\left\{J(W) : \|W\|_{L^2} = 1, W \in H_0^1(\Omega)\right\} < R - \lambda < 0.$$

PROOF: For W as above we have that

$$W(\xi,\eta) = \begin{cases} \sum_j a_j(\xi)\psi_j^+(\eta), & \xi = 0 \\ \sum_j a_j(\xi)\psi_j^-(\eta), & \xi < 0, \end{cases}$$

where

$$a_j(\xi) = \begin{cases} (W(\xi,\eta), \psi_j^+)_{L^2[0,1]}, & \xi > 0 \\ (W(\xi,\eta), \psi_j^-)_{L^2[0,1]}, & \xi < 0 \end{cases}$$

and $\sum_j a_j^2 = 1$; (we normalize $\psi_j^{\pm}$ to have unit L^2-norm). Then

$$\begin{aligned} J(W) &= \int_{\xi<0} \sum_j \left(-\dot{a}_j^2 + \rho_j^- a_j^2\right) \, d\xi + \\ &+ \int_{\xi>0} \sum_j \left(-\dot{a}_j^2 + \rho_j^+ a_j^2\right) \, d\xi - \lambda \\ &\leq \max(\rho_1^-, \rho_j^+) \int_{-\infty}^{\infty} a_j^2 \, d\xi - \lambda \\ &= R - \lambda. \end{aligned}$$

It immediately follows from Lemma A.2 and the Lax-Milgram lemma that (A.2) is uniquely solvable for $G_1 \in L^2(\Omega_1)$.

Department of Mathematics and Statistics
University of Massachusetts, Amherst MA 01003

The author is partially supported by NSF Grant DMS 8320845

Research at MSRI supported in part by NSF Grant DMS 812079-05.

References

1. Amick, C. and L.E. Fraenkel, *Steady solutions of the Navier-Stokes equations representing plane flow in channels of variaous types*, Acta Math. **144** (1980), 83–152.
2. Fife, P., *Mathematical Aspects of Reacting and Diffusing Systems*, Springer Lecture Notes in Math. (1979).
3. Fife, P. and J.B. McLeod, *The approach of solutions on nonlinear diffusion equations to travelling front solutions*, Arch. Rat. Mech. Anal. (1975).
4. Fife, P. and L.A. Peletier, *Clines by variables selection and migration*, Proc. R. Soc. Lond. **B 214** (1981), 99–123.
5. Fife, P. and L.A. Peletier, *Nonlinear diffusion in population genetics*, Arch. Rat. Mech. Anal. **64** (1977), 93–109.
6. Gardner, R., *Existence of multidimensional travelling wave solutions of an initial-boundary value problem*, J. Diff. Eq. **61** (1986), 335–379.
7. Henry, D., *Geometric theory of semilinear parabolic equations*, Springer Lecture Notes in Math. **840** (1981).
8. Schaaf, R., *A class of Hamiltonian systems with increasing periods*, J. Reine Angew. Math. **363** (1985), 96–109.
9. Smoller, J. and A. Wasserman, *Global bifurcation of steady state solutions*, J. Diff. Eq. **39** (1981), 269–290.

A Strong Form of the Mountain Pass Theorem and Application

H. Hofer

1. Introduction.

Variational methods are a strong tool in proving existence of solutions of differential equations. In this paper we prove a strong form of the mountain pass theorem which was prompted by the need of a theorem which, besides an existence statement for critical points, gives in addition information about the "fine structure" of the functional near to them. The abstract theorems are based on results in [**H1,H2,H3**] and [**EH1,EH2**]. Related results have been also obtained in [**PS1,PS2**].

We denote by $(E, \| \ \|)$ a real Banach space and by $\Phi : E \to R$ a C^1-map. We put

$$\mathrm{Cr}(\Phi, d) = \{u \in E \mid D\Phi(u) = 0, \Phi(u) = d\}$$
$$\mathrm{Cr}(\Phi) = \bigcup_{d \in R} \mathrm{Cr}(\Phi, d)$$

and

$$\Phi^d = \Phi^{-1}((-\infty, d]), \quad \Phi_c = \Phi^{-1}([c, +\infty))$$
$$\dot{\Phi}^d = \Phi^{-1}((-\infty, d))$$

DEFINITION 1. *A mountain pass characterization for Φ (abbreviated to "mpc") is a triplet $(e_0, e_1, d) \quad E \times E \times \mathsf{R}$ such that*

$$d > \max\{\Phi(e_0), \Phi(e_1)\}$$
$$d = \inf_{\gamma \in \Gamma} \sup \Phi(|\gamma|)$$

where $\Gamma = \{\gamma \in C([0,1], E) | \quad \gamma(0) = e_0, \gamma(1) = e_1\}$, $|\gamma| = \gamma([0,1])$ and Φ satisfies the Palais-Smale Condition at the level d:

$(\mathrm{PS})_d$ *any sequence $(u_n) \subset E$ such that $D\Phi(u_n) \to 0$ in E^* and $\Phi(u_n) \to d$ must be precompact.*

The Ambrosetti-Rabinowitz mountain pass theorem states the following:

THEOREM 1 (Ambrosetti-Rabinowitz). *If (e_0, e_1, d) is an mpc for Φ then* $\mathrm{Cr}(\Phi, d) \neq 0$.

We are interested in this paper to obtain some additional information about the fine structure of the critical points in $\mathrm{Cr}(\Phi, d)$.

DEFINITION 2. *A subset $\mathcal{P} \subset \mathrm{Cr}(\Phi, d)$ has the mountain pass property if it connected and, for every open neighborhood U of $\mathcal{P}$ in E the set $\{u \in U \mid \Phi(u) < d\}$ is neither empty nor path connected. We call $\mathcal{P}$ a mountain pass component if it has the mountain pass property, and there is no subset $\mathcal{P}' \subset \mathrm{Cr}(\Phi, d)$, $\mathcal{P}' \neq \mathcal{P}$, $\mathcal{P} \subset \mathcal{P}'$, with the mountain pass property.*

If $e \in E$ and $A \subset \mathrm{Cr}(\Phi, d)$ we shall write

$$e \underset{\Phi}{\rightarrow} A$$

provided $A \cap \mathrm{cl}(B) \neq \emptyset$, where B is the path component of e in $\{u \in E \mid \Phi(u) < d\}$, and $\mathrm{cl}(B)$ is the closure of B in E. If $A = \{u\}$ we write also $e \underset{\Phi}{\rightarrow} u$.

Let (e_0, e_1, d) be a mpc for Φ.

DEFINITION 3. *Take $u_0 \in \mathrm{Cr}(\Phi, d)$. We call u_0 mountain pass essential (abbreviated to "mpe") if there is a mountain pass component $\mathcal{P} \subset \mathrm{Cr}(\Phi, d)$ such that $u_0 \in \mathcal{P}$ and*

(i) $e_0 \underset{\Phi}{\rightarrow} u_0$

(ii) *Either u_0 has the mountain pass property, or $u_0 \in \mathrm{cl}(\mathcal{P}_M) \setminus \mathcal{P}_M$, where $\mathcal{P}_M$ denotes the set of $u \in \mathcal{P}$ which are local minima of Φ in E.*

In other words, if u_0 does not have the mountain pass property, then it is not a local minimum, but it is the limit of a sequence of local minima in $\mathcal{P}$. The main result is the following:

THEOREM 2. *Let (e_0, e_1, d) be an mpc for some $\Phi \in C^1(E, R)$. Then $\mathrm{Cr}(\Phi, d)$ contains a mountain pass essential point u_0 such that $e_0 \underset{\Phi}{\rightarrow} u_0$.*

As simple examples in $E = R$ show the result is in a far-reaching sense optimal.

2. Proof of the Main Result.

First we study the set of all $u \in \mathrm{Cr}(\Phi, d)$ which belong to a mountain pass component. We denote it by $\mathrm{Cr}(\Phi, d)_{mp}$. We assume $(\mathrm{PS})_d$.

LEMMA 1. *$\mathrm{Cr}(\Phi, d)_{mp}$ is compact.*

PROOF: $\mathrm{Cr}(\Phi, d)$ is compact since $(\mathrm{PS})_d$ holds. Hence it is enough to show that $\mathrm{Cr}(\Phi, d)_{mp}$ is closed. Pick a sequence $(u_n) \subset \mathrm{Cr}_{mp} := \mathrm{Cr}(\Phi, d)_{mp}$ converging to some $u_0 \in \mathrm{Cr} := \mathrm{Cr}(\Phi, d)$ and let $\mathcal{P}_0$ be the connected component containing u_0. We argue indirectly. We find an open neighborhood U of $\mathcal{P}_0$ such that $U \cap \dot{\Phi}^d = \emptyset$ or $U \cap \dot{\Phi}^d \neq \emptyset$ and path connected. We shall show that for some large n_0 we have $\mathcal{P}_{n_0} \subset U$ giving a contradiction. Here $\mathcal{P}_n$ denotes the mountain pass component containing u_n. Assume $\mathcal{P}_n \not\subset U$ for all $n \in N$. Then, eventually taking a subsequence we find $\hat{u}_n \in \mathcal{P}_n \setminus U$ such that $\hat{u}_n \to \hat{u}_0 \notin U$. We show that $\hat{u}_0 \in \mathcal{P}_0$. Namely if $\hat{u}_0 \notin \mathcal{P}_0$ we can apply Whyburn's Lemma in Cr to find closed subsets A and B of Cr such that

$$\mathrm{Cr} = A \cup B$$
$$A \cap B = \emptyset$$
$$A \supset \mathcal{P}_0, \quad B \supset \{u_0\}.$$

Clearly A and B are both open in Cr. Now using the fact that $u_n \to u_0 \in \mathcal{P}_0 \supset A$ and $\hat{u}_n \to \hat{u}_0 \in B$ and $u_n, \hat{u}_n \in \mathcal{P}_n$ we infer for some n_0 large

$$\mathcal{P}_n \cap A \neq \emptyset, \quad \mathcal{P}_n \cap B \neq \emptyset$$

which is impossible since $\mathcal{P}_n \subset \mathrm{Cr}$ is connected.

□

We need the following form of the standard deformation lemma.

LEMMA 2. *Let $\Phi \in C^1(E, R)$ satisfy $(PS)_d$. Given $\bar{\varepsilon} > 0$ and open neighborhoods W and V of $\mathrm{Cr}(\Phi, d)$ such that $\mathrm{cl}(W) \subset V$ and $\mathrm{dist}(\partial V, W) > 0$ there exist $\varepsilon \in (0, \bar{\varepsilon})$ and a continuous map $T : [0,1] \times E \to E$ such that*

(i) $T(0, \cdot) = Id$

(ii) *$t \to \Phi(T(t, u))$ is non-increasing for all $u \in E$*

(iii) $T(\{1\} \times (\Phi^{d+\varepsilon} \setminus W)) \subset \Phi^{d+\varepsilon}$

(iv) $T([0,1] \times \mathrm{cl}(W)) \subset V$

(v) *$T(t, u) = u$ for all $(t, u) \in [0,1] \times (\Phi^{d-\bar{\varepsilon}} \cup \Phi_{d+\bar{\varepsilon}})$*

PROOF OF THEOREM 2: Let $\bar{\varepsilon} = \frac{1}{2}(d - \max\{\Phi(e_0), \Phi(e_1)\})$. For $\delta > 0$ we put $V^\delta = B_\delta(\mathrm{Cr})$ and $W^\delta = B_{\frac{1}{2}\delta}(\mathrm{Cr})$ and find by Lemma 2 $\varepsilon_\delta \in (0, \bar{\varepsilon})$ and T^δ such that (i)–(v) hold. Put

$$\Gamma = \{\gamma \in \mathrm{Cr}([0,1], E) \mid \gamma(0) = e_0, \gamma(1) = e_1\}$$

and pick $\gamma_\delta \in \Gamma$ with

$$|\gamma_\delta| \subset \Phi^{d+\varepsilon_\delta}.$$

We define $\hat{\gamma}_\delta \in \Gamma$ by

$$\hat{\gamma}_\delta(t) = T^\delta(1, \gamma_\delta(t)), \quad t \in [0,1].$$

We denote by ς the path component of e_0 in $\dot{\Phi}^d$ and define $t_0^\delta \in [0,1]$ by

$$t_0^\delta = \sup\{t \in [0,1] \mid \gamma_\delta(t) \in \varsigma\}.$$

By the definition of d we must have $0 < t_0^\delta < 1$ and

$$\Phi(\hat{\gamma}_\delta(t)) = d.$$

By the construction of $\hat{\gamma}_\delta$ this implies

$$\mathrm{dist}(\hat{\gamma}_\delta(t_0^\delta), \mathrm{Cr}) \leq \delta.$$

We pick a sequence (δ_n) which converges monotonically to 0 such that $u_n := \hat{\gamma}_{\delta_n}(t_0^{\delta_n})$ converges (eventually after picking a subsequence) to some $u_0 \in \mathrm{Cr}$ (Cr is compact). We shall show that $u_0 \in \mathrm{Cr}_{mp} \neq \emptyset$. By the definition of t_0^δ it is clear that

$$u_0 \in \mathrm{cl}(\varsigma) \cap \mathrm{Cr}$$

so that u_0 cannot be a local minimum. Assume $u_0 \in \mathrm{Cr}_{mp}$. Denote by $\mathcal{P}$ the maximal connected component in Cr containing u_0. We find an open neighborhood U of u_0 containing $\mathcal{P}$ such that $U \cap \dot{\Phi}^d \neq 0$ and path connected. Applying Whyburn's Lemma to $A = \mathcal{P}$ and $B = \mathrm{Cr} \setminus U$ in Cr ($B = \emptyset$ is possible) we can construct disjoint open neighborhoods in E.

$$V_1 \supset A, V_2 \supset B$$
$$\mathrm{Cr} \subset V_1 \cup V_2$$
$$V_1 \cap V_2 = \emptyset.$$

Using the fact that $\operatorname{dist}(\partial(V_1 \cup V_2), \mathrm{Cr}) > 0$ we have for n_0 large enough

$$V^{\delta_{n_0}} \subset V_1 \cup V_2, \quad u_{n_0} \in V_1.$$

Consequently there exists some $t' \in (0, t_0^{\delta_{n_0}})$ close to $t_0^{\delta_{n_0}}$ such that

$$\hat{\gamma}_{\delta_{n_0}}(t') \in V_1 \cap \varsigma \subset U \cap \varsigma.$$

Consider $t' = \sup\{t \in [t_0^{\delta_{n_0}}, 1] \mid \hat{\gamma}_{\delta_{n_0}}(t) \in V_1\}$, clearly $t'' > t_0^{\delta_{n_0}}$. If $t'' = 1$ we infer that $\hat{\gamma}_{\delta_{n_0}}(t'') \in \dot{\Phi}^d \cap V_1 \subset U \cap \dot{\Phi}^d$ for a suitable $t''' < 1$ arbitrarily close to $t'' = 1$ giving the contradiction $e_1 \in \varsigma$. Hence $t'' < 1$. Therefore $\hat{\gamma}_{\delta_{n_0}}(t'') \in \partial V_1$. Hence

$$\hat{\gamma}_{\delta_{n_0}}(t'') \notin V^{\delta_{n_0}}$$

which implies $\hat{\gamma}_{\delta_{n_0}}(t'') \in \dot{\Phi}^d$. Consequently there exists some $t''' \in (t_0^{\delta_{n_0}}, t'')$ with

$$\hat{\gamma}_{\delta_{n_0}}(t''') \in \dot{\Phi}^d \cap V_1 \subset \dot{\Phi}^d \cap U$$

implying that $\hat{\gamma}_{\delta_{n_0}}(t''') \in \varsigma$ by the construction of U. This however contradicts the definition of $t_0^{\delta_{n_0}}$.

Summing up, we have shown the existence of a mountain pass component $\mathcal{P} \subset \mathrm{Cr}(\Phi, d)$ containing a point $u_0 \in \mathrm{cl}(\varsigma) \cap \mathrm{Cr}$. In other words

$$e_0 \underset{\Phi}{\rightarrow} \mathcal{P}, \quad \mathcal{P} \text{ a mountain pass component in } \mathrm{Cr}(\Phi, d).$$

Define $\mathcal{P}_{mp} = \{u \in \mathcal{P} \mid$ for every open neighborhood U of u the set $U \cap \dot{\Phi}^d$ is nonempty and not path connected $\}$. If there exists $\hat{u} \in \mathcal{P}_{mp}$ with $e_0 \underset{\Phi}{\rightarrow} \hat{u}$ we are done. Let us assume that $e_0 \underset{\Phi}{\rightarrow} \hat{u}$ does not hold for every $\hat{u} \in \mathcal{P}_{mp}$. Define $\mathcal{P}_M = \{u \in \mathcal{P} \mid u \text{ is a local minimum of } \Phi\}$. We show that $\mathcal{P}_M \neq \emptyset$. Arguing indirectly, assume $\mathcal{P}_M = \emptyset$. We find consequently for every $u \in \mathcal{P} \setminus \mathcal{P}_{mp}$ an open neighborhood U_u in E such that $U_u \cap \dot{\Phi}^d \neq \emptyset$ and path connected. Since by our assumption $\mathcal{P}_M = \emptyset$ we conclude $\mathcal{P} \subset \mathrm{cl}(\dot{\Phi}^d)$. Hence

$$\mathcal{P}_c = \{u \in \mathcal{P} \setminus \mathcal{P}_{mp} \mid e_0 \underset{\Phi}{\rightarrow} u\}$$

is open and nonempty. Trivially $\mathcal{P}_c$ is closed. Since $\mathcal{P}$ is connected we have $\mathcal{P} = \mathcal{P}_c$. Hence $\mathcal{P}_{mp} = \emptyset$ by our assumption that $e_0 \underset{\Phi}{\rightarrow} u$ does not hold for every $u \in \mathcal{P}_{mp}$. Define now, still assuming $\mathcal{P}_M = \emptyset$,

$$U = \bigcup_{u \in \mathcal{P}} U_u.$$

Then U is an open neighborhood of $\mathcal{P}$ such that $U \cap \dot{\Phi}^d \neq \emptyset$ and path connected. Consequently $\mathcal{P}$ cannot be a mountain pass component contradicting our assumption. Hence $\mathcal{P}_M \neq \emptyset$ if we assume that $e_0 \underset{\Phi}{\rightarrow} u$ does not hold for every $u \in \mathcal{P}_{mp}$. As we have already seen the set $\mathcal{P}_c$ is closed and nonempty. By our assumption $\mathcal{P}_{mp} = \emptyset$. Since $\mathcal{P}$ is connected and $\mathcal{P}_c = \{u \in \mathcal{P} \mid e_0 \underset{\Phi}{\rightarrow} u\}$ is trivially closed and nonempty we must have

$$\mathcal{P}_c \cap \mathrm{cl}(\mathcal{P}_M) \neq \emptyset.$$

In fact, by our assumption that $e_0 \underset{\Phi}{\rightarrow} u$ does not hold for any $u \in \mathcal{P}_{mp}$ we inferred that $\mathcal{P}_{mp} = \emptyset$. If now $\mathrm{cl}(\mathcal{P}_M) \cap \mathcal{P}_c = \emptyset$ we find a sequence $(u_n) \subset \mathcal{P} \setminus (\mathcal{P}_M \cup \mathcal{P}_c)$ which converges to some $u_0 \in \mathcal{P}_c$. u_0 has an open neighborhood U_0 such that $U_0 \cap \dot{\Phi}^d$ is nonempty and path connected. For some n_0 large we have $u_{n_0} \in U_0$. Since $u_{n_0} \notin \mathcal{P}_M$ and $\dot{\Phi}^d \cap U_0$ is path connected we infer that $u_{n_0} \in \mathrm{cl}(\varsigma)$ since $e_0 \underset{\Phi}{\rightarrow} u_0$. Hence in other words

$$e_0 \underset{\Phi}{\rightarrow} u_{n_0}$$

giving that $u_{n_0} \in \mathcal{P}_c$ contradicting the assumption $u_{n_0} \notin \mathcal{P}_c$. Consequently we have shown that

$$\mathcal{P}_c \cap \mathrm{cl}(\mathcal{P}_M) \neq \emptyset.$$

Since $\mathcal{P}_M \cap \mathcal{P}_c = \emptyset$ we find $\hat{u} \in \mathcal{P}_c \cap (\mathrm{cl}(\mathcal{P}_M) \setminus \mathcal{P}_M)$ which completes the proof of our theorem.

□

3. Application.

The first application we want to describe is the following. Assume E is a real Hilbert space and $\Phi \in C^2(E, R)$ having a gradient of the form Identity-compact. Suppose

(Φ) For all $u \in \mathrm{Cr}(\Phi)$ the first (smallest) eigenvalue λ_1 of the linearization $\Phi''(u_0) \in \mathcal{L}(E)$ at u_0 is simple provided $\lambda_1 = 0$.

We have the following:

THEOREM 3. *Let (e_0, e_1, d) be an mpc for Φ. Suppose (Φ) holds and $\mathrm{Cr}(\Phi, d)$ consists only of isolated critical points. Then there exists $u_0 \in \mathrm{Cr}(\Phi, d)$ such that*

$$\deg_{\mathrm{loc}}(\Phi', u_0, 0) = -1.$$

Here $\deg_{\mathrm{loc}}$ denotes the local Leray-Schauder degree.

PROOF: By Theorem 2 there exists a mountain pass essential critical point $u_0 \in \mathrm{Cr}(\Phi, d)$. Since $\# \mathrm{Cr}(\Phi, d) < +\infty$ by $(\mathrm{PS})_d$ we find that $\{u_0\}$ is a mountain pass component or in the terminology of [**H1,H2,H3**] of mountain pass type. As proved in [**H2,H3**] the hypothesis (Φ) now implies our assertions.

□

Theorem 3 is applicable to second order elliptic boundary value problems. Consider the differential equation

$$(E) \qquad \begin{aligned} -\Delta u &= f(x,u) && \text{in } \Omega \\ u &= 0 && \text{on } \partial\Omega \end{aligned}$$

where $\Omega \subset \mathbb{R}^N$, $N \geq 3$, is a bounded domain with smooth boundary $\partial\Omega$. Assume $f \in C^1(\bar{\Omega} \times \mathbb{R}, \mathbb{R})$ (the regularity assumption of the x-dependence can be considerably weakened) and $f'(x,s) = D_2 f(x,s)$ satisfies the growth condition

$$|f'(x,s)| \leq C(1 + |s|^{\sigma-1})$$

for all $(x,s) \in \bar{\Omega} \times \mathbb{R}$ where $C > 0$ and $\sigma \in [1, (N+2)/(N-2))$. Under this hypothesis the solutions of (E) can be found as critical points of the C^2-functional $\Phi \in C^2(E,R)$, where $E = H_0^1(\Omega)$ with norm $\|u\|^2 = \int_\Omega \langle \nabla u, \nabla v \rangle$ given by

$$\Phi(u) = \tfrac{1}{2}\|u\|^2 - \int_\Omega \hat{f}(x, u(x))\, dx.$$

Here $\hat{f}(x,s) = \int_0^s f(x,t)\, dt$. As a consequence of the results of Manes-Micheletti [**MM**] or Hess-Kato [**HK**] Φ satisfies (Φ). Theorem 3 is the key ingredient in the proof of the following result. Assume

$$(f) \qquad \begin{aligned} & f \in C^1(R,R), |f'(s)| \leq C(1 + |s|^{\sigma-1} \\ & s \in [1, (N+2)/(N-2)) \\ & \lim_{s \to +\infty} f'(s) \in (\lambda_k, \lambda_{k+1}) \quad k \geq 2 \\ & \limsup_{s \to -\infty} f(s)/s < \lambda_1. \end{aligned}$$

Here $\lambda_1 < \lambda_2 \leq \lambda_3$ is the sequence of eigenvalues of $-\Delta u = \lambda u$ in Ω, $u = 0$ on $\partial\Omega$, with multiplicity $m_k = m(\lambda_k)$. It is well-known that $m_1 = 1$. Denote by $I > 0$, $\int_\Omega I^2 = 1$, the eigenfunction belonging to λ_1.

THEOREM 4. *There exist numbers $s^* \leq s^{**}$ such that the following holds. The parameter dependent problem*

$$(E_s) \qquad \begin{aligned} -\Delta u &= f(u) + sI \quad &\text{in } \Omega \\ u &= 0 &\text{on } \partial\Omega \end{aligned}$$

satisfies

(i) *(E_s) has no solution for $s > s^{**}$*
(ii) *(E_s) has at least one solution for $s = s^{**}$*
(iii) *(E_s) has at least two solutions for $s \in [s^*, s^{**})$*
(iv) *(E_s) has at least four solutions for $s < s^*$.*

SKETCH OF THE PROOF: We restrict ourselves to (iv). (E_s) possesses for all $s < s^{**}$ a pair of strict sub- and supersolutions, say $\underline{u} < \bar{u}$. It is not difficult to construct a local maximum u_1 of Φ with $\underline{u} < u_1 < \bar{u}$. If now $s < s^*$ for a suitable $s^* \leq s^{**}$ we find a solution u^2 close to the function $\frac{s}{\lambda_1 - \lambda} I$ where $\lambda = \lim_{s \to \infty} f'(s)$. Further u_2 is a nondegenerate critical point of Φ with Morse index $m_k \geq 2$ and local Leray-Schauder degree $d_2 = (-1)^{m_k}$. Using some the fact that (E_s) is not solvable for $s > s^{**}$ together with estimates one sees that $d := \deg(\Phi', B_R, 0) = 0$ for every large radius R. Next using that $\lim_{\sigma \to \infty} \Phi(\sigma I) = -\infty$ we obtain a mountain pass characterization (u_1, e_1, d) which provides us with a critical point u_3 having local degree $d_3 = -1$ (we may assume (E_s) has only a finite number of solutions). We must have clearly $u_3 \neq u_1$. But also $u_3 \neq u_2$ since $m_k \geq 2$ implies that u_2 cannot be mountain pass essential. The local degree at u_1 is by well-known results $d_1 = 1$. Assuming now that we have only three solutions we must have by the additivity of the degree

$$d = d_1 + d_2 + d_3$$

or more precisely

$$0 = 1 - 1 + (-1)^{m_k} \neq 0.$$

This contradiction proves our claim.

□

More details of the proof and concerning former research on this problem can be found in [**H4**].

Next we describe briefly an application to the minimal period problem in the theory of Hamiltonian systems. See [**EH1**] for complete details. Consider the Hamiltonian system

$$(HS) \qquad \dot{x} = JH'(x)$$

Asymptotic Behaviour of Solutions of the Porous Media Equation with Absorption

S. KAMIN

Let $u(x,t)$ be a solution of the Cauchy problem for the porous medium equation

$$\left.\begin{aligned} &(1) \qquad u_t = \Delta u^m, \quad x \in \mathbb{R}^n, \quad t > 0 \\ &(2) \qquad u(x,0) = \varphi(x) \end{aligned}\right\} \quad \text{(PM)}$$

We are interested in the behaviour of $u(x,t)$ as $t \to \infty$. The problem will also be studied for the porous medium equation with absorption:

$$\left.\begin{aligned} &(3) \qquad v_t = \Delta v^m - v^p \\ &(4) \qquad v(x,0) = \varphi(x). \end{aligned}\right\} \quad \text{(PMA)}$$

Assume that $m \geq 1$, $p > m$ and let $\varphi(x)$ be a bounded nonnegative function. The existence and uniqueness of nonnegative solutions of (1), (2) and of (3), (4) are known. These solutions are continuous for $t > 0$. It was shown recently ([**1,6–11**]) that the large time behaviour of u and v depends on the behaviour of $\varphi(x)$ for large $|x|$. Let us assume that

(5) $$\lim_{|x|\to\infty} |x|^\alpha \varphi(x) = A > 0$$

The reason we are beginning with the power function is that in this case one representative of the class of solutions is known, namely, the exact similarity solution with initial datum a power of $|x|$.

We first consider problem (PM). It was proved in [**7**] that, if $\alpha > n$, then

(6) $$\lim_{t\to\infty} t^\ell |u(x,t) - E_M(x,t)| = 0$$

uniformly on $\{x, t : |x| \leq Rt^{\frac{\ell}{n}}\}$, where $R > 0$,

(7) $$\ell = \frac{n}{2 + n(m-1)}$$

and $E_M(x,t)$ is the Barenblatt-Pattle (B.-P.) solution of Eq. (1) with mass $M = \| \varphi \|_{L_1}$. If $0 < \alpha < n$ then, as proved in [1],

$$\lim_{t\to\infty} t^{\lambda}|u(x,t) - W_{\alpha,A}(x,t)| \to 0 \tag{8}$$

uniformly on $\{x,t : |x| \leq Rt^{\frac{\lambda}{\alpha}}\}$, where

$$\lambda = \frac{\alpha}{2+\alpha(m-1)} \tag{9}$$

and $W_{\alpha,A}(x,t)$ is the solution of (PM) with initial datum $A|x|^{-\alpha}$.

Results (6) and (8) mean that the large time behaviour of $u(x,t)$ is described by a similarity solution.

Now let us consider the borderline case $\alpha = n$. We show that in this case asymptotic behaviour of $u(x,t)$ is different from its behaviour in the two previous cases. Namely, on the assumption that

$$|x|^n \varphi(x) \to A, \tag{10}$$

we prove that

$$\lim_{t\to\infty} \tau^{\ell} \left| \frac{u(x,t)}{B \ln \tau} - E(x,\tau) \right| = 0, \tag{11}$$

where $t \cong B^{1-m} \frac{\tau}{(\ln \tau)^{m-1}}$, B is some constant depending on n, m and A, and E is the B.-P. solution with $M = 1$.

It follows from (11) that for large t

$$u(x,t) \cong B \ln \tau E(x,\tau).$$

We shall refer to $B \ln \tau E(x,\tau)$ as a "*reconstructed* similarity solution". This means here that the similarity solution E is evaluated at a different instant of time and also mutiplied by a certain function. The relationship (11) follows from a more general result, Theorem 1, which be formulated later. Theorem 1 can also be used to handle other cases, e.g., $\varphi(x) \sim A|x|^{-\alpha} \ln|x|$ or $\varphi(x) \sim A|x|^{-\alpha} \frac{1}{\ln|x|}$, $(o < \alpha < n)$ for large x, thus generalizing the results of [**1**].

To explain the meaning of the reconstructed similarity solution, we begin with a simple example. Let $u(x,t)$ be the solution of the heat equation

$$u_t = u_{xx}\,, \qquad x \in \mathbb{R}^1\,, \qquad t > 0 \tag{12}$$

satisfying the initial conditions $u(x,0) = \varphi(x)$. Assume that

(13) $$\varphi(x)|x| \to A \quad \text{as} \quad |x| \to \infty$$

Define $u_k(x,t) = f(k)u(kx, k^2t)$, $k \geq 1$, where

$$f(k) = \frac{k}{\int_{|y|<k} \varphi(y)dy} = \frac{1}{\int_{|x|<1} \varphi(kx)dx}.$$

Then $u_k(x,t)$ is a solution of eq.(12) with initial data

$$u_k(x,0) = \varphi_k(x) = f(k)\varphi(kx).$$

It follows from (13) that, for large k,

(14) $$\int_{|y|<k} \varphi(y)dy \cong 2A \ln k$$

and

$$\varphi(kx) \cong \frac{A}{k|x|}.$$

Therefore

$$\varphi_k(x) \cong \frac{kA}{2A|x| \ln k} = \frac{1}{2|x| \ln k} \to 0$$

as $k \to \infty$, $\forall x \neq 0$. On the other hand,

$$\int_{|x|<1} \varphi_k(x)dx = 1.$$

Therefore, $\varphi_k(x) \to \delta(x)$ in D' as $k \to \infty$. Thus, we may expect that the sequence of solutions $u_k(x,t)$ will converge to the fundamental solution $E(x,t)$ of (12). Then

$$f(k)u(kx, k^2t) \to E(x,t).$$

In particular, for $t = 1$,

(15) $$f(k)u(kx, k^2) \to E(x,1) = kE(kx, k^2).$$

Transforming $k^2 \to t$, $kx \to x$, we obtain

$$f(\sqrt{t})u(x,t) - \sqrt{t}E(x,t) \to 0 \quad \text{as} \quad t \to \infty.$$

By (14)

$$f(\sqrt{t}) \cong \frac{\sqrt{t}}{A \ln t},$$

and we get

$$\frac{\sqrt{t}}{A \ln t} u(x,t) - \sqrt{t} E(x,t) \to 0.$$

The last relation can be rewritten as

$$u(x,t) \cong A \ln t E(x,t).$$

In this case, therefore, the solution $u(x,t)$ is close to $A \ln t E(x,t)$, which is not a solution of the heat equation. This result may be obtained directly from the Poisson integral formula, but the above formal approach seems easier. The convergence in (15) can be proved, and thus the formal considerations are justified.

To be able to formulate the results we need some definitions.

We shall use two families of similarity solutions of the porous media equation (1):

1. The source-type solution $E(x,t)$ with mass 1. This is the B.-P. solution for $m > 1$ and the fundamental solution of the heat equation for $m = 1$.

Thus, for $m > 1$,

$$E(x,t) = t^{-\ell} \left\{ \left[a^2 - \frac{(m-1)\ell |x|^2}{2mnt^{2\ell/n}} \right]_+ \right\}^{1/(m-1)}$$

where ℓ is defined by (17) and a is a constant chosen so that $\int_{\mathbf{R}^n} E(x,t)dx = 1$. For $m = 1$

$$E(x,t) = \frac{1}{(4\pi t)^{n/2}} \exp\left(-\frac{|x|^2}{4t}\right)$$

Note that

$$E_M(x,t) = ME(x, M^{m-1}t)$$

where $E_M(x,t)$ is a solution of (1) with $E_M(x,0) = M\delta(x)$.

2. The solution $W_\alpha(x,t)$ of (1) with initial data

$$W_\alpha(x,0) = \frac{n-\alpha}{S_1}|x|^{-\alpha} \qquad (0 < \alpha < n)$$

where S_1 is the surface area of the unit sphere in $\mathbf{R}^n$. The function $W_\alpha(x,t)$ has the form

$$W_\alpha(x,t) = t^{-\lambda} g_\alpha(|x| t^{-\frac{\lambda}{\alpha}})$$

where λ is defined by (9) and $g = g_\alpha(\eta)$ is a solution of the problem

$$(g^m)'' + \frac{n-1}{\eta}(g^m)' \frac{\lambda}{\alpha}\eta g' + \lambda g = 0,$$
$$\eta > 0, \quad g'(0) = 0, \quad \lim_{\eta\to\infty} \eta^\alpha g(\eta) = \frac{n-\alpha}{S_1}.$$

Note that, for every A,

$$W_{\alpha,A}(x,t) = A\frac{S_1}{n-\alpha} W_\alpha\left(x, \left(\frac{AS_1}{n-\alpha}\right)^{m-1} t\right)$$

where $W_{\alpha,A}$ satisfies (1) and $W_{\alpha,A}(x,0) = A|x|^{-\alpha}$.

Define

$$U_\alpha(x,t) = \begin{cases} W_\alpha(x,t) & \text{if } \alpha < n \\ E(x,t) & \text{if } \alpha = n \end{cases}$$

Then

$$U_\alpha(x,0) = \Phi_\alpha(x),$$

where

$$\Phi_\alpha(x) = \begin{cases} \frac{n-\alpha}{S_1}|x|^{-\alpha} & \text{if } \alpha < n \\ \delta(x) & \text{if } \alpha = n. \end{cases} \tag{16}$$

Note that, $\forall k > 0$,

$$U_\alpha(x,t) = k^\alpha U_\alpha(kx, k^{2+\alpha(m-1)}t), \quad 0 < \alpha \le n.$$

Now let $u(x,t)$ be a solution of problem (PM). For $k > 0$, define

$$u_k(x,t) = f(k)u(kx, k^2(f(k))^{m-1}t).$$

It is readily seen that, for any function $f(k) > 0$, u_k is a solution of (PM) with initial data

$$u_k(x,0) = \varphi_k(x) = f(k)\varphi(kx).$$

Choose

$$f(k) = k^n \Big/ \int_{|x|<k} \varphi(x)dx. \tag{17}$$

Consider the sequence of solutions $\{u_k(x,t)\}$ and the corresponding sequence of initial conditions $\varphi_k(x)$. Our main assumption is that

$$\lim_{k\to\infty} \varphi_k(x) = \Phi_\alpha(x)$$

in the distributional sense. In other words,

H_1: There exists some $\alpha \in (0,n]$ such that, for any $\psi(x) \in C_0^\infty(\mathbb{R}^n)$,

$$\lim_{k\to\infty}\int_{\mathbb{R}^n}\psi(x)f(k)\varphi(kx)dx = \int_{\mathbb{R}^n}\psi(x)\Phi_\alpha(x)dx,$$

where $f(k)$ and Φ_α are defined by (17) and (16), respectively.

REMARK 1: Assumption H_1 generalizes condition (5). In fact, if $|x|^\beta\varphi(x) \to A$, $\beta \in (0,n]$, then H_1 holds with $\alpha = \beta$. Moreover, let

$$\varphi(x) = \frac{A}{1+|x|^\beta|\ln|x||}, \qquad \beta \in (0,n].$$

Then H_1 holds with $\alpha = \beta$, but (5) is not satisfied.

REMARK 2: The function $f(k)$ defined in (17) depends on $\varphi(x)$ and is different from the function k^α figuring in previous papers. Nevertheless, if k is large, $f(k)$ behaves like k^α, provided $\varphi(x)$ satisfies appropriate conditions. Thus the transformation employed here is a generalization of those used previously and enables us to consider a larger class of initial data.

We also add the assumption

H_2: There exists a positive constant b such that

$$0 \le f(k)\varphi(kx) \le b, \qquad k \ge 1,\ |x| \ge 1$$

where $f(k)$ is defined by (17).

THEOREM 1. *If assumptions H_1, H_2 hold for some $\alpha \in (0,n]$, then the solution $u(x,t)$ of (PM) satisfies the relationship*

$$\text{(18)}\quad \lim_{\tau\to\infty}\left| f\left(\tau^{\frac{1}{2+\alpha(m-1)}}\right) u\left(x, \tau^{\frac{2}{2+\alpha(m-1)}} f^{m-1}\left(\tau^{\frac{1}{2+\alpha(m-1)}}\right)\right) - \tau^{\frac{\alpha}{2+\alpha(m-1)}} U_\alpha(x,\tau)\right| = 0$$

uniformly on the sets $P_R = \{x,\tau : |x| \le R\tau^{\frac{1}{2+\alpha(m-1)}}\}$, $R > 0$, where f is defined by (17).

To prove Theorem 1, we first prove that $\{u_k(x,t)\}$ is convergent as $k \to \infty$. As in [**1**], we use the estimates from [**2,3**] and the uniqueness results of [**3,12**]. This gives:

$$u_k(x,t) \to U_\alpha(x,t) \quad \text{as} \quad k \to \infty.$$

Applications.

We now present some consequences of Theorem 1.

Rewrite (18) as follows:

$$u(x,t) \cong \frac{\tau^{\frac{\alpha}{2+\alpha(m-1)}}}{f(\tau^{\frac{1}{2+\alpha(m-1)}})} U_\alpha(x,\tau), \qquad 0 < \alpha \le n \qquad (t \to \infty). \tag{19}$$

where $t = \tau^{\frac{2}{2+\alpha(m-1)}} f^{m-1}(\tau^{\frac{1}{2+\alpha(m-1)}})$. This shows that, for large t, $u(x,t)$ is close to a reconstructed similarity solution, which is evaluated at a different time if $m > 1$.

Example 1. $\varphi(x)|x|^n \to A$ as $|x| \to \infty$. Then, by (17),

$$f(k) \cong \frac{k^n}{AS_1 \ln k} \qquad \text{as } k \to \infty,$$

and by (19),

$$u(x,t) \cong \frac{AS_1 \ln \tau}{2 + n(m-1)} E(x,\tau),$$

$$t \cong \left[\frac{2 + n(m-1)}{AS_1}\right]^{m-1} \frac{\tau}{(\ln \tau)^{m-1}}.$$

In particular, if $m = 1$, this gives $t \cong \tau$; hence, for the heat equation,

$$u(x,t) \cong \frac{AS_1 \ln t}{2} E(x,t).$$

Example 2. $\varphi(x)|x|^\alpha \to A$ as $|x| \to \infty$ for some $\alpha \in (0,n)$. Then

$$f(k) \cong \frac{n-\alpha}{AS_1} k^\alpha \qquad (k \to \infty)$$

$$t \cong \left(\frac{n-\alpha}{AS_1}\right)^{m-1} \tau$$

and

$$u(x,t) \cong \frac{AS_1}{n-\alpha} W_\alpha \left(x, \left(\frac{AS_1}{n-\alpha}\right)^{m-1} t\right) = W_{\alpha,A}(x,t).$$

Thus we obtain (8).

Now consider the problem (PMA) with $p > m + \frac{2}{n}$ and $\alpha = n$. In addition to H_1 (with $\alpha = n$), we shall assume here that H_2 holds, and also:

H_3: $\varphi \notin L_1(\mathbb{R}^n)$

H_4: There exists $\mu \in \left(\frac{2+mn}{p}, n\right)$ such that $|x|^\mu \varphi(x) \le B < \infty$, $\forall x \in \mathbb{R}^n$.

REMARK 3: If H_3 is false, then $\varphi \in L_1$ and the result of [**7**] is applicable.

REMARK 4: If $\varphi(x)|x|^n \to A$ as $|x| \to \infty$, then H_4 holds with $\mu = n - \varepsilon$ for any small $\varepsilon > 0$.

THEOREM 2. *If assumptions $H_1 - H_4$ hold for $\alpha = n$ and $v(x,t)$ is a solution of (PMA), then*

$$\lim_{\tau\to\infty}\left|f(\tau^{\frac{1}{2+n(m-1)}})v(x,\tau^{\frac{2}{2+n(m-1)}}f^{m-1}(\tau^{\frac{1}{2+n(m-1)}})-\tau^{\frac{n}{2+n(m-1)}}E(x,t)\right|=0$$

uniformly on the sets $P_R = \{x,\tau : |x| \le R\tau^{\frac{1}{2+n(m-1)}}\}$, $R > 0$, where f is defined by (17).

REMARK 5: It follows from Theorem 2 that, in the case considered ($p > m + \frac{2}{n}$, $\alpha = n$), the influence of the absorption term becomes negligible as $t \to \infty$. This phenomenon has already been observed in several cases ([**8-11**]).

Remark 5 implies that Example 1 is also valid for (PMA). Thus, if $|x|^n\varphi(x) \to A$, then

$$v(x,t) \cong \frac{AS_1 \ln\tau}{2+n(m-1)}E(x,\tau)$$

$$t \cong \left[\frac{2+n(m-1)}{AS_1}\right]^{m-1}\frac{\tau}{(\ln\tau)^{m-1}}.$$

Tel-Aviv University, Tel-Aviv, ISRAEL

The results presented here were obtained in a joint paper of the author and M. Ughi [**12**].

REFERENCES

1. N. D. Alikakos and R. Rostamian, *On the uniformization of the solutions of the porous medium equation in* $\mathbb{R}^n$, Israel J. Math. **47** (1984), 270-290.
2. D. G. Aronson and L. A. Caffarelli, *The initial trace of a solution of the porous medium equation*, Trans. AMS **280** (1983), 351-366.
3. P. Bénilan, M. G. Crandall and M. Pierre, *Solutions of the porous medium equation in* $\mathbb{R}^n$ *under optimal conditions on initial values*, Indiana Univ. Math. J. **33** (1984), 51-87.
4. M. Bertsch, R. Kersner and L. A. Peletier, *Positivity versus localization in degenerate diffusion equations*, Nonlinear Anal. TMA **9** (1985), 987-1008.
5. E. Di Benedetto, *Continuity of weak solution to a general porous media equation*, Indiana Univ. Math. J. **32** (1983), 83-118.
6. M. Escobedo and O. Kavian, *Asymptotic behaviour of positive solution of a nonlinear heat equation*, to appear.
7. A. Friedman and S. Kamin, *The asymptotic behaviour of gas in an n-dimensional porous medium.*, Trans. AMS **262** (1980), 551-563.
8. V. A. Galaktionov, S. P. Kurdjumov and A. A. Samarskii, *On asymptotic "eigenfunctions" of the Cauchy problem for a nonlinear parabolic equation*, Mat. Sbornik **126** (1985), 435-472. (in Russian).
9. L. Gmira and L. Véron, *Large time behaviour of the solutions of a semilinear parabolic equation in* $\mathbb{R}^n$, J. Diff Equ. **53** (1984), 258-276.
10. S. Kamin and L. A. Peletier, *Large time behaviour of solutions of the heat equation with absorption*, Ann. Sc. Norm. Sup. Pisa Ser. IV **XII** (1985), 393-408.
11. S. Kamin and L. A. Peletier, *Large time behaviour of solutions of the porous media equation with absorption*, Israel J. Math. **55** (1986), 129-146.
12. S. Kamin and M. Ughi, *On the behaviour as* $t \to \infty$ *of the solutions of the Cauchy problem for certain nonlinear parabolic equations*, J. Math. Anal. Appl., to appear.
13. M. Pierre, *Uniqueness of the solutions of* $u_t - \Delta\phi(u) = 0$ *with initial datum a measure*, Nonlinear Anal. TMA **6** (1982), 175-187.
14. E. S. Sabininia, *On the Cauchy problem for the equation of non-stationary gas filtration in several space variables*, Dokl. Akad. Nauk S.S.S.R. **136** (1961), 1034-1037. (in Russian).